Preface

During the academic year 1983/84, the Science and Engineering Research Council of the United Kingdom gave generous financial support for two symposia, at the Universities of Warwick and Durham, on hyperbolic geometry, Kleinian groups and 3-dimensional topology. The symposium at Durham was also sponsored by the London Mathematical Society. I would like to express my thanks to both the SERC and the LMS for their help and support. It is a pleasure to acknowledge the help of my co-organizer at Durham, Peter Scott, who was also an unofficial co-organizer at Warwick. He made an essential contribution to the great success of the symposia.

The world's foremost contributors to this very active area were all invited, and nearly all of them came. The activity centred on the University of Warwick, and climaxed with a two week long intensive meeting at the University of Durham during the first two weeks of July 1984. There was earlier a period of intense activity during the Easter vacation of 1984, when a number of short introductory lectures were given. The text of the most important of these series of lectures, by S.J. Patterson, is published in these Proceedings.

The papers published here are the result of an invitation to all those attending the two Symposia to submit papers. Not all the papers submitted were the subject of talks given during the Symposia – the contents of the Proceedings are based on their relevance to the subject, and not on their accuracy as documents recording the events of the Symposia. Also, a number of important contributions to the Symposia are not published here, having been previously promised elsewhere.

One of the few advantages of being an editor is that one can confer certain rights and privileges on oneself. I have taken the opportunity of accepting as suitable for publication several rather large papers of which I was the the author or co-author, and which have a substantial element of exposition. This is a field which is expanding very quickly, mainly under Thurston's influence, and more material of an expository nature is sorely needed, as many of those attempting to penetrate the area will testify. I hope that my own efforts in this direction will be of some help.

Because of the amount of material submitted, it has been necessary to publish two separate books. The division was done on the basis of first sorting into five different fairly narrowly defined subject areas, and then trying to balance the sizes of the two books. The books are entitled "Analytical and Geometric Aspects of Hyperbolic Space" and "Low-Dimensional Topology and Kleinian Groups".

At an early stage I made the decision to set the books by computer, with the advantage that the entire process would be under my own control. This has been an interesting experience. I will content myself with the comment that computer typesetting is not the joy and wonder that I once thought it would be; the considerable delay in publication has been largely due to the unforeseen difficulties encountered in this process.

My particular thanks go to Russell Quin, without whose help the typesetting difficulties would never have been overcome. I am grateful to Kay Dekker, who has spent many hours creating fonts of special characters needed for this work. I would like to thank the University of Warwick Computer Unit for the use of their facilities for the printing.

Finally I must thank the contributors for their patience and forbearing during the long delay before publication.

D.B.A.Epstein
Mathematics Institute,
University of Warwick,
Coventry, CV4 7AL,
ENGLAND.

12 July 1986

LONDON MATHEMATICAL SOCIETY LECTURE NOTE SERIES

Managing Editor: Professor J.W.S. Cassels, Department of Pure Mathematics and Mathematical Statistics, 16 Mill Lane, Cambridge CB2 1SB, England

The books in the series listed below are available from booksellers, or, in case of difficulty, from Cambridge University Press.

4 Algebraic topology, J.F.ADAMS
5 Commutative algebra, J.T.KNIGHT
8 Integration and harmonic analysis on compact groups, R.E.EDWARDS
11 New developments in topology, G.SEGAL (ed)
12 Symposium on complex analysis, J.CLUNIE & W.K.HAYMAN (eds)
13 Combinatorics, T.P.McDONOUGH & V.C.MAVRON (eds)
16 Topics in finite groups, T.M.GAGEN
17 Differential germs and catastrophes, Th.BROCKER & L.LANDER
18 A geometric approach to homology theory, S.BUONCRISTIANO, C.P.ROURKE & B.J.SANDERSON
20 Sheaf theory, B.R.TENNISON
21 Automatic continuity of linear operators, A.M.SINCLAIR
23 Parallelisms of complete designs, P.J.CAMERON
24 The topology of Stiefel manifolds, I.M.JAMES
25 Lie groups and compact groups, J.F.PRICE
26 Transformation groups, C.KOSNIOWSKI (ed)
27 Skew field constructions, P.M.COHN
29 Pontryagin duality and the structure of LCA groups, S.A.MORRIS
30 Interaction models, N.L.BIGGS
31 Continuous crossed products and type III von Neumann algebras, A.VAN DAELE
32 Uniform algebras and Jensen measures, T.W.GAMELIN
34 Representation theory of Lie groups, M.F. ATIYAH et al.
35 Trace ideals and their applications, B.SIMON
36 Homological group theory, C.T.C.WALL (ed)
37 Partially ordered rings and semi-algebraic geometry, G.W.BRUMFIEL
38 Surveys in combinatorics, B.BOLLOBAS (ed)
39 Affine sets and affine groups, D.G.NORTHCOTT
40 Introduction to Hp spaces, P.J.KOOSIS
41 Theory and applications of Hopf bifurcation, B.D.HASSARD, N.D.KAZARINOFF & Y-H.WAN
42 Topics in the theory of group presentations, D.L.JOHNSON
43 Graphs, codes and designs, P.J.CAMERON & J.H.VAN LINT
44 Z/2-homotopy theory, M.C.CRABB
45 Recursion theory: its generalisations and applications, F.R.DRAKE & S.S.WAINER (eds)
46 p-adic analysis: a short course on recent work, N.KOBLITZ
47 Coding the Universe, A.BELLER, R.JENSEN & P.WELCH
48 Low-dimensional topology, R.BROWN & T.L.THICKSTUN (eds)
49 Finite geometries and designs, P.CAMERON, J.W.P.HIRSCHFELD & D.R.HUGHES (ed
50 Commutator calculus and groups of homotopy classes, H.J.BAUES
51 Synthetic differential geometry, A.KOCK
52 Combinatorics, H.N.V.TEMPERLEY (ed)
54 Markov process and related problems of analysis, E.B.DYNKIN
55 Ordered permutation groups, A.M.W.GLASS
56 Journées arithmêtiques, J.V.ARMITAGE (ed)
57 Techniques of geometric topology, R.A.FENN
58 Singularities of smooth functions and maps, J.A.MARTINET
59 Applicable differential geometry, M.CRAMPIN & F.A.E.PIRANI
60 Integrable systems, S.P.NOVIKOV et al
61 The core model, A.DODD
62 Economics for mathematicians, J.W.S.CASSELS
63 Continuous semigroups in Banach algebras, A.M.SINCLAIR

64 Basic concepts of enriched category theory, G.M.KELLY
65 Several complex variables and complex manifolds I, M.J.FIELD
66 Several complex variables and complex manifolds II, M.J.FIELD
67 Classification problems in ergodic theory, W.PARRY & S.TUNCEL
68 Complex algebraic surfaces, A.BEAUVILLE
69 Representation theory, I.M.GELFAND *et al.*
70 Stochastic differential equations on manifolds, K.D.ELWORTHY
71 Groups - St Andrews 1981, C.M.CAMPBELL & E.F.ROBERTSON (eds)
72 Commutative algebra: Durham 1981, R.Y.SHARP (ed)
73 Riemann surfaces: a view towards several complex variables,A.T.HUCKLEBERRY
74 Symmetric designs: an algebraic approach, E.S.LANDER
75 New geometric splittings of classical knots, L.SIEBENMANN & F.BONAHON
76 Linear differential operators, H.O.CORDES
77 Isolated singular points on complete intersections, E.J.N.LOOIJENGA
78 A primer on Riemann surfaces, A.F.BEARDON
79 Probability, statistics and analysis, J.F.C.KINGMAN & G.E.H.REUTER (eds)
80 Introduction to the representation theory of compact and locally compact groups, A.ROBERT
81 Skew fields, P.K.DRAXL
82 Surveys in combinatorics, E.K.LLOYD (ed)
83 Homogeneous structures on Riemannian manifolds, F.TRICERRI & L.VANHECKE
84 Finite group algebras and their modules, P.LANDROCK
85 Solitons, P.G.DRAZIN
86 Topological topics, I.M.JAMES (ed)
87 Surveys in set theory, A.R.D.MATHIAS (ed)
88 FPF ring theory, C.FAITH & S.PAGE
89 An F-space sampler, N.J.KALTON, N.T.PECK & J.W.ROBERTS
90 Polytopes and symmetry, S.A.ROBERTSON
91 Classgroups of group rings, M.J.TAYLOR
92 Representation of rings over skew fields, A.H.SCHOFIELD
93 Aspects of topology, I.M.JAMES & E.H.KRONHEIMER (eds)
94 Representations of general linear groups, G.D.JAMES
95 Low-dimensional topology 1982, R.A.FENN (ed)
96 Diophantine equations over function fields, R.C.MASON
97 Varieties of constructive mathematics, D.S.BRIDGES & F.RICHMAN
98 Localization in Noetherian rings, A.V.JATEGAONKAR
99 Methods of differential geometry in algebraic topology, M.KAROUBI & C.LERUSTE
100 Stopping time techniques for analysts and probabilists, L.EGGHE
101 Groups and geometry, ROGER C.LYNDON
102 Topology of the automorphism group of a free group, S.M.GERSTEN
103 Surveys in combinatorics 1985, I.ANDERSEN (ed)
104 Elliptic structures on 3-manifolds, C.B.THOMAS
105 A local spectral theory for closed operators, I.ERDELYI & WANG SHENGWANG
106 Syzygies, E.G.EVANS & P.GRIFFITH
107 Compactification of Siegel moduli schemes, C-L.CHAI
108 Some topics in graph theory, H.P.YAP
109 Diophantine analysis, J.LOXTON & A.VAN DER POORTEN (eds)
110 An introduction to surreal numbers, H.GONSHOR
111 Analytical and geometric aspects of hyperbolic space, D.B.A.EPSTEIN (ed)
112 Low dimensional topology and Kleinian groups, D.B.A.EPSTEIN (ed)
113 Lectures on the asymptotic theory of ideals, D.REES
114 Lectures on Bochner-Riesz means, P.A.TOMAS & Y-C.CHANG
115 An introduction to independence for analysts, H.G.DALES & W.H.WOODIN
116 Representations of algebras, P.J.WEBB (ed)
117 Homotopy theory, E.REES & J.D.S.JONES (eds)
118 Skew linear groups, M.SHIRVANI & B.WEHRFRITZ

London Mathematical Society Lecture Note Series. 112

Low Dimensional Topology and Kleinian Groups

Warwick and Durham 1984

Edited by D.B.A. EPSTEIN
The Mathematics Institute, University of Warwick

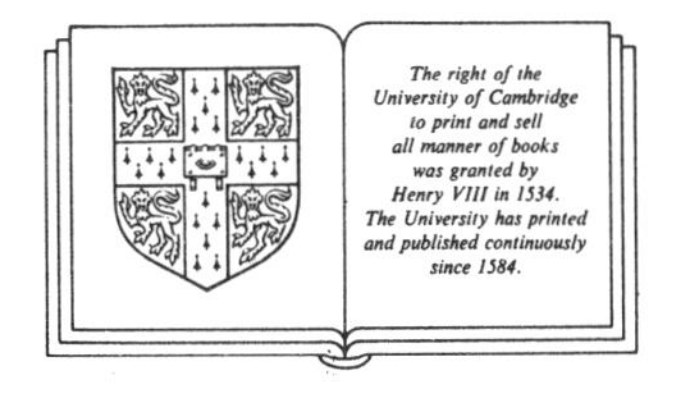

CAMBRIDGE UNIVERSITY PRESS
Cambridge
London New York
Melbourne Sydney

Published by the Press Syndicate of the University of Cambridge
The Pitt Building, Trumpington Street, Cambridge CB2 1RP
32 East 57th Street, New York, NY 10022, USA
10, Stamford Road, Oakleigh, Melbourne 3166, Australia

First published 1986

Printed in Great Britain at the University Press, Cambridge

Library of Congress cataloging in publication data applied for

British Library cataloguing in publication data

Low dimensional topology and Kleinian groups:
Warwick and Durham 1984. ---(London Mathematical Society
lecture note series, ISSN 0076-0052; 112)
1. Low dimensional topology
I. Epstein, D.B.A. II. Series
514'.3 QA612.14

ISBN 0 521 33905 7

Contents

Part A: Surfaces

Geodesics with multiple self-intersections and symmetries on Riemann surfaces

Joan S. Birman

Caroline Series

Irwin Kra has raised the question as to whether, if M is a surface of constant negative curvature, one can find points $\mathbf{v} \in M$ such that a closed smooth geodesic passes through $\mathbf{v}$ three or more times. John Stillwell (private communication) has raised a different question: if M is a closed surface of genus g, and if $\mathbf{v} \in M$ is the image on M of the $4g$ vertices of a rotationally symmetric geodesic polygon which is a fundamental domain in the hyperbolic disc for M, how can one tell whether the geodesic loop corresponding to a word in $\pi_1(M)$ passes through $\mathbf{v}$ or not?

Troels Jørgensen (private communication) has answered Kra's question in the affirmative for certain Schottky groups which are constructed specifically so as to force multiple intersection points. In this note we answer Kra's question in a different way, by applying a lemma of Jakob Nielsen to show that geodesics with multiple self-intersections occur naturally on Riemann surfaces which admit certain symmetries. Using the group-theoretic analogue of this lemma we are able to construct many examples and to describe the free homotopy classes of the loops in question as words in $\pi_1(M)$. Choosing the multiple intersection point to be Stillwell's vertex $\mathbf{v}$, we also solve Stillwell's problem. The words corresponding to loops passing through $\mathbf{v}$ contain "half the defining relator" the same number of times as they pass through $\mathbf{v}$, and in addition satisfy certain symmetries which we describe below.

We think of M as the quotient D/Γ of the Poincaré disc by a Fuchsian group Γ. We denote points of M in bold type $\mathbf{v}$, while points in D appear in standard type v. Thus $\mathbf{v}$ denotes the image of v in M.

1. Nielsen's Lemma

Let Γ be a Fuchsian group acting on the Poincaré disc $\mathbb{D}$, let R be a finite sided geodesic polygon which is a fundamental domain for the action, and let N be the net of translates of ∂R. Assume further that N is a union of complete geodesics. (This last condition is not satisfied for arbitrary Γ, R, however it holds for example in the standard tiling for closed surface groups by symmetrical $4g$-gons; for the standard tiling for $SL(2, \mathbb{Z})$, and for the usual fundamental regions for Schottky groups).

If $v \in \mathbb{D}$, let Ω_v be the hyperbolic isometry which rotates by π about v. Let

$$V = V(\Gamma,R) = \{v \in \mathbb{D} \mid \Omega_v(N) = N\}.$$

Assume that Γ, R are chosen so that $V \neq \emptyset$. Then V is infinite, for if $v \in V$, then $\gamma(v) \in V$ for each $\gamma \in \Gamma$. Nielsen's lemma, proved in [Nielsen, 1927] for very special choices of Γ,R, and points in V, but valid more generally, is as follows:

Lemma 1. (Section 5 of [Nielsen, 1927]). *Let $v,w \in V$. Let A be the geodesic through w and v and let ρ be the hyperbolic translation with axis A which carries w to v. Then $\rho^k \in \Gamma$ for some $k \leq 4r$, where $2r$ is the number of sides of R.*

Proof. (Nielsen): Let Γ_R be the finite set of elements which pair corresponding sides of R. Let $X = \{x, x^{-1} \mid x \in \Gamma_R\}$. Label each edge of R which does not cover part of ∂M by the corresponding element of X, i.e. if s, $s' \in \partial R$ and $x \in X$, $x(s) = s'$, assign the label x to the side of s interior to R. This labelling extends to a labelling on all oriented edges of N, by assigning to each edge the label on its pre-image in R. (Thus each edge of N has two mutually inverse labels, one on each side.)

If γ is an isometry of $\mathbb{D}$ which preserves N, then γR will be a polygon in the tiling. Hence γ induces a map $\gamma_\star : X \to X$, defined by sending the label on an edge of R to the label on its image under γ. Since the cyclic order of the labels in each polygon in the tiling agrees with that in R, it follows that if there are $2r$ elements in X then $\gamma_\star$ is a $2r$ cycle so that $\gamma_\star^k =$ identity for some k with $1 \leq k \leq 2r$. Clearly $\gamma \in \Gamma$ if and only if $\gamma(N) = N$ and $\gamma_\star =$ identity.

Let v, w, A, ρ be as in the statement of the lemma. We may as well assume that $\rho \notin \Gamma$. We claim that $\rho^2(N) = N$. First, note that $\Omega_v(A) = A$ and $\Omega_v(N) = N$, also by symmetry $\Omega_v(w) = w' \in V$. Thus if n is an edge of N which intersects A transversally, the angle between n and A is the same as that between $\Omega_v(n)$ and A. Choose such an edge which intersects A at a distance $d \geq 0$ measured along A from w, on the side opposite to v. Then $\Omega_v(n)$ intersects A at a point distance d from w', measured away from v, and $\Omega_{w'}\Omega_v(n)$ intersects A at distance d from w' measured towards v, making the same angle as n with A. Clearly $\rho^2(n) = \Omega_{w'}\Omega_v(n)$, and ρ^2 preserves all distances and angles measured along n. Thus $\rho^2(N) = N$, and the result follows by applying the remarks above with $\rho^2 = \gamma$.

□

2. Geodesics with multiple self-intersections

In this section we show how Nielsen's lemma may be applied to answer the question of Kra.

With the notation of Lemma 1, let $v, w \in A$ be chosen so that $v, w \in V$ and so that there are no points equivalent to v on A between v and w. If $\rho \in \Gamma$, it is clear that A projects to a simple loop on Γ. More generally, if k is the smallest non-negative integer with $\rho^k \in \Gamma$, and if r of the points $w = \rho(v), \rho^2(v), \ldots, \rho^{k-1}(v)$ are equivalent to v, then the projection of A will pass through the image $\mathbf{v}$ of v on M exactly $r+1$ times. Thus in order to answer Kra's question we need only look for cases in which $r > 1$. That this situation indeed occurs is illustrated by the example of a closed surface of odd genus, as follows.

Let M be such a closed surface of a genus g represented as $\mathbb{D}/\Gamma$, where Γ has fundamental domain a symmetrical $4g$-gon of vertex angle $\frac{\pi}{2g}$ centred on $0 \in \mathbb{D}$. Choose the side pairing so that if $X = \{x_1, x_1^{-1}, \ldots, x_{2g}, x_{2g}^{-1}\}$, then the order of the labels on R is $x_1, x_2^{-1}, x_1^{-1}, x_2, \ldots, x_{2g-1}, x_{2g}^{-1}, x_{2g-1}^{-1}, x_{2g}$. (With our conventions this implies that the defining relation is

$$\prod_{i=1}^{g} \left(x_{2i-1} x_{2i} x_{2i-1}^{-1} x_{2i}^{-1} \right)).$$

Clearly 0, the centre of $\mathbb{D}$, belongs to V.

If g is even, rotation Ω_0 about 0 preserves the labelling on the edges of N, on the other hand, if g is odd, it does not. This means that for g odd, Ω_0 does not induce a topological mapping of M. If the points v, w above are chosen to be 0, $x0$ where $x \in X$, so that v, w are in adjacent polygons, then one sees easily that $\rho \notin \Gamma$ and that if g is even, $\rho^2 \in \Gamma$ while if g is odd, $\rho^2 \notin \Gamma$. Moreover the points $0, \rho(0), \rho^2(0), \ldots, \rho^{k-1}(0)$ are all equivalent under Γ. Thus if g is odd the geodesic A joining 0 to $x0$ projects to a geodesic $\mathbf{A}$ on M which passes through $\mathbf{0}$ at least three times. In the case $g = 3$, the multiplicity is exactly 3.

3. Multiply palindromic words in $\pi_1(M)$

In the example given in Section 2, it is not immediately clear how to compute the number of equivalent points on A between v and $\rho^k(v)$ when v,w are far apart. In this section we overcome this difficulty and at the same time find the free homotopy class of the loop $\mathbf{A}$, a problem of interest in its own right because of the striking patterns obtained. For simplicity of exposition, we assume throughout this section that Γ is the symmetric closed surface group of Section 2, with R the symmetric $4g$-gon as before. We do not place any restriction on the edge pairings other than that they determine a closed surface of genus g.

An oriented geodesic A in $\mathbb{D}$ determines a bi-infinite *edge path sequence*

$$S = \ldots e_{-2}e_{-1}e_1e_2\ldots, \quad e_j \in X,$$

which records the sequence of labels on the far side of edges of N crossed by A. (If A coincides with a net edge, push it off the net slightly to one side; if it passes through a vertex deform it slightly round the vertex, as in Figure 1.) The geodesic A projects to a smooth closed geodesic $\mathbf{A}$ on M if and only if S is periodic. If S is periodic of fundamental period U, then U represents the free homotopy class of $\mathbf{A}$. These assertions are well known and are developed in detail in, for example, [Birman-Series, 1984] for the easiest case $\partial M \neq \emptyset$, and in [Birman-Series, 1985] in full generality.

Conversely, any word W in the symbols of X determines a bi-infinite sequence $S = \ldots WWW \ldots$, which in turn gives an edge-path in $\mathbb{D}$ defined by joining the centres of adjacent polygons in the tiling so that the edge cuts correspond to S. One has:

Lemma 2. *Let W be a word in the symbols of X, $W \neq$ id in Γ. Then the edge-path sequence $S = \ldots WWW \ldots$ has limit points $S_\infty = \lim_{n \to \infty} W^n 0$, $S_{-\infty} = \lim_{n \to \infty} W^{-n} 0$, and the geodesic A joining S_∞ to $S_{-\infty}$ projects to a*

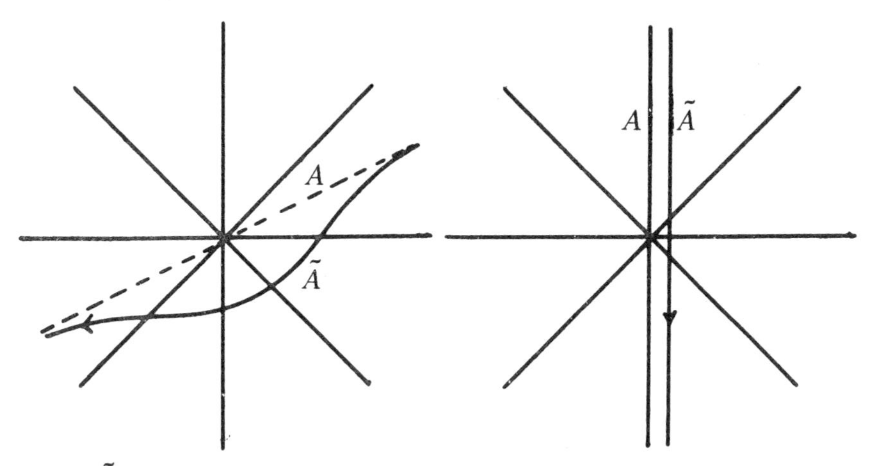

Figure 1. *$\tilde{A}$ is the deformation of A described in the text.*

smooth closed geodesic $\mathbf{A}$ *on* M *having the free homotopy class of* W.

Proof. This follows by the method of [Series, 1981], Proposition 4.2. The assumption that W be a shortest word is unnecessary; all we need is that the points $W^n 0$ tend to $\partial \mathbb{D}$ as $n \to \infty$, and this follows from the lemma on noting that W is certainly hyperbolic, so that only finitely many points $W^n 0$ lie in any compact ball in $\mathbb{D}$.

$\square$

With these preliminaries, our first goal is to find necessary and sufficient conditions on a periodic sequence $S = \ldots WWW \ldots$ to ensure that the axis of the element of Γ defined by W is a geodesic passing through $v \in V$. The idea is that for such an axis A, the fact that $\Omega_v(A) = A$ forces S to be invariant under a rule determined by Ω_v. Let $\psi : X \to X$ be the $2r$ cycle sending $x \in X$ to the label next to x in anticlockwise order around ∂R, and let $\tau : X \to X$ be the involution $\tau(x) = x^{-1}$. As we shall see, our Ω_v-invariant sequences can be characterised combinatorially in terms of ψ and τ.

The easiest case to treat is that of the example in Section 2, where v is taken to be the origin $0 \in \mathbb{D}$. Rewrite the sequence S as $\ldots f_3 f_2 f_1 e_1 e_2 e_3 \ldots$, $e_i, f_j \in X$, where 0 lies in the region R between the crossings f_1 and e_1, as in Figure 2.

For each adjacent pair $f_j f_{j-1}$, there is a unique integer n_j, $1 \le n_j \le 2g$, such that $f_j = \psi^{n_j} \tau(f_{j-1})$. The rotated half sequence $e_1 e_2 e_3 \ldots$ is then determined by the conditions:

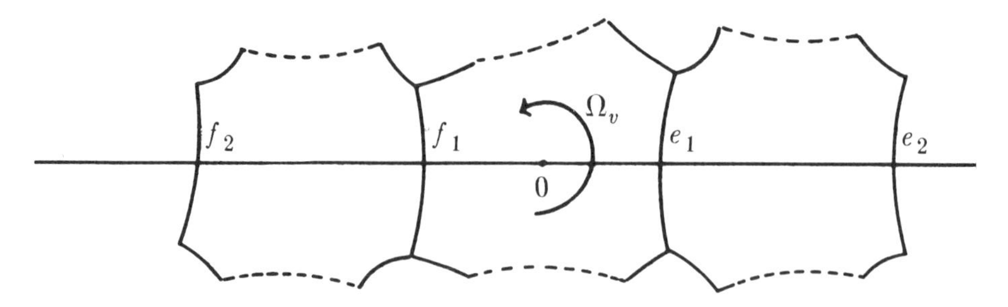

Figure 2.

(1) $\quad e_1 = \tau\psi^{2g}(f_1)$

(2) $\quad e_j = \tau\psi^{n_j}(e_{j-1})$ where $f_j = \psi^{n_j}\tau(f_{j-1}),\ j \geq 2.$

This holds because as A is rotated about 0 the edge with label f_j goes over to the edge with label e_j^{-1}, also since A is a diameter of $\mathbb{D}$ and R a symmetric $4g$-gon, the edges $f_1 e_1^{-1}$ are diametrically opposite on R, and so related by ψ^{2g}, since R has $4g$ edges.

We introduce the shorthand $E = F^\star$ to indicate that a finite or infinite word E is obtained from a finite or infinite word F by rules (1) and (2). If $S = FE$ and $E = F^\star$, we write $S = F \star E$ and say that S is *palindromic* (*rel* $X, \psi, 0$) at the interface between F and E.

Lemma 3. *Let $S = \ldots WWW \ldots$ be periodic with fundamental period W. Then $S = \ldots \star W \star W \star W \star \ldots$ if and only if the geodesic representative of the free homotopy class of W passes through the point* $\mathbf{0} \in M$.

Proof. The necessity has already been proved in the discussion following the conditions (1) and (2).

Assume conversely that $S = \ldots \star W \star W \star W \star \ldots$. Since S is palindromic at the interface between adjacent copies of W, we have $W = W^\star$, and therefore $S = \ldots WWW \ldots = \ldots \star W \star W \star W \star \ldots$. let S_∞, $S_{-\infty}$ be the limit points of $\ldots WWW \ldots$ on $\partial\mathbb{D}$, and let A be the geodesic joining them. Clearly S_∞, $S_{-\infty}$ are also the limit points of $\ldots \star W \star W \star W \star \ldots$. But then Ω_0 maps A to itself (reversing orientation), and so A passes through 0 and $\mathbf{A}$ passes through $\mathbf{0}$. By Lemma 2, the free homotopy class of $\mathbf{A}$ is $W = W^\star$.

$\square$

We are now ready to solve the problem posed at the beginning of this section, for our special group Γ and fundamental domain R, when $v = 0$ and $w \in \Gamma 0$.

Proposition 4. *Let A be the geodesic in $\mathbb{D}$ which joins $w = f_n^{-1}\dots f_1^{-1}0$ to 0, $f_i \in X$, $1 \leq i \leq n$. Assume that $f_1\dots f_n$ cannot be written as $B\star C$ for any words B, C in X. Let $F_1 = f_n\dots f_1$, $F_{j+1} = F_j^\star$, $j \in \mathbb{Z}$. (Notice that this makes sense since $F^{\star\star} = F$.) Let $S = \dots F_{-2}F_{-1}F_0F_1F_2\dots$. Then there is $k \in \mathbb{N}$ such that*

$$S = \dots\star F_1\star F_2\star\dots\star F_k\star F_1\star\dots\star F_k\star F_1\star\dots$$

i.e. S is periodic, with fundamental period $W = F_1\dots F_k$ and palindromic (rel $X, \psi, 0$) at each interface between adjacent F_j's. Also, W represents the free homotopy class of **A**, *a smooth closed geodesic which passes through* **0** *exactly k times.*

Proof. By Theorem 2.8 of [Birman-Series, 1985], the sequence $f_n\dots f_2f_1$ may be taken to be the sequence of edges of N cut by the geodesic arc joining w to 0. By construction, it follows that S is the edge-path sequence for the geodesic A, because S is obtained by repeatedly rotating the arc about about its end points. By Lemma 1, A is the axis of an element of Γ and hence S is periodic. (This could also be proved by noting that only a finite number of distinct blocks F_i can occur.) The fundamental period cannot be a proper subword of F_1 for if it were there would necessarily be a translate of 0 on A between w and 0, implying that $F_1 = F_{11}\star F_{12}$.

Since $F_{j+1} = F_j^\star$, $j \in Z$, it follows that $F_{j+1}F_{j+2} = (F_{j-1}F_j)^\star$ and more generally that

$$F_{j+1}\dots F_{j+r} = (F_{j-r+1}\dots F_j)^\star,\ r > 0,$$

hence S is palindromic at the interface between F_j and F_{j+1} for each j. Therefore the fundamental period W of S must be a multiple of the length of F_1, implying that $W = F_1\dots F_k$ for some $k \geq 1$.

Let $W_i = F_iF_{i+1}\dots F_kF_1\dots F_{i-1}$. Then $S = \dots\star W_i\star W_i\star\dots$, so by Lemma 3 the geodesic representative of W_i passes through **0**. Since the W_i's are all conjugate and therefore all represent the same geodesic loop, namely **A**, it follows that **A** has a k-fold multiple self-intersection at **0**. □

Example We give an example to show how Proposition 4 enables one to compute the integer k by means of equations (1) and (2). Let M have genus 3, and let $\pi_1(M)$ be presented by

$$x_1, \dots, x_6\ ;\ \prod_{i=1}^{3} x_{2i-1}x_{2i}x_{2i-1}^{-1}x_{2i}^{-1}.$$

Choose $F_1 = x_1x_2^{-1}$, so that w is the centre of a polygon which is twice-removed from R. Then $f_2 = x_1$, $f_1 = x_2^{-1}$ and $f_2 = \psi^{n_2}(\tau f_1)$ gives $x_1 = \psi^{n_2}(x_2)$ and hence $n_2 = 3$. Using (1) and (2) we find $F_2 = x_4^{-1}x_3$, F_3 $x_5x_6^{-1}$, $F_4 = x_2^{-1}x_1$, $F_5 = x_3x_4^{-1}$, $F_6 = x_6^{-1}x_5$, and $F_6^\star = x_1x_2^{-1} = F_1$ so that the cycle repeats. Hence the geodesic joining $x_1^{-1}x_2 0$ to 0 has free homotopy class

$$x_1x_2^{-1}x_4^{-1}x_3\, x_5x_6^{-1}\, x_2x_1^{-1}\, x_3x_4^{-1}\, x_6^{-1}x_5$$

and passes through **0** exactly six times.

Generalizations

The point 0 used in Proposition 4 is very special, however analogous methods work equally well for other points of V and indeed for other groups. For example, if R is the same symmetric $4g$-gon as above, then both the vertices of R and the mid-points of the sides of R are in V. Let V_0, V_N, V_M be respectively the sets of Γ translates of $\mathbf{0} \in \mathbb{D}$, the vertices of R and the mid-points of sides of R, and let $V' = V_0 \cup V_N \cup V_M \subseteq V$. Choose $v \in V'$ and suppose that S is the edge path sequence of a geodesic A passing through v. Just as before, the fact that $\Omega_v(A) = A$ forces a symmetry on S. To describe it, observe that if $v \in V_N \cup V_M$ then v is the mid-point of a geodesic arc joining the centres 0_1, 0_2 of adjacent polygons R_1, R_2, as in Figure 3.

The portion of the edge-path S arising from this arc will be a cycle (half the defining relator) when $v \in V_N$ or a symbol in X when $v \in V_M$. Thus S has the form

$$S = \ldots f_3f_2f_1\, C\, e_1e_2e_3 \ldots$$

where

$$\begin{aligned} C &= \emptyset \text{ if } v \in V_0 \\ &= \text{half the defining relator if } v \in V_N \\ &= x \in X \text{ if } v \in V_M. \end{aligned}$$

The e_j's are related to the f_j's in exactly the same way as before (using relation (2)). With these observations the reader should have no difficulties in adapting Proposition 4 to the more general case.

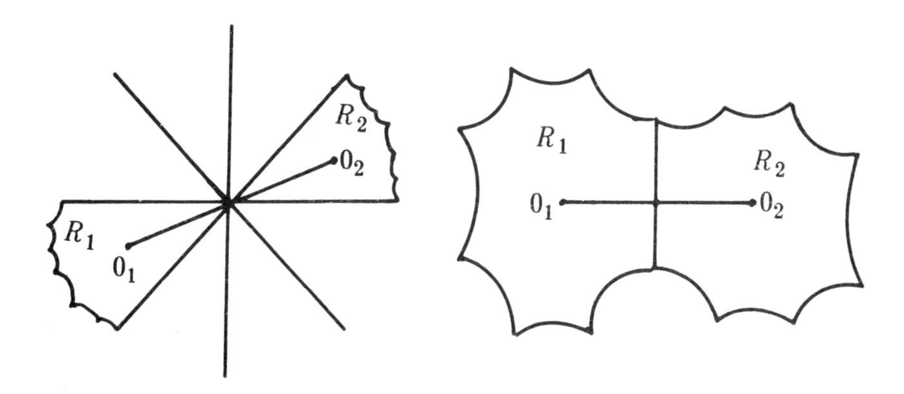

Figure 3.

References

Birman-Series, 1984.
J. S. Birman and C. Series, "An algorithm for simple curves on surfaces," *J.L.M.S. (2)*, **29**, pp. 331-342, 1984.

Birman-Series, 1985.
J. S. Birman and C. Series, "Dehn's algorithm revisited, with application to simple curves on closed surfaces," *Proc. Conf. on Combinatorial Group Theory 1984*, Alta, Utah, 1985.

Nielsen, 1927.
J. Nielsen, "Untersuchungen zur Topologie der geschlossenen Zweiseitigen Flächen," *Acta Math.*, **50**, pp. 189-358, 1927.

Series, 1981.
C. Series, "The infinite word problem and limit sets in Fuchsian groups," *Ergod. Thy and Dyn. Systs.*, **1**, pp. 337-360, 1981.

J.S. Birman,
Columbia University,
N.Y. 10027,
U.S.A.

C. Series,
University of Warwick,
Coventry, CV4 7AL,
England.

The Classification of Pseudo-Anosovs

Lee Mosher

Introduction

The classification of conjugacy classes in mapping class groups of surfaces began with the work of Nielsen [Nielsen, 1944].
He analysed a mapping class according to its actions on the circle at infinity of hyperbolic space. Using this method, he was able to give a complete classification for finite order mapping classes. In addition, he was able to classify those mapping classes now known (in Thurston's terminology) as reducible with only finite order components. Nielsen also recognized certain phenomena which have come to be associated with pseudo-Anosov mapping classes, defined by Thurston.

Nielsen did not extend his work far enough to, say, give a complete classification for pseudo-Anosov mapping classes. His work has recently been extended by Gilman [Gilman, 1981] to give a proof of Thurston's trichotomy; and by Hemion [Hemion, 1979] to give an algorithm for the conjugacy problem for mapping class groups of surfaces. Also, correspondences between Nielsen's and Thurston's theories have been given by Miller [Miller, 1982] and Casson [Casson, 1983].

It still remained, however, to give an explicit, complete classification of all conjugacy classes, and in particular, of pseudo-Anosov conjugacy classes. This is the problem taken up here, where we describe a method which uses Thurston's theory of measured foliations and train tracks to give a classification of pseudo-Anosov conjugacy classes. The methods we develop here are not quite strong enough to give a complete classification; we shall have to be content with a classification up to powers. The complete classification requires further refinement of techniques; this is the subject of current work in progress [Mosher, 1986].

The idea is as follows: a pseudo-Anosov is described, essentially up to a power, by its stable foliation; this is described in turn by an infinite sequence of train tracks, connected by elementary combinatorial operations called "splittings". This is in analogy to a real number being described by its continued fraction expansion. And just as a quadratic irrational has an eventually periodic continued fraction expansion, so is the "train track" expansion of a pseudo-Anosov stable foliation eventually periodic, up to the action of the mapping class group.

We do not exactly use train tracks, however, instead preferring to develop an essentially equivalent combinatorial tool called a "cell-division with distinguished prong", or CDP. The reason is that a CDP carries less combinatorial data than a train track, and is therefore basically easier to compute with. In Sections 0, we define enough of the main combinatorial tools to state Theorem 1, which describes explicitly the periodic nature of the above mentioned train track expansion of a pseudo-Anosov. We also state Theorem 2, the classification theorem, in terms of these expansions. Section 1 gives the connection with the theory of measured foliations and train tracks. The proof of Theorem 1 divides neatly into an existence part and a uniqueness part; these occupy Section 2 and 3. The heart of the combinatorial analysis is in Section 3.1, where the main connection is given between the analytic and the combinatorial properties of a measured foliation. Section 4 gives the proof of theorem 2, which is quite short after all the previous development. An appendix is included to make precise certain intuitive notions that cropped up in the writing, concerning the action of a mapping class on the "sectors" of a measured foliation.

This material is essentially contained in my thesis [Mosher, 1983], though it has also been recast into a more thoroughly structured form; it has been extended from the case of a punctured surface to include the case of a closed surface, an extension which involved no new conceptual difficulties.

Nothing is included here on how to actually compute the combinatorial invariants described herein. This computational problem is solved in [Mosher, 1983], in the case of a punctured surface; it will also be recast, in a future work. The computational problem for a closed surface does involve some new concepts, which will hopefully be developed in another future work.

0. Main Definitions and Theorems

0.1. Measured Foliations. A *punctured surface* is a pair (S, P) where S is a compact, connected, orientable surface and P is a finite, possibly empty, set of points on S, called the *punctures*. $S-P$ is often called the *interior* of (S, P). The only surfaces not allowed here are the sphere with 0, 1, 2, or 3 punctures, and the non-punctured torus.

A *measured foliation* on (S, P) is a codimension 1 singular foliation, equipped with a transverse invariant measure. The singularities are n-pronged singularities: for interior singularities, $n \geq 3$; for puncture singularities, $n \geq 1$. By convention, a puncture is always considered to be a singularity, even when $n = 2$. A *transverse invariant measure* assigns a positive number to each arc embedded transversely to the foliation; this number is invariant under isotopy along leaves, and adds correctly.

There is an equivalence relation on measured foliations generated by isotopy and Whitehead moves, where a *Whitehead move* collapses to a point a leaf segment l joining two singularities, at least one of which is not a puncture. $MF(S, P)$, or just MF, denotes the collection of equivalence classes of measured foliations. The class of a measured foliation f will be denoted $\{f\}$.

Let $\mathcal{S}$ denote the set of isotopy classes rel P of simple closed curves in $S-P$ that do not bound a disc with zero or one puncture. Let $\mathbb{R}_+$ denote the non-negative reals. MF maps to $\mathbb{R}_+^{\mathcal{S}}$ in the following manner: if $\{f\} \in MF$ and $C \in \mathcal{S}$, $\langle\{f\}, C\rangle$ is defined to be the infimum of the measures of simple closed curves representing C, measured with respect to foliations representing $\{f\}$.

Theorem (Thurston). *This gives a 1-1 mapping of MF into $\mathbb{R}_+^{\mathcal{S}}$, and under this inclusion, MF is the cone on a sphere of dimension*

$$6 \cdot g(S) + 2 \cdot |P| - 7 .$$

For every positive number λ, we say that $\{f\}$ is *projectively equivalent* to $\{\lambda \cdot f\}$. The space of projective equivalence classes of measured foliations, PF, is therefore naturally included in $P\mathbb{R}_+^{\mathcal{S}}$ as a sphere of the above dimension. The projection mapping will be denoted $\pi: MF \to PF$; thus the projective class of $\{f\}$ is $\pi\{f\}$.

0.2. Mapping Classes. The *mapping class group* of (S, P), denoted $M(S, P)$, is defined to be $\pi_0(\text{Homeo}^+(S, P))$, where $\text{Homeo}^+(S, P)$ is the group of orientation preserving homeomorphisms of S fixing P as a set. $M(S,P)$ acts naturally on MF and PF.

Notation: Elements of $\text{Homeo}^+(S, P)$ and $M(S, P)$ will be represented (resp.) by lower case Greek (e.g. ϕ or ψ) and upper case Greek (e.g. Φ or Ψ).

If f is a measured foliation, f is *arational* if there are no closed leaf cycles of f rel P. This property remains invariant under isotopy and Whitehead moves, so we can properly speak of an arational element of MF or PF.

A mapping class Φ is *pseudo-Anosov* if the action of Φ on PF has an arational fixed point. Thurston shows that in this case, Φ has two arational fixed points, and Φ acts as a discrete flow from one fixed point to the other.

A mapping class Φ is *reducible* if Φ permutes a set of disjoint elements of S.

Thurston's Classification Theorem. *If Φ is infinite order and irreducible, then Φ is pseudo-Anosov. Moreover, Φ is represented by a pseudo-Anosov homeomorphism ϕ.*

This means that the stable and unstable points in PF are represented (resp.) by a pair of transverse measure foliations f_s and f_u, such that for some $\lambda > 1$, $\phi(f_s) = \lambda \cdot f_s$ and $\phi(f_u) = \lambda^{-1} \cdot f_u$. These equations hold exactly, not just up to equivalence. Qualitatively, ϕ stretches the leaves of f_s and squashes the leaves of f_u, both by a factor of λ. It follows that f_s and f_u are *canonical models* for their respective classes, meaning that there are no leaf segments connecting singularities. We also say that such an arational measured foliation is a *canonical* arational measured foliation.

Up to topological conjugacy, there is a unique pseudo-Anosov homeomorphism in each pseudo-Anosov mapping class [Fathi-Laudenbach-Poenaru, 1979, Exp. 12]. Thus we can say that two pseudo-Anosov homeomophisms on the same (punctured) surface are topologically conjugate if and only if their respective mapping classes are conjugate in the mapping class group. Thus, the main result of this paper can be viewed as a classification of topological conjugacy classes of pseudo-Anosov homeomophisms, or of conjugacy classes in $M(S, P)$ of pseudo-Anosov mapping classes.

0.3. Cell-Divisions. Throughout this paper a *cell-division* of (S, P) will mean exclusively a cell division δ of S with a single vertex v, located either at a puncture or in the interior. The following conditions must also hold:

1) an arc of δ does not contain any puncture in its interior.

2) an n-gon of δ has no more than one puncture in its interior.

3) any 1-gon or 2-gon of δ has exactly one puncture in its interior.

In general, an n-gon is a quotient space of an n-sided polygon, identifying *all* vertices, and with (perhaps) some side pairings. Thus, an arc of δ may bound the same n-gon of δ on both sides. Throughout this paper, two cell-divisions which are isotopic rel P will be considered identical.

Cell divisions will be the main tool for studying pseudo-Anosovs. They will have extra structure attached to them, and various operations performed on them, all with the goal of reaching a combinatorial understanding of pseudo-Anosovs. They will play the role usually played in this subject by train tracks. This is done because of the restrictive combinatorial class of train tracks that we will be dealing with; also because, in the case of a punctured surface, the number of combinatorial possibilities is smaller for cell-divisions than for train tracks, by a factor which is linear in the genus of S and the size of P. This computational simplification is bought at the price of having inelegant definitions which depend on whether the vertex of a cell-division is at a puncture or in the interior.

Given a cell-division δ, a *prong* of δ is an ordered pair (e, e') of arc ends in δ, with e' adjacent to e in the positive direction around v. The symbols $\star$ and # will often denote a prong.

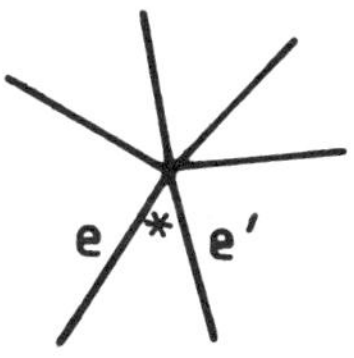

Figure 1.

Let δ be a cell-division of (S, P). We shall define a *cell division with distinguished prong*, or *CDP* whose *underlying cell-division* is δ. The definition depends on whether δ has a puncture vertex or an interior vertex.

First suppose δ has a puncture vertex. Let $\star$ be a prong of δ which is located in a non-punctured 3-gon T of δ. The pair $(\delta, \star)$ is a cell-division with distinguished prong, with underlying cell-division δ. It is often denoted with a Δ. It is easy to see that a 3-gon can have no side identifications, so the three sides of the distinguished triangle T are distinct arcs of δ. These three arcs are denoted h^L,h^R, and h^O: the arcs to the Left, Right, and Opposite from the distinguished prong.

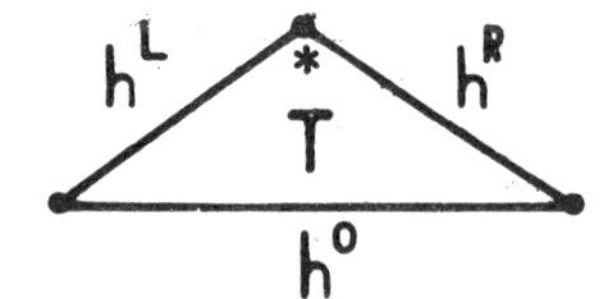

Figure 2.

Now suppose the vertex of δ is in the interior. A CDP Δ whose underlying cell-division is δ still has a distinguished prong $\star$, satisfying all the conditions of the previous paragraph. But there is an additional prong #, called the *singularity prong* of Δ, which is not allowed to lie in the distinguished triangle. Formally, Δ is the triple $(\delta,\star,\#)$. There are two other conditions imposed on the choice of #. The first, called the *interior singularity condition*, is that # is located in a non-punctured n-gon, where $n \geq 4$. The other condition is called the *recurrence condition*; to define this, first note that $\star$ and # separate the set of arc ends of δ, which have a naturally cyclic ordering, into two disjoint subsets. The recurrence condition says that if there is an arc with both of its ends in one of these subsets, then there is another arc with both of its ends in the other subset. For motivation of these conditions, see Section 1.1. Thus, unless otherwise stated, given a CDP with interior vertex, the singularity prong is required to satisfy the interior singularity condition and the recurrence condition.

Note: It is clear from the conventions adopted that a CDP Δ "has a singularity prong" if and only if the vertex of the underlying cell-division of Δ is in the interior of (S,P).

Let Δ, Δ' be CDPs with underlying cell-divisions δ, δ'. We say that Δ and Δ' are *combinatorially equivalent*, or that they have the same *combinatorial type*, if there is a mapping class Φ such that $\Phi(\delta)=\delta'$, Φ takes the distinguished prong of Δ to that of Δ', and if Δ, Δ' have interior vertices, Φ takes the singularity prong of Δ to that of Δ'.

Note: Proposition A in the appendix assures that the action of Φ on prongs is well-defined.

We give here a more intrinsic and formal description of combinatorial type. Let Δ be a CDP with underlying cell-division δ. Let E be a set in 1-1 correspondence with the arc ends of δ. Let σ be the permutation of E which takes $e \in E$, corresponding to some arc end, to $\sigma(e)$, corresponding to the next end around v in the positive direction. Note that σ has a single cycle. Let τ be the transposition of E which pairs e_0, $e_1 \in E$ when e_0, e_1 corresponds to two ends of the same arc in δ. The n-gons of δ correspond in an obvious way to cycles of the permutation $\tau \circ \sigma$, as indicated by the following figure:

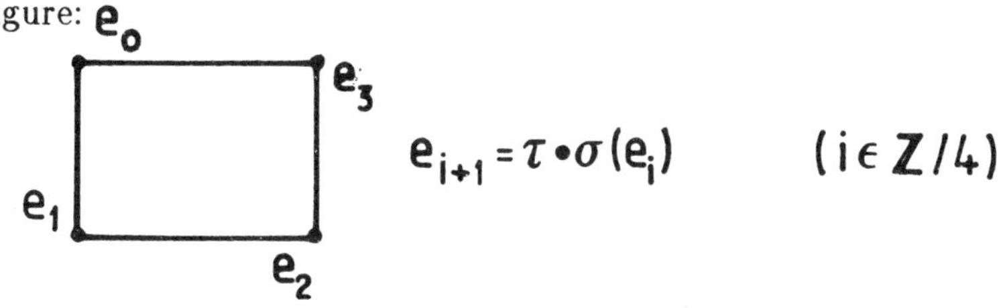

e_0 e_3 e_1 e_2

$$e_{i+1} = \tau \bullet \sigma(e_i) \qquad (i \in \mathbb{Z}/4)$$

Figure 3.

The n-gons of δ which contain a puncture then correspond to a certain subset Q of the cycles of $\tau \circ \sigma$. The distinguished prong of Δ corresponds to a pair $\star = (e, \sigma(e))$ such that e is contained in a 3-cycle of $\tau \circ \sigma$ which is not contained in Q. The singularity prong $\# = (f, \sigma(f))$, if it exists, is such that f does not lie in the same cycle of $\tau \circ \sigma$ containing e; one can easily write down the interior singularity condition and the recurrence condition for # in this language. The *combinatorial type* **E** *of* Δ is formally defined to be the sextuple $(E, \sigma, \tau, Q, \star, \#)$, where # is empty if Δ has a puncture vertex. More generally, a *combinatorial type* is defined to be any sextuple $(E, \sigma, \tau, Q, \star, \#)$ satisfying the conditions herein described.

Two combinatorial types **E** and **E**$'$ are *equivalent* if there is a 1-1 mapping from E to E' preserving all the structure.

The proof of the following fact is left to the reader; consult [Mosher, 1983, Section 2] for the flavour, but beware of differences in definitions.

Fact. *Two CDPs are combinatorially equivalent if and only if their combinatorial types are equivalent.*

The combinatorial type of Δ will be denoted $[\Delta]$.

The main reason for introducing the formal concept of a combinatorial type is to show that the results of this paper are effective. This formal notion can be easily implemented using a well-structured computer language such as C. I am in fact presently engaged in such a program.

Throughout this paper, we will be dealing with many different classes of objects constructed on (S, P), and $M(S, P)$ will act on all the objects in a given class. We will often say that a given construction is *combinatorial* meaning that it is invariant under the action of $M(S, P)$, and thus has a *combinatorial type*. The term *natural* will occasionally be used as a synonym for combinatorial. Combinatorial types can always be described by augmenting sextuples $\mathbf{E}=(E,\sigma,\tau,Q,\star,\#)$ with more data. For instance, if we wish to distinguish a certain arc of a CDP Δ, and to describe the combinatorial type of this arc, it can be done by augmenting the combinatorial type $\mathbf{E}$ of Δ with appropriate cycle of τ. We shall avoid continuing this formal discussion, however.

In lieu of this, it is often helpful to have a more pictorial description of combinatorial types. E and σ can be pictured as a finite set of points arranged on a circle, and can be represented by drawing chords of the circle, connecting corresponding points. For instance from the triangulation of a surface of genus 2 of Figure 4,

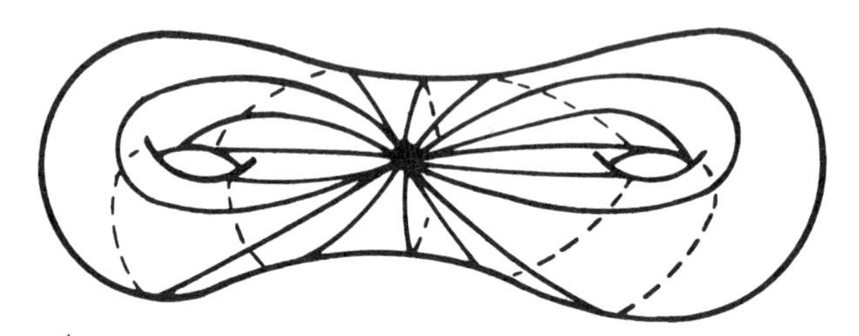

Figure 4.

one arrives at the picture of Figure 5:

Figure 5.

Q is clumsy to deal with, so examples are best taken from cell-divisions with no punctures in any n-gons. The distinguished prong can be represented by drawing the symbol $\star$ adjacent to the appropriate gap in the circle (see Figure 6).

Learning to manipulate these pictures seems to be a necessary prerequisite to doing any actual calculations by hand. The interested reader may refer to [Mosher, 1983], which is profusely illustrated with many such pictorial calculations.

0.4. Splittings. Let Δ be a CDP. The CDP Δ^L, obtained from Δ by a *Left splitting*, is defined by the following process. First, remove the arc h^L from Δ, joining the distinguished triangle with some k-gon to form a $k+1$-gon γ. Now insert a new arc, which will be h^O for Δ^L; this arc is the image of h^L under a $1/k+1$ counter-clockwise rotation of γ. This is made clear by a few examples:

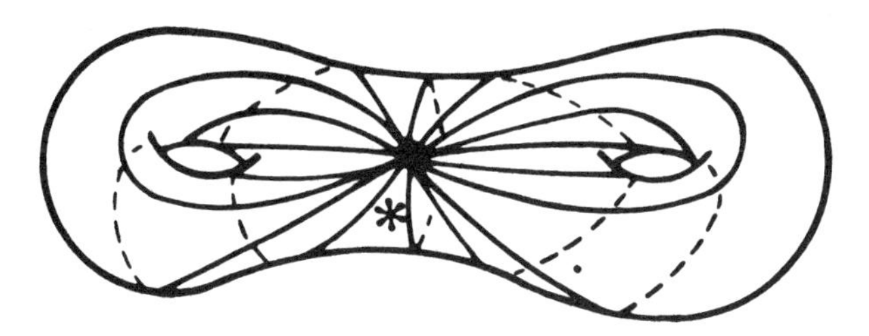

Figure 6.

Figure 7.

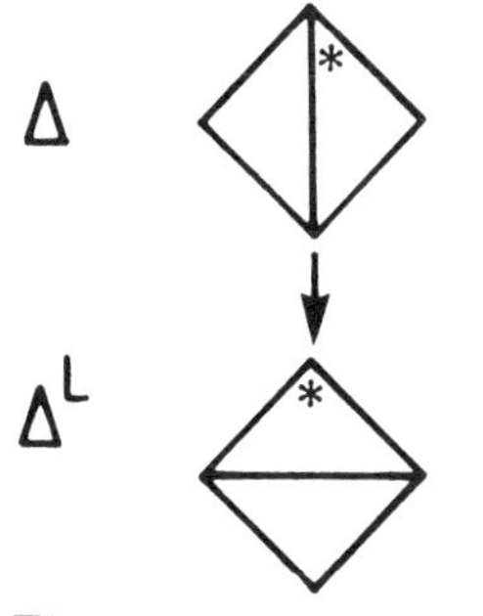

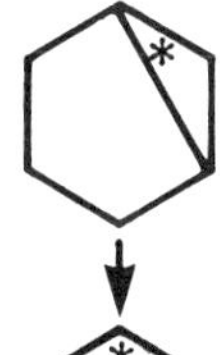

Figure 8.

The distinguished prong of Δ^L is the prong of Δ^L which "contains" the distinguished prong of Δ, as indicated in the above diagrams. If, in addition, Δ has a singularity prong located in some n-gon $\gamma^\#$, then several things can happen. If $\gamma^\#$ is not adjacent to h^L, then $\gamma^\#$ is also in the cell-division Δ^L; so the singularity prong of Δ is still a prong of Δ^L, and is thus chosen as the singularity prong of Δ^L. But if $\gamma^\#$ is adjacent to h^L, let γ' be the k-gon of Δ^L adjacent to h^O.

We set up a 1-1 correspondence between the prongs of $\gamma^\#$ and those of γ', which will tell how to transfer the singularity prong of Δ to one in Δ^L. Let $\#_0$ be the prong of $\gamma^\#$ adjacent to $\star$. Let $\#'_0$ be the prong of γ' adjacent to h^R. Now extend this correspondence to a cyclic correspondence. For clarity, here is the correspondence when $\gamma^\#$ is 1, 2, or 3-gon (which cannot actually occur if # satisfies the interior singularity condition).

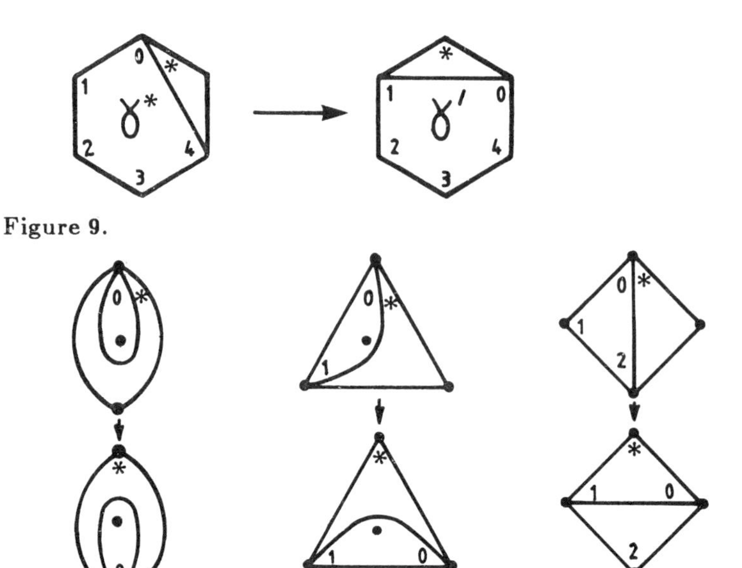

Figure 9.

Figure 10.

Note that, from the recurrence condition, $\#_0$ could never occur as the singularity prong.

There is also a CDP Δ^R, obtained from Δ by a *Right splitting*. Δ^R is defined by reversing the orientation of the definition for Δ^L.

Notation: We will often say that $\Delta \to \Delta'$ *is a splitting*, meaning that Δ' is obtained from Δ by a (Right or Left) splitting.

It is clear that a splitting is a combinatorial operation in the sense that if $\Delta \to \Delta'$ is a splitting, then so is $\Phi(\Delta) \to \Phi(\Delta')$, for every $\Phi \in M(S, P)$. Thus we can speak of the combinatorial type of a splitting. The combinatorial type of $\Delta \to \Delta'$ is often denoted $[\Delta] \to [\Delta]$.

0.5 Digraphs.

This section will describe a directed graph $\tilde{\mathcal{G}}(S, P)$ that $M(S, P)$ acts freely on. The quotient graph $\mathcal{G}(S, P)$ will also be discussed and analyzed, allowing us to make a precise statement of the classification theorem.

The vertices of $\tilde{\mathcal{G}}(S, P)$ are the CDPs of (S, P). The directed edges of $\tilde{\mathcal{G}}(S, P)$ are splittings.

Note that a CDP with interior vertex whose singularity prong does not satisfy the recurrence condition is not a vertex of $\tilde{\mathcal{G}}(S, P)$. It is possible that a splitting on a CDP that does satisfy the recurrence condition results in a CDP that does not. Thus, not every vertex of $\tilde{\mathcal{G}}(S,P)$ has both a Left and a Right outgoing edge. Similarly, not every vertex has both a Left and a Right incoming edge. It is evident, though, that if a CDP has a punctured vertex, then it represents a vertex of $\tilde{\mathcal{G}}(S,P)$ having two outgoing and two incoming edges.

$M(S,P)$ acts on $\tilde{\mathcal{G}}(S,P)$, and the following lemma can be used to prove that the action is free:

Lemma 0. *Let δ be a cell-division of (S,P), $\star$ a prong of δ, and Φ an element of $M(S,P)$. If $\Phi(\delta,\star) = (\delta,\star)$, then Φ is the identity.*

Proof. The proof is to show that Φ fixes the puncture set P, and that Φ is isotopic to the identity on the arcs of δ. From this it follows easily that Φ is isotopic to the identity on the n-gons of δ.

Evidently Φ acts as a permutation on the punctures, arcs and arc ends of δ. It suffices to show that this permutation is the identity. First note that the ends of h^L and h^R adjacent to the distinguished prong are fixed. Continuing around the vertex, it follows that each arc end is fixed. Thus, Φ can be isotoped to the identity on the arcs of δ.

Consider a puncture p, contained in some n-gon γ. Since Φ fixes arcs, Φ maps the interior of γ to itself. Since p is the only puncture in γ, $\Phi(p) = p$.

Lemma 0

The quotient of $\tilde{\mathcal{G}}(S,P)$ under the action of $M(S,P)$ is therefore a digraph $\mathcal{G}(S,P)$. The vertices of $\mathcal{G}(S,P)$ consist of the combinatorial types of CDPs on (S,P). The directed edges of $\mathcal{G}(S,P)$ are the combinatorial types of splittings.

The graph $\mathcal{G}(S,P)$ is not connected: it has many different components. We will label each component of $\mathcal{G}(S,P)$ by defining a combinatorial property of a CDP Δ called the *polygon type* of Δ.

The polygon type of Δ consists of two sequences of integers,

$$p_1, p_2, p_2, \ldots$$

$$q_3, q_4, q_5, \ldots$$

plus a sequence s which is either empty or has one integer. p_m for $m \geq 1$ is the number of punctured m-gons in Δ. q_n for $n \geq 3$ is the number of non-punctured n-gons in Δ, not counting the distinguished triangle. If Δ has an interior vertex, s is the size of the n-gon containing the singularity prong. Otherwise, s is empty. The polygon type will usually be denoted

$$(p_1, p_2, \ldots; q_3, q_4, \ldots; s)$$

The possibilities for the polygon type are limited by two equations, one coming from the standard Euler characteristic equation,

$$2g = 3/2 + \sum_{m \geq 1} (m/2 - 1)p_m + \sum_{n \geq 3} (n/2 - 1)q_n$$

and the other by counting punctures:

$$|p| = \sum_{m \geq 1} p_m + \begin{cases} 1 \text{ if } s \text{ is empty} \\ 0 \text{ otherwise} \end{cases}$$

Note that the polygon type of Δ is a combinatorial invariant. Also, the polygon type is invariant under a splitting. From this it follows that all the combinatorial CDPs in a given component of $\mathcal{G}(S, P)$ have the same polygon type. we can therefore speak of the polygon type of a component of $\mathcal{G}(S, P)$.

A *loop* is defined to be a closed, directed edge path in $\mathcal{G}(S, P)$, i.e.

$$[\Delta_0] \to [\Delta_1] \to [\Delta_2] \to \cdots \to [\Delta_n] = [\Delta_0].$$

A loop comes from a single component of $\mathcal{G}(S, P)$ and so has a well-defined polygon type.

It is evident that a loop represents a conjugacy class in $M(S, P)$; but it is worthwhile to make this correspondence explicit.

Let $\mathcal{L}$ be a loop in $\mathcal{G}(S, P)$, such as

$$[\Delta_0] \to [\Delta_1] \to \cdots \to [\Delta_n] = [\Delta_0]$$

Note that there is no natural starting point for a loop. Thus the above representation of $\mathcal{L}$ includes a choice of starting point $[\Delta_0]$ for $\mathcal{L}$.

Now choose a CDP Δ_0 representing $[\Delta_0]$. This determines a sequence of splittings

$$\Delta_0 \to \Delta_1 \to \cdots \to \Delta_n$$

whose combinatorial types give the loop $\mathcal{L}$. Since Δ_n has the same combinatorial type as Δ_0, there is a $\Phi \in M(S, P)$ such that $\Phi(\Delta_0) = \Delta_n$.

Given the choice of starting point for the loop (at $[\Delta_0]$), plus the choice of CDP representing $[\Delta_0]$, Φ is uniquely determined.

Suppose a different CDP Δ'_0 is chosen to represent $[\Delta_0]$. Then there is a $\Psi \in M(S,P)$ such that $\Psi(\Delta_0) = (\Delta'_0)$. It follows that $\Psi(\Delta_1) = \Delta'_1$ and, by induction, $\Psi(\Delta_n) = \Delta'_n$. Thus, if $\Phi'(\Delta'_0) = \Delta'_n$,

$$\Phi' = \Psi \circ \Phi \circ \Psi^{-1}$$

Suppose the loop is started at $[\Delta_i]$. The sequence of splittings

$$\Delta_0 \to \cdots \to \Delta_i \to \cdots \to \Delta_n$$

can be extended to

$$\Delta_0 \to \cdots \to \Delta_i \to \cdots \to \Delta_n \to \cdots \to \Delta_{n+1}$$

in such a way that

$$\Phi(\Delta_j) = \Delta_{n+j} \quad \text{for } j = 1, \ldots, i$$

Thus, the conjugacy class determined by $\mathcal{L}$ does not depend on where a starting point for $\mathcal{L}$ is chosen, or what representative CDP is chosen for that starting point.

Here is a useful definition: Given a loop $\mathcal{L}$, and $\Phi \in M(S,P)$ whose conjugacy class is represented by $\mathcal{L}$, let $\Delta_0 \to \cdots \to \Delta_M$ be a sequence of splittings such that $\Phi(\Delta_0) = \Delta_M$, and such that $\mathcal{L}$ is $[\Delta_0] \to \cdots \to [\Delta_M] = [\Delta_0]$. Then we say that the sequence $\Delta_0 \to \cdots \to \Delta_M$ is a *representative sequence of splittings* for $\mathcal{L}$ and Φ.

We are now ready for a statement of the main results. Let $\Phi \in M(S,P)$ be pseudo-Anosov. Let $[f]$ be the stable foliation of Φ. Φ has a *singularity type*, which is a pair of sequences

$$r_1, r_2, \ldots$$

$$s_3, s_4, \ldots$$

where r_m is the number of punctures at which f has m-pronged singularities, and s_n is the number of interior n-pronged singularities of f. This is well-defined because a canonical model for an arational measured foliation class is unique up to isotopy.

The singularity type of Φ is limited by equations similar to those for the polygon type of a CDP. From the Euler-Poincaré formula,

$$2g = 2 + \sum_{m \geq 1} (m/2 - 1) \cdot r_m + \sum_{n \geq 3} (n/2 - 1) \cdot s_n$$

and by counting punctures,

$$|p| = \sum_{m \geq 1} r_m$$

Theorem 1. *Let Φ be a pseudo-Anosov of singularity type $(r_1, r_2, \ldots; s_3, s_4, \ldots)$ and suppose that Φ fixes all its singularities and does not rotate the prongs at any singularity. Then for each $m \geq 1$ the conjugacy class of Φ is represented by exactly $m \cdot r_m$ distinct loops of polygon type $(p_1, p_2, \ldots; q_3, q_4, \ldots; \emptyset)$ where if $m = 1$, then*

$$p_1 = r_1 - 1 \qquad \text{1)}$$

$$p_i = r_i \qquad i \geq 2$$

$$q_j = s_j \qquad j \geq 3$$

and if $m \geq 2$, then

$$p_m = r_m - 1 \qquad \text{2)}$$

$$p_i = r_i \qquad i \neq m$$

$$q_{m+1} = s_{m+1} + 1$$

$$q_j = s_j \qquad j \neq m+1$$

and for each $n \geq 3$ the conjugacy class of Φ is represented by exactly $n \cdot s_n$ distinct loops of polygon type $(p_1, p_2, \ldots; q_3, q_4, \ldots; (n+1))$ where

$$p_i = r_i \qquad i \geq 1 \qquad \text{3)}$$

$$q_n = s_n - 1$$

$$q_{n+1} = s_{n+1} + 1$$

$$q_j = s_j \qquad j \neq n, n+1$$

In each case, loops are counted with multiplicity, where the multiplicities are determined as follows: Let N_f be the subgroup of $M(S, P)$ fixing $\pi[f]$ (N_f is a finite extension of the cyclic group generated by Φ, acting by permutation on the sectors at singularities of f). The multiplicities for each $m \geq 1$ and $n \geq 3$ are the sizes of the orbits of the action of N_f on sectors at m-pronged puncture singularities and n-pronged interior singularities.

Note: Suppose Φ satisfies the conditions of Theorem 1. Theorem 1 does not say that there are no other loops representing the conjugacy class of Φ. When Φ has the simplest possible singularity type, namely $r_m = 0$ and $q_n = 0$ for $m > 1$ and $n > 3$, then it is true that the only loops representing the conjugacy class of Φ are the loops mentioned in Theorem 1. But if Φ has a more complicated singularity type, there can be a plethora of loops representing Φ. For instance, suppose the singularity type of Φ is as above, with the exception of a single $q_n = 1$ for some $n > 3$. Then the number of loops representing Φ is greater than the number of loops from Theorem 1 by a *multiplicative factor*, which is itself greater than the number of different ways of adding a set of disjoint interior edges to an n-sided polygon.

Thus, the uniqueness half of the proof of Theorem 1 will involve a combinatorial characterization of the loops constructed in the existence half of the proof. Also, an effective procedure will be given to check whether a loop is pseudo-Anosov, and to check whether it has the combinatorial property which says that it is one of the loops of Theorem 1.

We can now state our classification theorem. If Φ is a pseudo-Anosov in $M(S,P)$ which fixes all its singularities and does not rotate the singular prongs at any singularity, we shall refer to the collection of loops (with multiplicities) described in Theorem 1 as the *pseudo-Anosov invariants* of Φ. In general, for an arbitrary pseudo-Anosov, let $n(\Phi)$(be the lowest positive integer such that $\Phi^{n(\Phi)}$ fixes all its singularities and singular prongs; this definition makes sense, since there are only finitely many singular prongs.

Theorem 2. *If Φ, $\Psi \in M(S,P)$ are pseudo-Anosov, and if $n(\Phi) = n(\Psi) = 1$, then Φ and Ψ are conjugate if and only if they have identical pseudo-Anosov invariants.*

More generally, if Φ, $\Psi \in M(S,P)$ are pseudo-Anosov, then Φ^m and Ψ^m are conjugate for some $m \geq 1$ if and only if there exists $n \geq 1$ such that Φ^n and Ψ^n fix all singularities and singular prongs, and Φ^n, Φ^n have identical pseudo-Anosov invariants.

For remarks on a more complete classification of pseudo-Anosov, refer to the end of Theorem 2.

Comment: note that the above theorems contain implicitly the statement that Φ, which is a pseudo-Anosov *mapping class*, acts in a well-defined manner on the sectors of its stable foliation. For a justification of this, see the Appendix.

1. CDPs and Measured Foliations

1.1 Parameterised Families of Measured Foliations

Given a CDP Δ, we will define $W(\Delta)$, the space of *transverse weights* on Δ; and also an embedding $L: W(\Delta) \to MF$. The definitions will depend on whether Δ has a puncture vertex. The definition of L will use the language of train tracks; for a reference, see [Thurston, 1979 Chapters 8-9], [Mosher, 1983], or [Harer-Penner, 1986].

Case: Δ has a puncture vertex.

Then $W(\Delta)$ is the space of functions μ which assign, to each arc h of Δ except h^O, a positive number $\mu(h)$.

Given $\mu \in W(\Delta)$, we seek to define a measured foliation f_μ, whose class $L(\mu)$ is well-defined.

Consider the set of arc ends of Δ. Let e^L and e^R be the ends of h^L and h^R adjacent to the distinguished prong. Starting with e^R, enumerate the ends in order around v:

$$e^R = e_1, \ldots, e_k = e^L$$

skipping over the ends of h^O.

Given $0 \neq \mu \in W(\Delta)$, μ can be considered as an assignment of a positive weight to each arc end e_i. Then one of the following happens:
Either:

1) there is a unique end e_{i_0} and a $t \in (0,1)$ such that

$$\sum_{i=1}^{i_0} \mu(e_i) + t\cdot\mu(e_{i_0}) = (1-t)\cdot\mu(e_{i_0}) + \sum_{i=i_0+1}^{k} \mu(e_i)$$

Or:

2) there is a unique end e_{i_0} $(i_0 < k)$ such that

$$\sum_{i=1}^{i_0} \mu(e_i) = \sum_{i=i_0+1}^{k} \mu(e_i)$$

We now construct a train track τ, depending on whether μ satisfies 1) or 2).

In case 1), split the arc containing e_{i_0} into two isotopic arcs. Move the puncture v into the region between the split ends of e_{i_0}. Throw away the arc h^O. Then smooth the intersection of all arc ends into a train track switch, by flattening the distinguished prong, and flattening the prong between the split ends of e_{i_0}. This gives the train track τ.

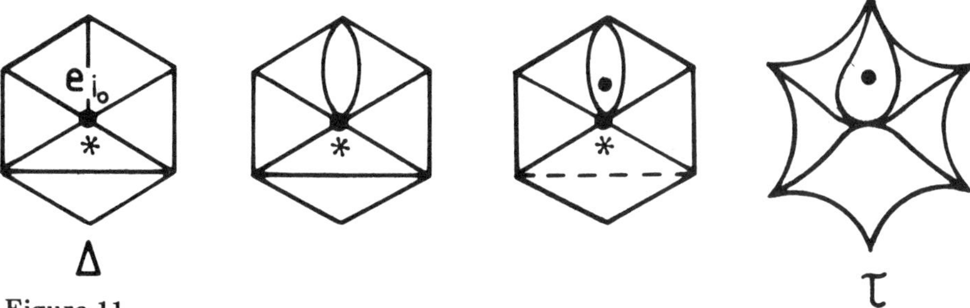

Figure 11.

μ can be made into a transverse weight on τ by assigning weights $t \cdot \mu(e_{i_0})$ and $(1-t) \cdot \mu(e_{i_0})$ to the two branches around v.

In case 2), the construction of τ is more straightforward. τ is constructed by moving the puncture v into the region between the ends (e_{i_0}, e_{i_0+1}); throwing away h^O; then forming a train track by flattening the distinguished prong and the prong (e_{i_0}, e_{i_0+1}). μ then becomes a transverse measure on τ with no fudging necessary.

The measured foliation f_μ is now constructed from τ and μ by Thurston's "highways" construction, which we give here for completeness.

For each branch b, replace b by a highway of width $\mu(b)$, foliated parallel to the traffic flow. The ends of all the highways are glued along an arc, and the gluing is done by isometries.

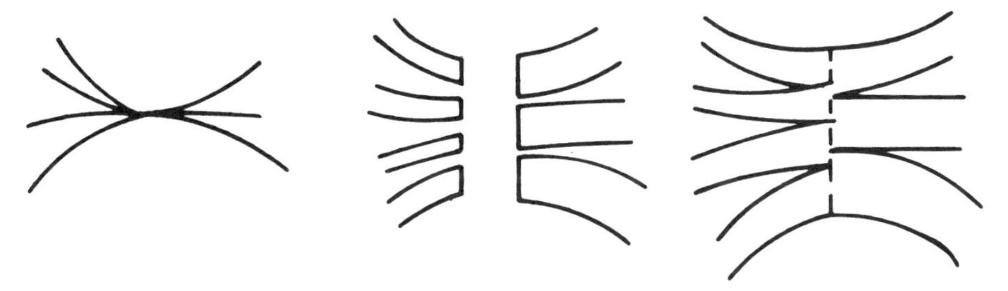

Figure 12.

Each complementary region is a disc with at most one puncture, and one or more cusps along the boundary. Choose a star-shaped spine for each region, having a single vertex (at the puncture if it exists), and edges going out to each cusp along the boundary. Now collapse each region to its spine. This defines the measured foliation f_μ. Note that, up to isotopy, the choice of a spine for each region is unique. Thus, f_μ is well-defined up to isotopy,

Figure 13.

so the class $L(\mu)$ of f_μ is well-defined.

For a proof of the following, the reader is referred to [Mosher, 1983], section 7.

Proposition. $L : W(\Delta) \to MF$ *is an embedding, preserving the homogeneous structure.*

Note that this does not follow directly from the corresponding theorem for train tracks, because the train track τ constructed above depends on the choice of μ. In the next case, this will not be a problem.

Note that there is a natural 1-1 correspondence from the n-gons of Δ (other than the distinguished triangle) onto the n-pronged singularities of f_μ, with the following exceptions:
in case 1) above, f_μ has an additional 1-pronged singularity at v;
in case 2) above, the $m+1$-gon γ containing the prong (e_{i_0}, e_{i_0+1}) will give rise to either

a) an m-pronged puncture singularity at v, if γ does not contain a puncture, or

b) a pair of puncture singularities (connected by an arc of f_μ), if γ contains a puncture. (In this case, it will not actually be possible to choose the spine to be star-shaped).

It is not necessary for f_μ to be a canonical model for an arational class, for there may be "accidental" joinings of singularities by leaf segments. However, If f_μ happens to be such a foliation, then the above discussion shows that the singularity type of f_μ is related to the polygon type of Δ as in Theorem 1. This discussion can be summarized in the following:

Proposition 3. *Suppose Δ is a CDP with a puncture vertex. Let $\mu \in W(\Delta)$. If f_μ is a canonical model for an arational measured foliation class, then the singularity type of f_μ and the polygon type of Δ are related as follows: if the singularity type of f_μ is $(r_i;s_j)$ then the polygon type of Δ is given by equation 1) of Theorem 1 in case 1) above, and by equation 2) of Theorem 1 in case 2a).*

Note. *case 2b) is ruled out by hypothesis, since in that case, f_μ is not arational.*

Case: Δ has an interior vertex.

Now suppose Δ is a CDP with an interior vertex. Thus, Δ has a singularity prong. As before, $W(\Delta)$ is a space of functions μ assigning positive weights to each arc of Δ except h^0. But now μ must satisfy a "switching condition": the distinguished prong and the singularity prong divide the arc ends of Δ into two subsets $e_1, \ldots, e_{i_0}$ and $e_{i_0+1}, \ldots, e_k$. The ends of h^0 are excluded from these subsets; but otherwise, both ends of each arc occur in the union of these subsets. Then μ must satisfy the equation

$$\sum_{i=1}^{i_0} \mu(e_i) = \sum_{i=i_0+1}^{k} \mu(e_i) \tag{*}$$

It is easy to see that the recurrence condition on Δ is equivalent to the existence of positive solutions to this equation. For the terms of the equation occur in pairs, and the two members of each pair are equal. So a positive solution exists if and only if, when one side of the equation contains both members of the same pair, the other side contains both members of some other pair.

Note: When the recurrence condition is satisfied, $W(\Delta)$ is defined to be the set of all positive functions on arcs(Δ) $-$ h^0 satisfying (*). This is a linear subspace of all positive functions on arcs(Δ)$-h^0$. What is the codimension of this subspace? It is tempting to say one, though this is not always true. For instance, equation (*) may be an identity, which occurs precisely when the distinguished prong and the singularity prong separate the two ends of each arc. In such a case, $W(\Delta)$ consists of all positive functions. Such an example is provided by the following picture, which is the pictorial representation of a combinatorial CDP of polygon type (;0,0,0,0,1;(7)) on a closed surface of genus 2. Note that for this example, choosing the singularity prong at any other place would result in a combinatorial CDP that did not satisfy the recurrence condition.

Figure 14.

Thus, when Δ has an interior vertex and satisfies the recurrence condition, $W(\Delta)$ is either all possible positive functions, or else a codimension 1 subspace of them.

The construction of f_μ proceeds exactly as in the previous case, except that the train track τ does not depend on μ. τ is constructed by flattening the distinguished prong and the singularity prong.

Given $\mu \in E(\Delta)$, the correspondence from n-gons of Δ (other than the distinguished triangle) to n-pronged singularities of f_μ is well-behaved, having the single exception that the (non-punctured) n-gon containing the singularity prong corresponds to an interior $n-1$ pronged singularity (where $n - 1 \geq 3$ is guaranteed by the interior singularity condition). It will be convenient to assume, as we can by isotoping f_μ, that this singularity coincides with the vertex of Δ.

The reason for the interior singularity condition is now clear. If the singularity prong were located in a punctured n-gon γ, then for every $\mu \in W(\Delta)$, the singularity of f_μ corresponding to γ would be at a puncture rather than in the interior. And if the singularity prong were located in a non-punctured 3-gon γ, then there would be no singularity at all corresponding to γ.

We can now state the analogous version of Proposition 3:

Proposition 4. *Suppose Δ is a CDP with interior vertex. Let $\mu \in W(\Delta)$. If f_μ is a canonical model for an arational measured foliation class, and if the singularity type of f_μ is $(r_i;s_j)$, then the polygon type of Δ is given by equation 3) of Theorem 1.*

As before, $L(\mu)$ is defined to be the class of f_μ. The fact that $L: W(\Delta) \to MF$ is an embedding follows directly from the analogous fact for $W(\tau)$, since in this case $W(\Delta)$ and $W(\tau)$ are the same space.

1.2 Transversals and rectangles. In this section the process of the previous section is reversed: starting from a measured foliation f, we will look for CDPs Δ such that $[f] \in L(W(\Delta))$.

Let s be a singularity of f. Let α be a transversal to f with one endpoint at s, called the *fixed endpoint* of α. The opposite end of α is assumed to lie at a regular point of f, called the *free endpoint* of α. α is said to be *based at* s. α is called *generic* if the leaf through the free endpoint of α intersects $\text{int}(\alpha)$ in both directions before intersecting the singular set of f. Otherwise, α is *non-generic*.

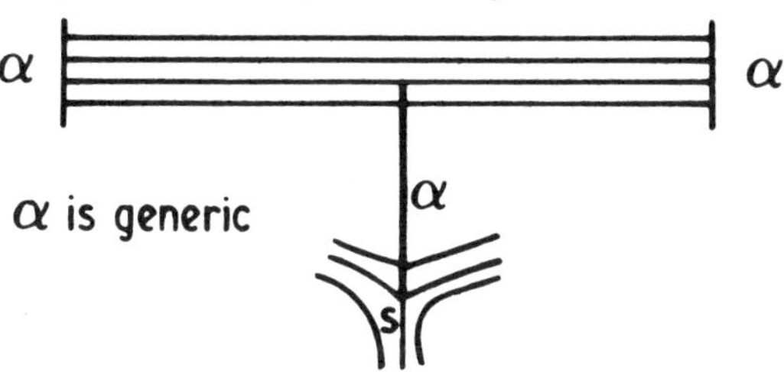

Figure 15. *For convenience, the picture is drawn in the universal cover $\tilde{S}$ of S, with all pertinent features lifted to $\tilde{S}$, such as the singularity s, the foliation f, the transversal α, etc. Many pictures hereafter will be drawn in $\tilde{S}$. Consequently, there will be, for instance, many different points labeled s, or transversals labeled α. Thus, the reader should be cautioned to view these different objects as being identified by some covering transformation.*

The definition of an f, α rectangle is adapted from [Fathi-Laudenbach-Poenaru, 1979 exposé 9]. An f, α *rectangle* is the image of an immersion with boundary folds $\mathbf{I} \times [0,t] \rightarrow S$ taking interior horizontal segments to non-singular leaf segments; taking the horizontal boundaries to finite unions of leaf segments and singularities; and taking vertical boundaries isometrically into α. Horizontal boundaries are required to either contain a singularity of f in the interior of the boundary, or to contain the free end of α. The difference with the definition in [Fathi-Laudenbach-Poenaru, 1979] is that a vertical side is allowed to fold around a possible 1-pronged singularity at the fixed end of α; similarly, horizontal sides are allowed to fold around other 1-pronged singularities.

With this definition, lemma 4 of exposé 9 of [Fathi-Laudenbach-Poenaru, 1979] still holds (see Figure 17).

Fact. *There is a unique collection of f, α rectangles such that the two sides of α are covered by vertical sides of rectangles. These rectangles have disjoint interiors.*

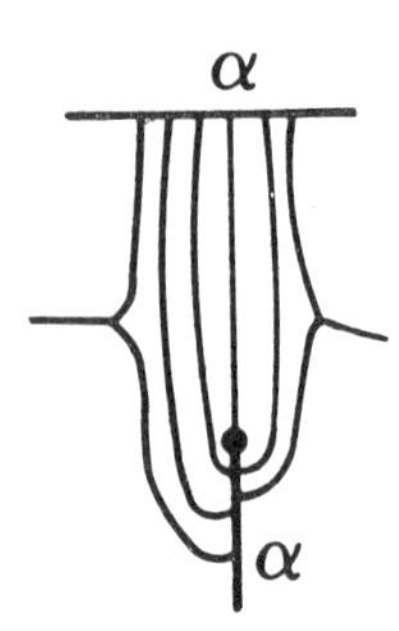

Figure 16.

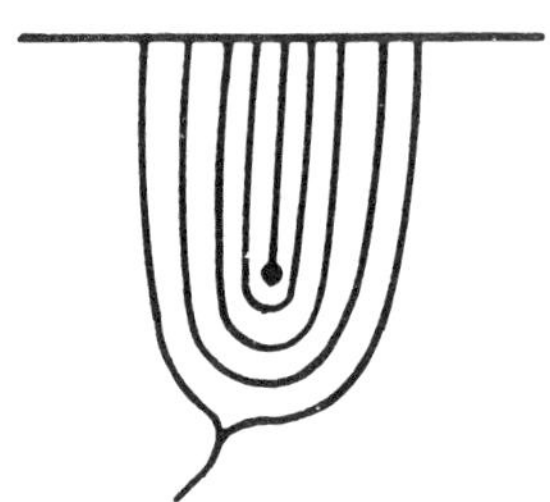

Figure 17.

Let $H_1, \ldots, H_n$ be the system of f, α rectangles. For each rectangle H_i, choose a single interior leaf $\mathbf{l}_i$, with ends on int(α). Now collapse α down to s, by homotoping the identity, through homeomorphisms fixed outside a neighbourhood of α, to a quotient map which identifies α to s, and is a homeomorphism away from α. Under this collapsing, the family of leaves $\mathbf{l}_i$ goes to a well-defined family of arcs h_i, each with endpoint at s.

Figure 18.

This family of arcs is not, in general, well-behaved. But when the arcs form a cell-division of (S, P), then we say that f and α *fill up* (S, P). In this case, we will assume that all possible Whitehead moves have been performed which are supported in the complement of int(α). In particular, this implies that when α is non-generic, the leaf intersecting the free end of α hits a singularity in only one direction; the other direction must hit int(α).

Note that when f and α fill up (S, P), the union of the f,α rectangles is all of S. The converse is not necessarily true, however.

When f and α fill up (S, P), and when α is generic, there is a natural choice for a distinguished prong. The distinguished triangle is the triangle whose three sides each correspond to a rectangle containing the free end of α on a horizontal side. Two of these rectangles contain the free end of α at a corner, and the distinguished prong is located between the corresponding arc ends.

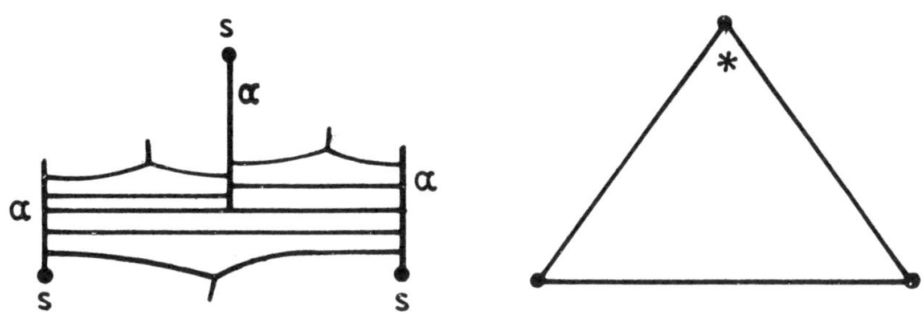

Figure 19.

If s is an interior k-pronged singularity, then the cell-division has a $k+1$-gon corresponding to s. The singularity prong is then located between the two ends of this $k+1$-gon corresponding to the two transverse

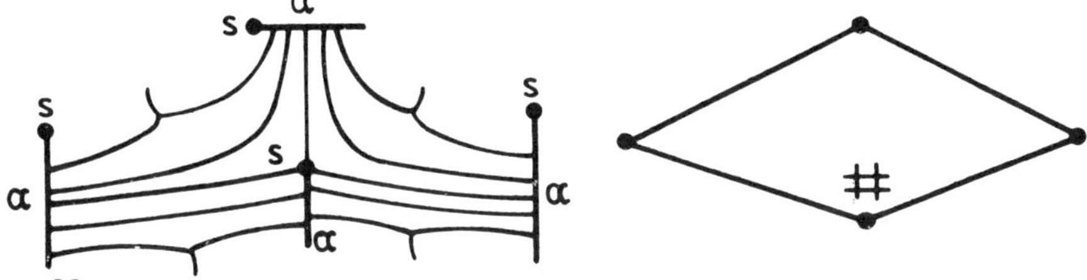

Figure 20.

sides of rectangles containing the fixed end of α at a corner. When α is generic, this defines a CDP which is denoted $\Delta(f, \alpha)$.

When f and α fill up (S, P) and α is non-generic, then every sub transversal α_- obtained by cutting off a sufficiently small amount of the free end of α is generic. In this case, $\Delta(f, \alpha)$ is by convention defined to be $\Delta(f, \alpha_-)$

The following key fact will be used at a crucial point of Proposition 7. It is an extension of lemma 5 of exposé 9 of [Fathi-Laudenbach-Poenaru, 1979].

Arationality Lemma. *When f is arational, then for any transversal α, f and α fill up (S, P).*

Proof. Suppose f and α do not fill up (S, P). Thus, the family of arcs $[h_i]$ does not form a cell-division; so there exists a complementary region R which contains a non-trivial singular cycle of S rel P. R corresponds, under the collapsing of α, to a complementary region R' of α union the family of leaves ℓ_i. R' is homotopy equivalent rel P to $\Sigma = \bigcup$ {singular leaf segments in R'}. Therefore Σ contains a non-trivial cycle of S rel P, so f is not arational.

Arationality Lemma

The following fact shows how the process of finding $\Delta(f,\alpha)$ is the reverse of the process which finds $L(\mu)$.

Proposition 5. *Let Δ be a CDP with vertex v, and let $\mu \in W(\Delta)$. Let f_μ be the measured foliation constructed from μ in Section 1.1. Then there exists a non-generic transversal α, based at v, such that f_μ and α fill up (S, P) and $\Delta(f_\mu, \alpha) = \Delta$.*

Suppose α' is another non-generic transverse to f_μ such that $\Delta(f_\mu,\alpha') = \Delta$. Then there exist a non-generic transversal α'' based at v, and a $\phi \in$ Homeo$^+(S, P)$ such that α'' is isotopic rel v along leaves to α', $\phi(f_\mu) = f_\mu$, $\phi(\alpha) = \alpha''$, and ϕ is in the mapping class of the identity.

Conversely, let f be a measured foliation, α a non-generic transversal to f based at some singularity s, and suppose that f and α fill up (S, P). Let $\Delta = \Delta(f, \alpha)$. Then there exists a transverse measure $\mu \in W(\Delta)$ such that $f_\mu = f$.

Comment: The Uniqueness statement given above for α and α' can be strengthened. The strongest possible statement is that α and α' are based at the same singularity, and are isotopic along leaves in f. We shall not need such a complete uniqueness statement, but we will formulate and use a related version in the proof of the multiplicity statement of Theorem 1 (Section 3.4).

Proof. The first half follows by a careful observation of the highways construction. Let τ be the train track determined by Δ and μ, as in Section 1.1.

f_μ is obtained by replacing each branch of τ with a foliated rectangle, gluing the rectangles along their transverse sides, and then collapsing the complementary regions. After collapsing, the transverse sides of the rectangles all line up to form a transversal α, with one endpoint at the vertex v. Evidently this α satisfies the conditions of the Proposition.

Now suppose that α' is another non-generic transversal to f_μ such that $\Delta(f_\mu,\alpha) = \Delta(f_\mu,\alpha') = \Delta$. Let $\mathbf{l}$ be the unique leaf segment connecting a singularity s to the free end of α; define $\mathbf{l}'$ and s' similarly with respect to α'. Evidently $\mathbf{l}$ intersects α from either the Left or the Right; similarly for $\mathbf{l}'$ and α'.

If $\mathbf{l}$, $\mathbf{l}'$ intersect α, α' from the same side, we choose $\alpha'' = \alpha'$. Otherwise, isotope α' along leaves so that the free end of α' passes the singularity s'; let α'' be the resulting transversal. Evidently, $\Delta(f_\mu, \alpha'') = \Delta(f_\mu, \alpha')$. It now follows, from the fact that $\Delta(f_\mu, \alpha) = \Delta(f_\mu, \alpha'')$, that the combinatorial pattern of gluing together the f, α rectangles is identical to the combinatorial pattern of gluing together the f, α'' rectangles. Also, since μ is well-defined, it follows that the f_μ width of a given f_μ, α rectangle is equal to the f_μ width of the corresponding f_μ, α'' rectangle. Therefore ϕ can be defined one rectangle at a time, in such a way that the transverse measure is preserved in each rectangle. ϕ is automatically well-defined along vertical sides of rectangles, but in order to match up ϕ along horizontal sides, ϕ may have to be changed by isotopy along leaves in each rectangle. The details are left to the reader. Evidently, $\phi(\Delta) = \Delta$, so by Lemma 0, ϕ is in the mapping class of the identity.

To prove the converse, consider the construction of Δ by finding the collection of f, α rectangles. Each rectangle corresponds to an arc of Δ, so the μ measure of an arc is just the f measure of a vertical side of the corresponding rectangle. The reader can check that, if s is an interior singularity, μ satisfies the switching condition.

Proposition 5

2. The Existence Half of Theorem 1

Let f be a measured foliation, α a non-generic transversal based at a singularity s of f, and suppose that f and α fill up (S, P). Let $\Delta = \Delta(f, \alpha)$. Let $\mathbf{l}$ be a singular leaf whose first intersection with $\mathrm{int}(\alpha)$ lies closest to the free endpoint of α, and suppose that $\mathbf{l}$ lies on the boundary of some f, α rectangle (this excludes the case where $\mathbf{l}$ comes from a 1-pronged singularity at the vertex of Δ). There are a few other cases to consider.

One case is that there is another singular leaf $\mathbf{l}'$ satisfying the same conditions as $\mathbf{l}$. Necessarily, $\mathbf{l}$ and $\mathbf{l}'$ hit α from opposite sides. In this case, we say that f *degenerates* at α.

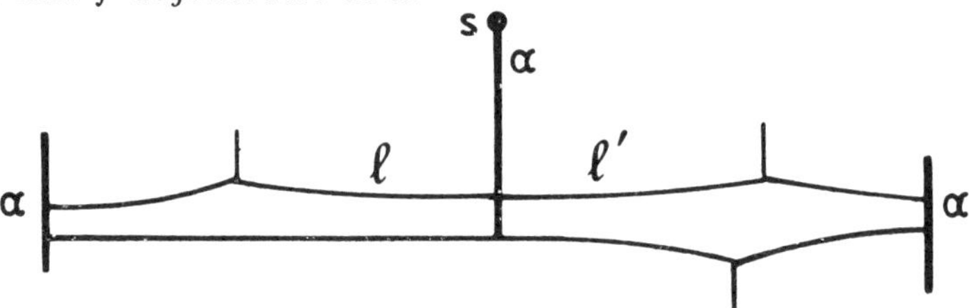

Figure 21.

Otherwise, $\mathbf{l}$ is uniquely determined, and it hits α either from the Left or the Right (where by convention the fixed end of α is always pointing up). Let α' be the subtransversal of α, based at s, whose free endpoint coincides precisely with the first intersection of $\mathbf{l}$ with $\text{int}(\alpha)$. α' is called the *first non-generic subtransversal of* α.

Proposition 6. *Suppose f does not degenerate at α, and let $\mathbf{l}$, α' be as above. Then f, α' fill up (S, P) and $\Delta(f, \alpha')$ is obtained from $\Delta(f, \alpha)$ by a Left or a Right splitting, according as $\mathbf{l}$ hits int(α) from the Left or Right.*

Proof. We shall consider only the case of the Right splitting.

Let s' be the singularity of f from which $\mathbf{l}$ emerges. The proof will be given in cases, each depending on the nature of the singularity s'. The proof for each case should be clear from the picture given. Recall that α_- is α with a small bit cut off of the free end, and $\Delta(f, \alpha)$ is defined to be $\Delta(f, \alpha_-)$; similarly for α'.

Case 1: $s' \neq s$. The picture is given only for the cases when $n(s')=3$, 2, or 1 ($n(s')=$ the number of singular leaves at s'). For $n(s') > 3$, the proof is similar. (See Figure 22)

Note: It is possible that the singular leaf $\mathbf{l}$, after crossing int(α), runs into a 1-pronged singularity at s. The pictures will all be slightly different in this case; for an example, the reader is referred to the proof of theorem 13.

Case 2: $s' = s$. The proof is also given only for the cases $n(s) = 1, 2$, or 3. The situation is more complicated now, however. Let H be the rectangle which contains the free end of α at an end of one horizontal side, and contains $\mathbf{l}$ in the other horizontal side. The latter horizontal side therefore contains the singularity s, so the interior of H contains transversals based at s, representing some sector at s. There will be several sub-cases, depending on how this sector is related to the sector at s which contains α.

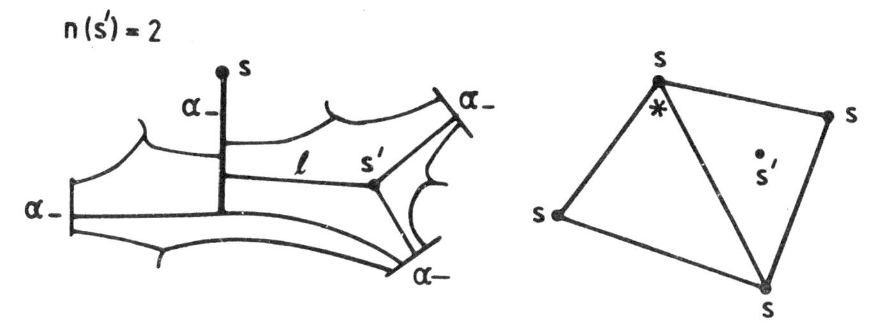

Figure 22. *$n(s')$ = 3. If s' is not a puncture singularity, the puncture in this picture can be ignored.*

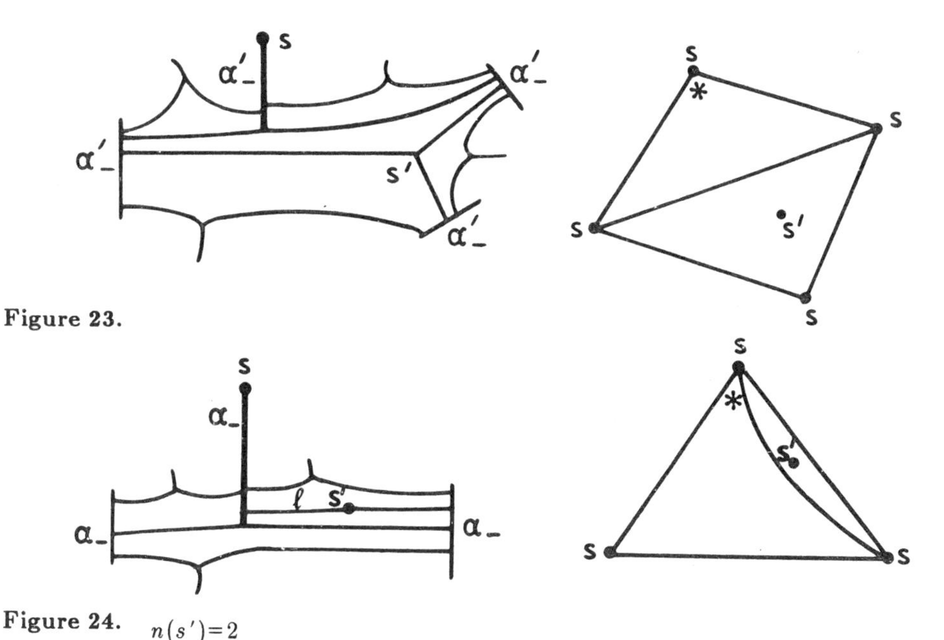

Figure 23.

Figure 24. $n(s')=2$

$n(s)$ = 1. This case does not occur, for the singular leaf segment l going from s to int(α) is in the interior of a rectangle which folds around the fixed end of α.

$n(s)$ = 2. There are two subcases:

a) H represents the sector at s containing α.

b) H represents the sector at s opposite to the sector containing α.

$n(s)$=3. In this case, it is possible that s is an interior singularity. If so, the pictures show that the singularity prong is passed on properly. If s is a

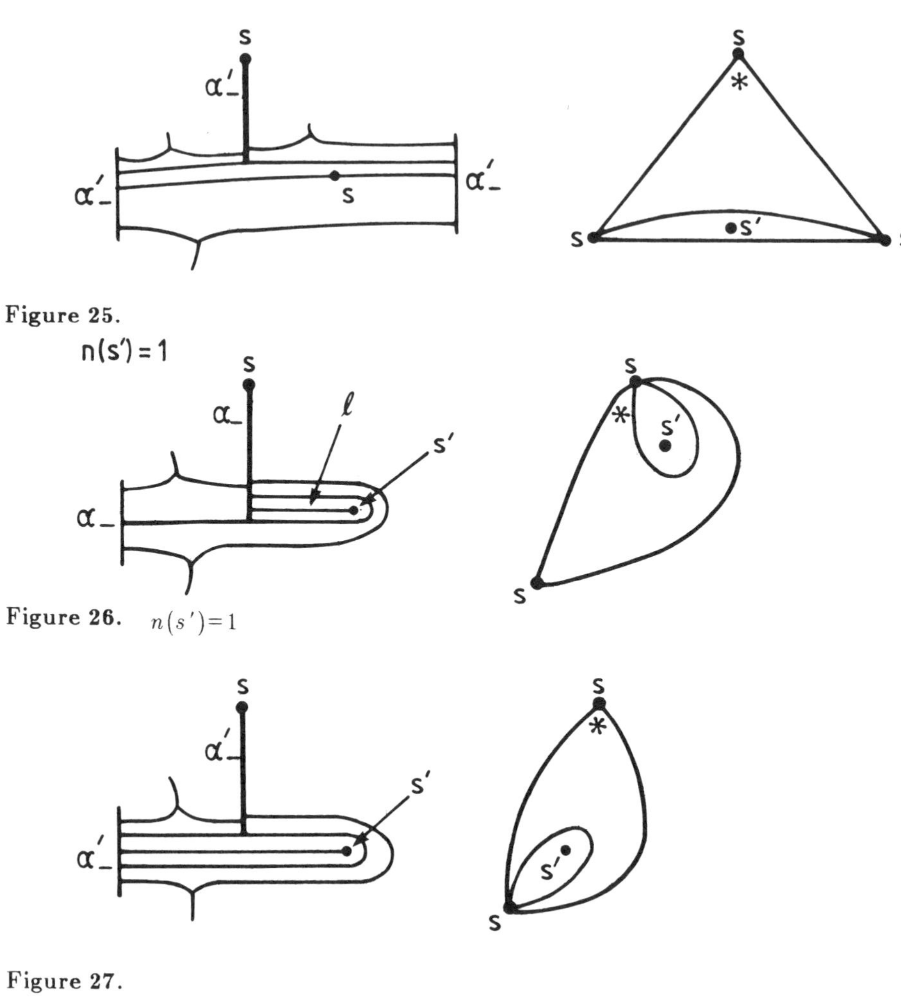

Figure 25.

Figure 26. $n(s')=1$

Figure 27.

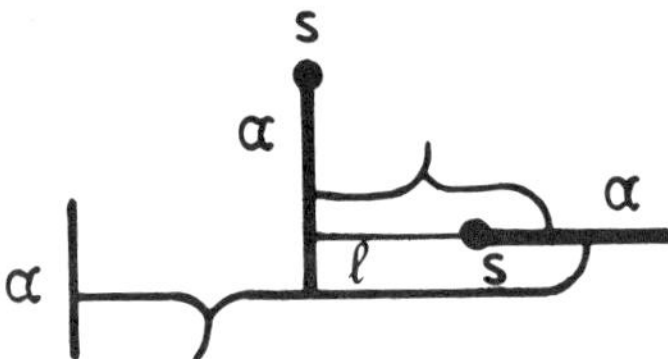

Figure 28.

puncture singularity, the singularity prong in the pictures can be ignored. There are three subcases:

a) H represents the sector at s containing α.

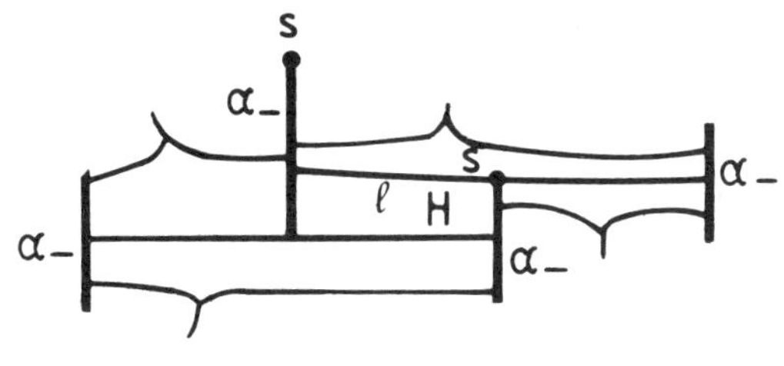

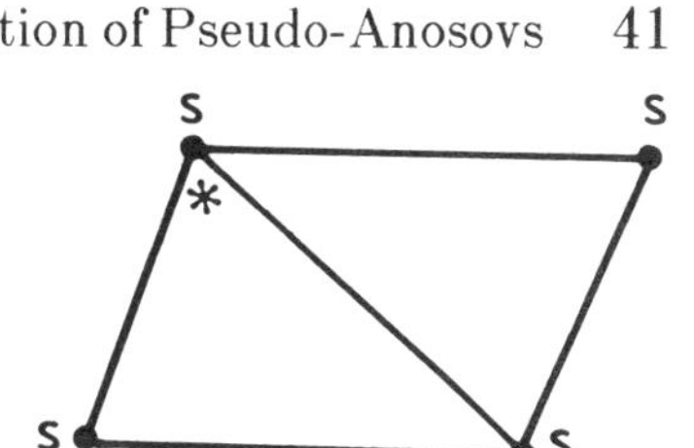

Figure 29.

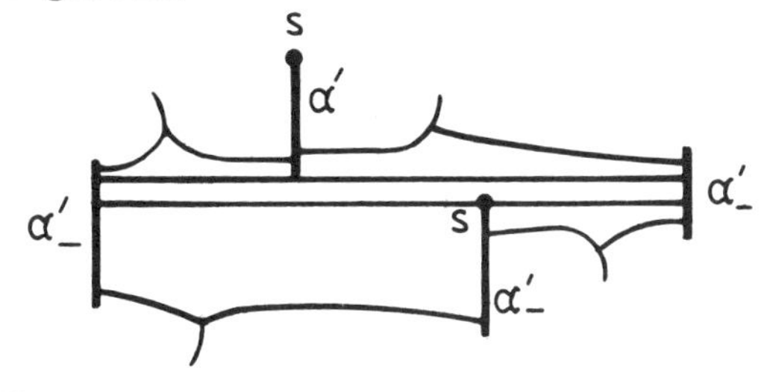

Figure 30.

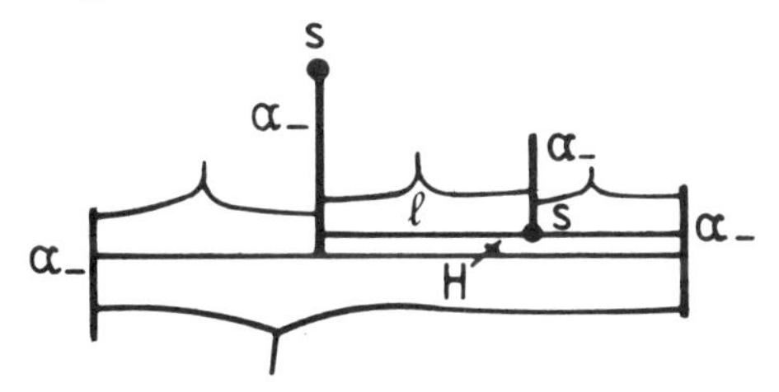

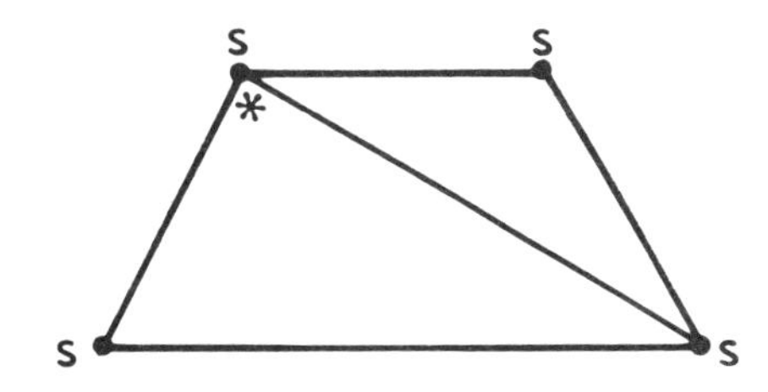

Figure 31.

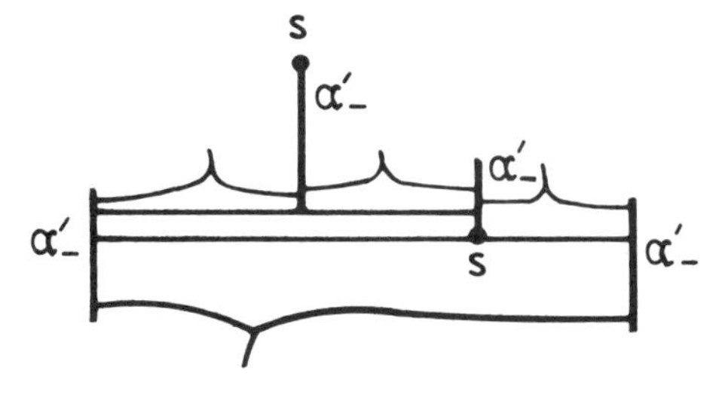

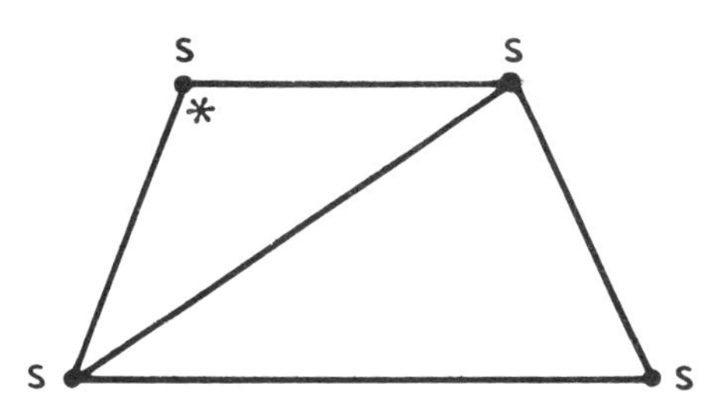

Figure 32.

b) H represents the sector at s which is 1/3 around s in the counter-clockwise direction from α.

c) H represents the sector at s which is 2/3 around s in the counter-clockwise direction from α.

Proposition 6

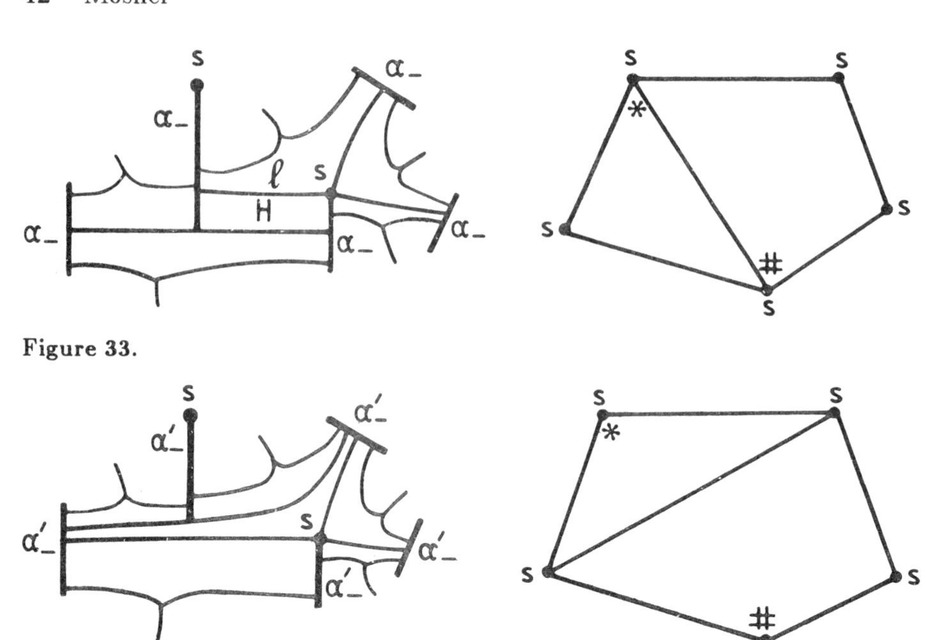

Figure 33.

Figure 34.

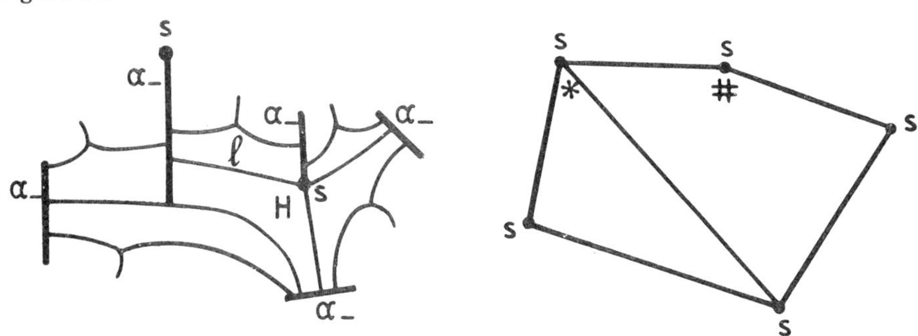

Figure 35.

These pictures justify the use of the term "splitting": after the transversal has been shortened, l is free to travel on to its next intersection with int(α), splitting one of the f, α rectangles along the way.

Now let α_0 be a non-generic transversal of f, based at a singularity of f. α_i is defined by induction: if f does not degenerate at α_i, define α_{i+1} to be the first non-generic sub-transversal of α_i.

Suppose that f is arational. Suppose also that f is a canonical model in its class. In this case, it follows that α_i is defined for all i, because if f degenerates at α_i, clearly a Whitehead move can be performed on f.

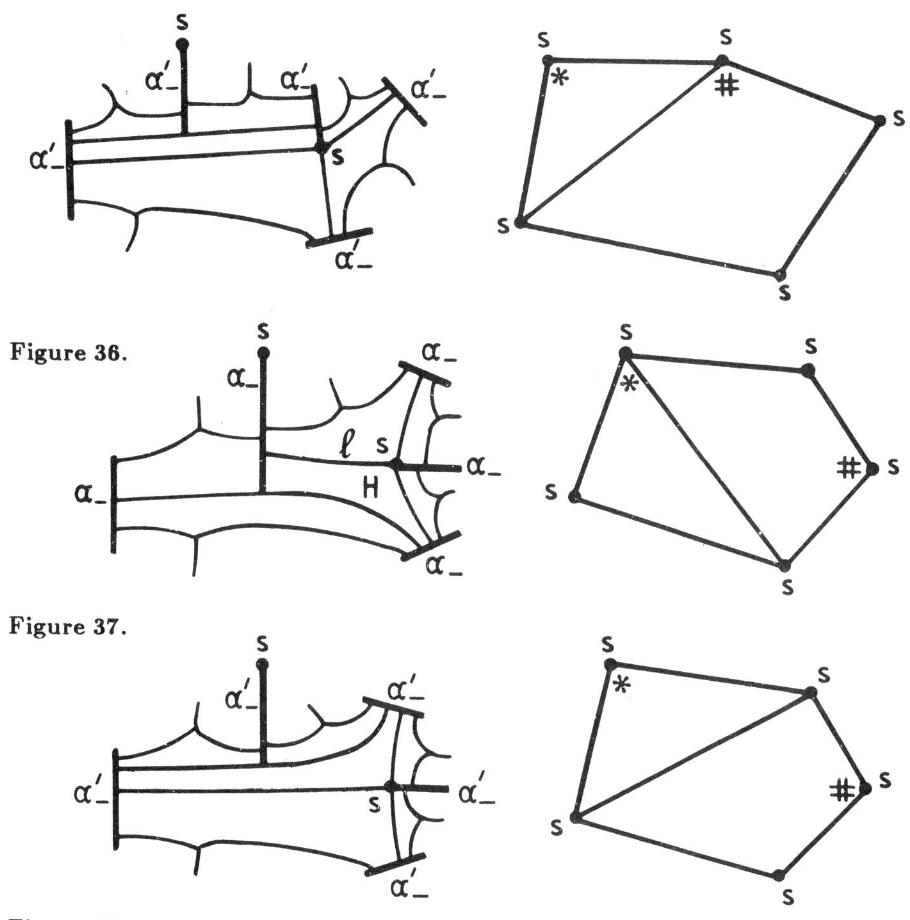

Figure 36.

Figure 37.

Figure 38.

Proposition 7. *If f is a canonical model for an arational measured foliation class, and α_0 is a transversal to f based at a singularity s of f, then*

$$\bigcap_i \alpha_i = s$$

Proof. The only way this could fail is if the sequence of free endpoints of α_i limited on some point $x \neq s$ of α_0.

Let α' be the subtransversal of α_0 ending at x. By the arationality lemma, since f is arational f and α' fill up (S, P). In particular, this implies that there is an f, α' rectangle H such that the free end x of α' is contained in the interior of a horizontal side of H.

Every subtransversal α'' of α_0 slightly larger than α' has its free end contained in int(H), and the leaf $\mathbf{l}$ containing this free end intersects int(α'')

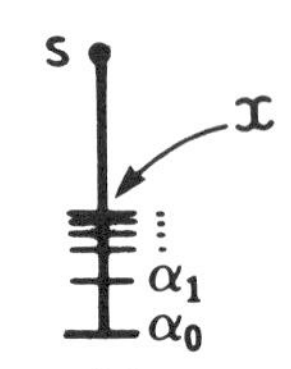

Figure 39.

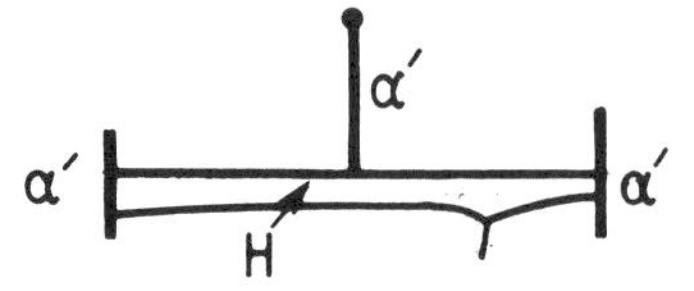

Figure 40.

(which is most of the vertical boundary of H) in both directions before hitting a singularity. Thus, α'' is generic.

However, for sufficiently large n, α_n is only slightly larger than α', but α_n is non-generic for every n. This contradiction proves the Proposition.

Proposition 7

To see more clearly what is going on, suppose f is not arational – suppose, for instance, that f has precisely two minimal closed subsets, bounded by a closed leaf cycle c; and suppose α_0 intersects c exactly once in $\text{int}(\alpha_0)$. Then $\alpha' = \bigcap_i \alpha_i$ will not be equal to s. In fact, the free end of α' will be the unique point in $c \cap \text{int}\,(\alpha_0)$. Note also that in this case, f and α' do not fill up (S, P).

Now let f, α_0 be as in Proposition 7, and consider the sequence of CDPs

$$\Delta_i = \Delta(f, \alpha_i) \qquad i = 0, 1, 2, \cdots$$

This sequence has the property that Δ_{i+1} is obtained from Δ_i by a Left or Right splitting. Such a sequence will be called a *string* of CDPs, starting at Δ_0, or an *R-L string*. In particular, the string defined above will be called the string *associated to f and α_0*. A string of CDPs will often be denoted by

$$\Delta_0 \to \Delta_1 \to \cdots \to \Delta_i \to \cdots$$

Such a string is completely determined by Δ_0 and the sequence of Lefts and Rights.

Now suppose that α_0 and α_0' are two non-generic transversals to f, based at the same singularity s. We say that α_0 and α_0' are in the same *sector* of f at s if, with respect to a coordinate chart for the foliation at s, α_0 and α_0' leave s between the same pair of adjacent singular leaves at s.

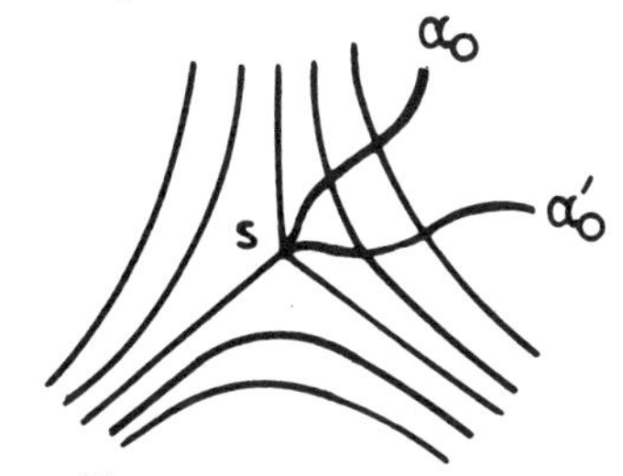

Figure 41.

If α_0 and α_0' are in the same sector at s, then α_0 and α_0' have a "common" subtransversal, that is, there is a transversal α'' based at s in the same sector as α_0 and α_0', such that α'' is isotopic along leaves to a subtransversal of α_0, and also to a subtransversal of α_0'. Thus we can isotope α_0 and α_0' along leaves so that α_0 and α_0' actually contain α'' as a common subtransversal. It is easy to see that isotopy along leaves does not affect the string associated to a transversal. It therefore follows from Proposition 7 that if f is a canonical model for an arational measured foliation class, there exist integers M, $N \geq 0$ such that α_M equals α'_N. In fact, for every integer $i \geq 0$, α_{M+i} is equal to α'_{N+i}. We have proven:

Stability Lemma 8. *Let f, α, α' be as shown above, and let Δ_i, Δ'_i be the strings of CDPs associated to α, α'. Then Δ_i and Δ'_i are* stably equivalent. *That is, there exist integers M, $N \geq 0$ such that*

$$\Delta_{M+i} = \Delta'_{N+i} \quad \textit{for every } i \geq 0.$$

The proof of the existence half of Theorem 1 now rests on the following:

Proposition 9. *Let ϕ be a pseudo-Anosov homeomorphism of (S, P) in the mapping class Φ. Let f be the stable foliation of ϕ, and suppose that ϕ fixes some singularity s of f, and does not rotate the sectors of s. Then to each sector of s there is a naturally associated loop representing the conjugacy class of Φ. If the singularity type of f is $(r_i;s_j)$, then the polygon type of the loop is given by equation 1), 2), or 3) of Theorem 1, according as the singularity s is a 1-pronged puncture singularity, an m-pronged puncture singularity $(m \geq 2)$, or an n-pronged interior singularity $(n \geq 3)$.*

("Natural" in this case means invariant under the action of any homeomorphism of (S, P).)

Proof. Let α_0 be a non-generic transversal to f based at s. Let

$$\Delta_0 \to \cdots \to \Delta_i \to \cdots$$

be the R-L string associated to f and α_0. Let $\alpha_0' = \phi(\alpha_0)$, and let

$$\Delta_0' \to \cdots \to \Delta_i' \to \cdots$$

be the R-L string associated to f and α_0'. Evidently,

$$\Delta_i' = \Phi(\Delta_i)$$

Moreover, since ϕ does not rotate the sectors at s, α_0 and α_0' are in the same sector at s. Thus, by the stability lemma, the two R-L strings are stably equivalent, so there exist integers M, $N \geq 0$ such that for all $i \geq 0$,

$$\Delta_{M+i} = \Delta'_{N+i}$$

But in fact a little more is true: if α'' is a common subtransversal, by the definition of a stable foliation, ϕ takes α'' to a proper subtransversal of itself. Therefore the shear number $M - N$ is positive, so for $j \geq N$,

$$\Phi(\Delta_j) = \Delta_j' = \Delta_{j+(M-N)}$$

This says that Δ_j and $\Delta_{j+(M-N)}$ have the same combinatorial type, where the mapping class taking Δ_j to $\Delta_{j+(M-N)}$ is Φ. Therefore,

$$[\Delta_n] \to [\Delta_{N+1}] \to \cdots \to [\Delta_M] = [\Delta_N]$$

is an R-L loop representing the conjugacy class of Φ.

The resulting loop does not depend on the original choice of α_0, for if a different transversal β_0 is chosen in the same sector, then by the stability lemma, there is a common subtransversal α'' to α_0, β_0, $\phi(\alpha_0)$, and $\phi(\beta_0)$.

This correspondence is natural, for if ψ is any homeomorphism of (S, P), then $\psi(\alpha_0)$ is a transversal to $\psi(f)$, in the sector of $\psi(f)$ corresponding to the sector of f containing α_0. If

$$\Delta_0 \to \cdots \to \Delta_i \to \cdots$$

is the string associated to f and α_0, then

$$\psi(\Delta_0) \to \cdots \to \psi(\Delta_i) \to \cdots$$

is the string associated to $\psi(f)$ and $\psi(\alpha_0)$, and it follows that the two loops are equal.

The statement about singularity types follows directly from Proposition 5, and Propositions 3 and 4.

Proposition 9

Let Φ be a pseudo-Anosov mapping class, ϕ a pseudo-Anosov homeomorphism representing Φ, and suppose that ϕ fixes all sectors of the stable foliation. Since ϕ is unique up to topological conjugacy, the property of fixing sectors is independent of the choice of ϕ; and if this property is satisfied, the collection of loops constructed in the proof of Proposition 9 is also independent of the choice of ϕ. Given such a loop $\mathcal{L}$, we shall say that $\mathcal{L}$ is a loop *associated to a sector of* ϕ.

We now prove the existence half of Theorem 1. Suppose Φ is a pseudo-Anosov mapping class which fixes all singularities of its stable foliation f_s, without rotating any of them. Let $(s_m; r_n)$ be the singularity type of f_s. Consider the collection of loops associated to the sectors of Φ. There are a total of $m \cdot r_m$ sectors located at m-pronged puncture singularities, and $n \cdot s_n$ sectors located at n-pronged interior singularities. This gives the count of loops made in Theorem 1.

Of course, there is no information as yet about whether or not these loops are all distinct. This will be taken up in Section 3.4, on counting multiplicities. And in Section 3.3 we will prove that there are no other loops representing the conjugacy class of Φ with polygon types as given in Theorem 1.

3. The Uniqueness Half of Theorem 1

3.1 A Characterisation of Those Loops Associated to Sectors of Pseudo-Anosovs.

In this section, we develop the combinatorial tools necessary to prove the uniqueness half of Theorem 1. Theorem 11, which gives a combinatorial characterisation of the loops constructed in Proposition 9, will be proven.

Let Δ, Δ' be CDPs on (S, P), where $\Delta \to \Delta'$ is a splitting. Let h, h' be arcs of Δ, Δ' (respectively), neither of which is the arc opposite the (respective) distinguished prong. h and h' are called *invariant* under the splitting if either h and h' are the same arc (meaning isotopic rel P), or, if $\Delta \to \Delta'$ is a D-splitting ($D = R$ or L), then $h = h^D$ and h' is the arc of Δ' which is the same as the arc h^0 of Δ, opposite the distinguished prong of Δ.

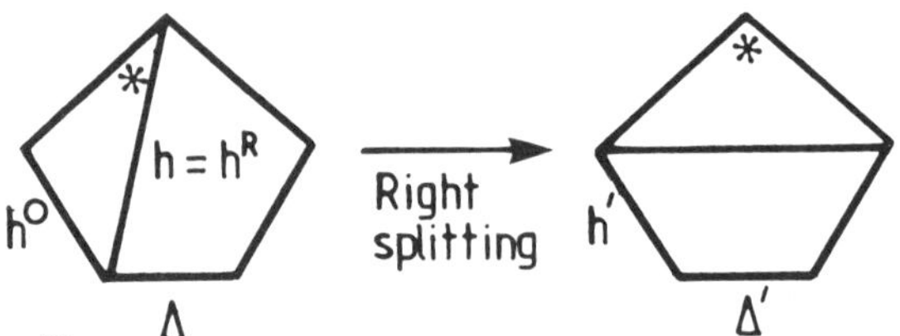

Figure 42.

Note that invariance is a well-defined 1-1 correspondence from arcs(Δ)$-h^O$ onto arcs(Δ')$-h^{O'}$. Thus, given subsets H, H' of the arcs of Δ, Δ' (respectively), H and H' are called *invariant* under the splitting if there is a 1-1 correspondence between H and H' such that the corresponding pairs are invariant.

Similarly, given a sequence of CDPs and splittings $\Delta_0 \to \cdots \to \Delta_M$ and subsets H_i of the arcs of Δ_i (resp.), we say that $\{H_i\}$ is *invariant* if H_i, H_{i+1} is invariant under $\Delta_i \to \Delta_{i+1}$ for each i. Note that invariance of $\{H_i\}$ is a combinatorial condition.

Now suppose that $\Delta \to \Delta'$ is a D-splitting ($D = R$ or L), and H, H' are invariant under the splitting. Let $\bar{D}$ be the complement of D. Consider the arc $h^{\bar{D}}$. This arc is in both CDPs Δ and Δ', and is adjacent to the distinguished prong of both.

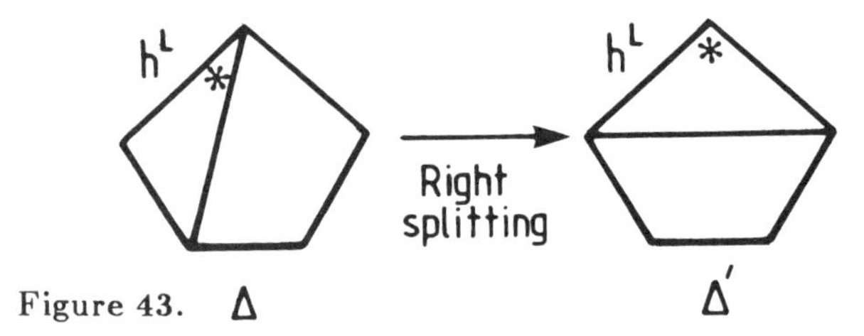

Figure 43.

We say that H and H' are *transparent* under the splitting if neither H nor, equivalently, H' contains the arc h^D.

Similarly, if $\{H_0, \ldots, H_M\}$ is invariant under the sequence $\Delta_0 \to \cdots \to \Delta_M$, then we can speak of $\{H_i\}$ being transparent. Evidently, transparency is also a combinatorial condition.

Therefore, given a combinatorial splitting $[\Delta] \to [\Delta']$, we can speak of sets of arcs of $[\Delta]$ and $[\Delta']$ transparent under the splitting.

Now suppose $[\Delta_0] \to [\Delta_1] \to \cdots \to [\Delta_M] = [\Delta_0]$ is an R-L loop $\mathcal{L}$, and H_i is a subset of the arcs of $[\Delta_i]$ such that, for each i, H_i, H_{i+1} are transparent under $[\Delta_i] \to [\Delta_{i+1}]$. Then we say that H_i is a *set of transparent arcs for* $\mathcal{L}$.

Given a loop $\mathcal{L}$, there is evidently an effective procedure for testing whether a given arc h is or is not contained in a set of transparent arcs for $\mathcal{L}$. Push h around the loop under the correspondence of invariance, perhaps several times around the loop, until either h comes back to itself (in which case h is contained in a transparent set), or h is adjacent to a distinguished prong without having the next splitting performed on it (in which case h is not in a transparent set). Since invariance is a 1-1 correspondence from the arcs of one *CDP* in the loop to the arcs of the next, and since there are only finitely many arcs in each CDP, one of these two events must eventually occur. In this manner we can associate to each loop $\mathcal{L}$ its *maximal transparent set* of arcs which contains all other transparent sets of arcs.

Next a simple lemma.

Lemma 10. *Let $\mathcal{L}$ be a loop, and suppose all splittings in* **L** *have the same parity. Then the conjugacy class represented by $\mathcal{L}$ is reducible, and the maximal transparent set of $\mathcal{L}$ is non-empty.*

Comment: In fact it can be shown that this conjugacy class is a fractional power of a Dehn twist.

Proof. Let Φ be a mapping class whose conjugacy class is represented by $\mathcal{L}$, and let $\Delta_0 \to \cdots \to \Delta_M$ be a representative sequence of splittings for $\mathcal{L}$ and Φ. Suppose, without loss of generality, that all splittings are Right. Then the same arc h lies to the left of the distinguished prong of every Δ_i, and $\Phi(h) = h$. Therefore Φ is isotopic to the identity on a regular neighbourhood of h. Moreover, any arc of $[\Delta_i]$ other than h^L is in the maximal transparent set of $\mathcal{L}$.

Simple Lemma

Theorem 11. *Given a loop $\mathcal{L}$ and $\Phi \in M(S, P)$ such that the conjugacy class of Φ is represented by $\mathcal{L}$, the following conditions are equivalent:*

1) *Φ is pseudo-Anosov, and $\mathcal{L}$ is associated to a sector of Φ.*

2) *The maximal transparent set of $\mathcal{L}$ is empty.*

Because of lemma 10, we can assume for the rest of the proof of Theorem 11 that $\mathcal{L}$ does not have constant parity.

To prove Theorem 11, we need to introduce coordinate systems for spaces of transverse weights and transition matrices for sequences of splittings.

Let Δ be a CDP, and let H be arcs$(\Delta) - h^O$. $W(\Delta)$ is a polygonal subcone of $\mathbf{R}^H$, of codimension 0 or 1. $\mathbf{R}^H$ is equipped with a standard basis. A *coordinate system* for $W(\Delta)$ is defined to be an enumeration of this basis.

Now let $\Delta \to \Delta'$ be a splitting. Recall that the relation of invariance gives a natural 1-1 correspondence from H to H'. Thus, given a coordinate system for $W(\Delta)$, there is a naturally induced coordinate system for $W(\Delta')$, which we say is *induced by the splitting*.

Given a CDP, define Δ^L and Δ^R to be the CDPs obtained by performing the appropriate splitting.

Splitting Lemma 12. *$W(\Delta^L)$ and $W(\Delta^R)$ are contained disjointly in $W(\Delta)$, as subspaces of MF. In fact, given $\mu \in W(\Delta)$,*

$$L(\mu) \in L(W(\Delta^L)) \Leftrightarrow \mu(h^L) < \mu(h^R)$$

$$L(\mu) \in L(W(\Delta^R)) \Leftrightarrow \mu(h^R) < \mu(h^L)$$

More specifically, in terms of the coordinate systems on $W(\Delta^L)$ and $W(\Delta^R)$ induced from a given coordinate system on $W(\Delta)$, the inclusion mappings are given by the following elementary matrices:

$$E_{i^R i^L} : W(\Delta^L) \to W(\Delta)$$

$$E_{i^L i^R} : W(\Delta^R) \to W(\Delta)$$

where i^L, i^R are the indices of the arcs h^L, h^R of Δ. (Such a matrix is called a transition matrix *of the splitting; it depends, of course, on the chosen coordinate system for $W(\Delta)$.)*

Comment 1: This statement contains implicitly the fact that, in the codimension 1 case, the transition matrix M, mapping all of $\mathbf{R}^{H'}$ to all of $\mathbf{R}^H$, takes $W(\Delta') \subset \mathbf{R}^{H'}$ to $W(\Delta) \subset \mathbf{R}^H$, when $\Delta \to \Delta'$ is a splitting.

Comment 2: Given $\mu \in W(\Delta)$, there is also a geometric interpretation for the case where $\mu(h^L) = \mu(h^R)$. If f_μ and α are chosen as in Proposition 5, then f_μ degenerates at α if and only if $\mu(h^L) = \mu(h^R)$.

Proof. Let $\Delta \to \Delta'$ be a splitting, which we will suppose without loss of generality to be a left splitting. Let M be the transition matrix. Let $\mu \in W(\Delta')$, and consider the transverse weight $M\mu \in W(\Delta)$. Let $f = f_{M\mu}$. Using Proposition 5, choose a non-generic transversal α to f, based at the vertex of Δ, such that $\Delta(f, \alpha) = \Delta$.

Let H^L, H^R be the f, α rectangles corresponding to the arcs h^L, h^R of Δ.

Let g^L, g^R be the arcs of Δ' such that h^L, g^L and h^R, g^R are both invariant under splitting.

By definition of M,

$$M\mu(h^L) = \mu(g^L)$$

$$M\mu(h^R) = \mu(g^L) + \mu(g^R)$$

Since $\mu \in W(\Delta')$, μ has positive weights, and it follows that

$$M\mu(h^L) < M\mu(h^R)$$

Therefore the f width of H^L is less than the f width of H^R. It follows that f does not degenerate at α, and moreover that if **l** is the unique singular leaf segment whose first intersection with $\mathrm{int}(\alpha)$ is closest to the free end of α, then **l** intersects α from the Left. Thus, if α' is the first non-generic subtransversal of α, then $\Delta(f, \alpha') = \Delta^L = \Delta'$.

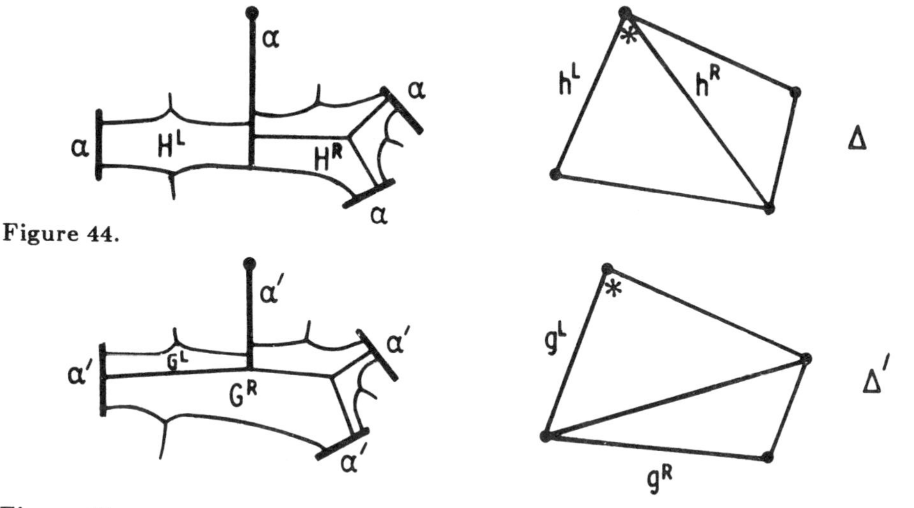

Figure 44.

Figure 45.

Moreover, if G^L, G^R are the f, α' rectangles corresponding to g^L, g^R, then the f widths of G^L, G^R are respectively equal to $\mu(g^L)$ and $\mu(g^R)$. Thus, the transverse weight on Δ' associated to f by Proposition 5 is μ.

Splitting Lemma 12

As an example of how this lemma might be used, consider a canonical arational measured foliation f, and a non-generic transversal α to f at a singularity s. Then the R-L string associated to f and α can be generated just from the numerical data of the transverse weight $\mu \in W(\Delta(f, \alpha))$ for which $L(\mu) = \{f\}$.

Now consider a more refined question: let Δ_0 be a CDP, $\mu \in W(\Delta_0)$. It is true that f_μ is a canonical arational measured foliation if and only if there exists a R-L string $\Delta_0 \to \cdots \to \Delta_i \to \cdots$ such that $L(\mu) \in \bigcap_i (W(\Delta_i))$?

The answer is almost true, but there is one other kind of measured foliation that f_μ can be. Suppose that Δ_0 has a puncture vertex. Recall Proposition 3, where the polygon type of Δ_0 is related to the singularity type of f_μ, *when* f_μ *is canonical*. Here is a slight restatement of that result: suppose the polygon type of Δ_0 is $(p_i;q_j;\emptyset)$. Then the singularity type $(r_i;s_j)$ of f_μ is given by either

$$r_1 = p_1 + 1$$

$$r_i = p_i \qquad i > 1$$

$$s_j = q_j \qquad j \geq 3$$

or, for some $m \geq 1$,

$$r_m = p_m + 1$$

$$r_i = p_i \qquad i \neq m$$

$$s_{m+1} = q_{m+1} - 1$$

$$s_j = q_j \qquad j \neq m+1$$

What this means is that Δ_0 can't tell whether there is a 1-pronged singularity at the vertex, or whether, for some m, there is an m-pronged singularity at the vertex, and one fewer $m+1$-pronged interior singularity than expected.

This discussion should serve as motivation for the following definition: given a CDP Δ with vertex v, and $\mu \in W(\Delta)$, we say that μ *has the same degeneracy as* Δ if, either there are no leaf segments connecting singularities of f_μ (i.e. f_μ is canonical arational), or there is at worst a leaf segment connecting a 1-pronged singularity at v (which is necessarily a puncture) to an interior singularity of f_μ.

Degeneracy Theorem 13. *Let Δ be a CDP with vertex v, $\mu \in W(\Delta)$. If μ has the same degeneracy as Δ, then there exists a unique infinite sequence D_i $(i = 1, 2, \ldots)$ of L s and R s, such that, if $\Delta_0 = \Delta$ and $\Delta_{i+1} = \Delta_i^{D_i}$, then $L(\mu) \in \bigcap_i L(W(\Delta_i))$. In this case, if in addition f_μ is a canonical model for $L(\mu)$, then there is a transversal α to f_μ based at v such that the R-L string associated to f_μ and α is $\Delta_0 \to \cdots \to \Delta_i \to \cdots$. Conversely, suppose $L(\mu) \in \bigcap_i L(W(\Delta_i))$ where $\Delta_0 \to \cdots \to \Delta_i \to \cdots$ is an R-L string, and let $\mu_i \in W(\Delta_i)$ be the unique transverse weight on Δ_i such that $L(\mu) = L(\mu_1)$ (μ_i is given by the splitting lemma). Then if*

$$\sup \mu_i(h) \to 0 \quad \textit{as } i \to \infty$$

where the sup is taken over the arcs of Δ_i, then μ has the same degeneracy as Δ_o.

Comment: This technical hypothesis is in fact necessary – there exist certain "reducible" CDPs, for which this hypothesis can be violated; and for such a CDP Δ, there are no arational measured foliations in $L(W(\Delta))$. For an example, see [Mosher, 1983, P. 98].

Proof. The proof of the uniqueness is easy – it follows directly from the fact that for any CDP Δ,

$$L(W(\Delta^L)) \cap L(W(\Delta^R)) \cap L(\textit{common boundary}) = \emptyset$$

The only case where is might not be clear that a string exists when f_μ has a leaf segment ℓ joining a 1-pronged singularity at v to an interior singularity s. Using Proposition 5, choose a non-generic transversal α based at v so that $\Delta(f, \alpha) = \Delta$. By construction of f_μ, α must intersect ℓ. The arguments for Proposition 6 and the Splitting Lemma go through with no trouble; *even in the case* that **l** intersects α in only one point x, which is the closest first intersection of a singular leaf with int(α) to the free end of α. Note that f_μ does not degenerate at α, since the part of **l** connecting x to v lies in the interior of an f_μ, α rectangle, one which folds around the 1-pronged singularity at v.

For completeness, we show how Proposition 6 works in this case. Replace α by the first non-generic sub-transversal α'. Note that the free end of α' is x.

Slide α' along **l** past the singularity s and perform a Whitehead move on **l**.

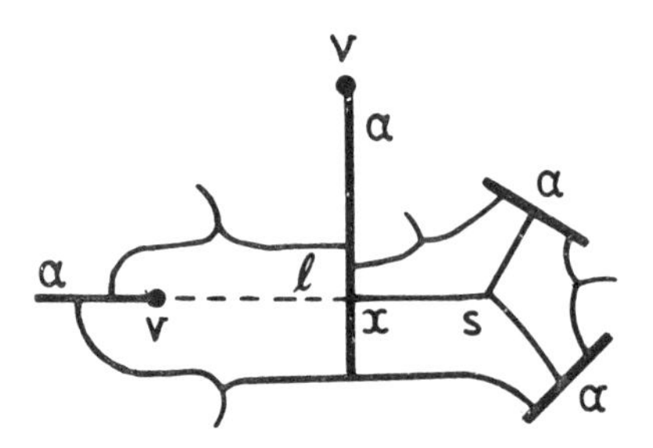

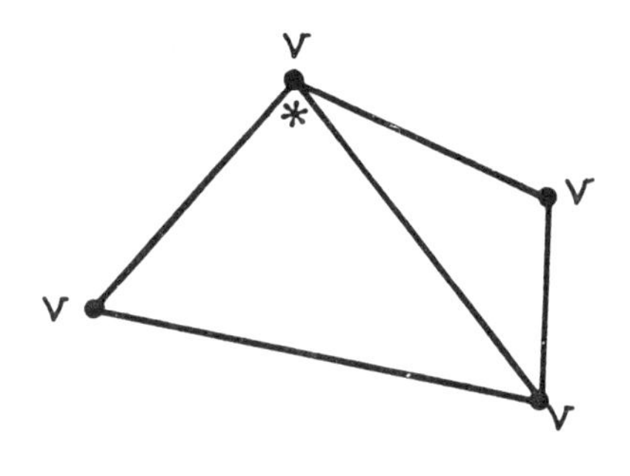

Figure 46.

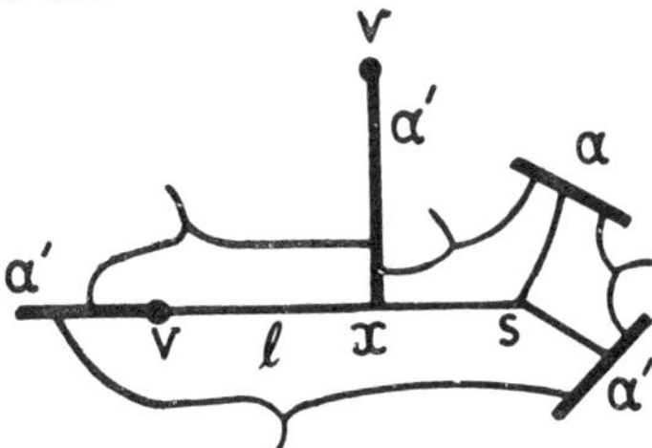

Figure 47.

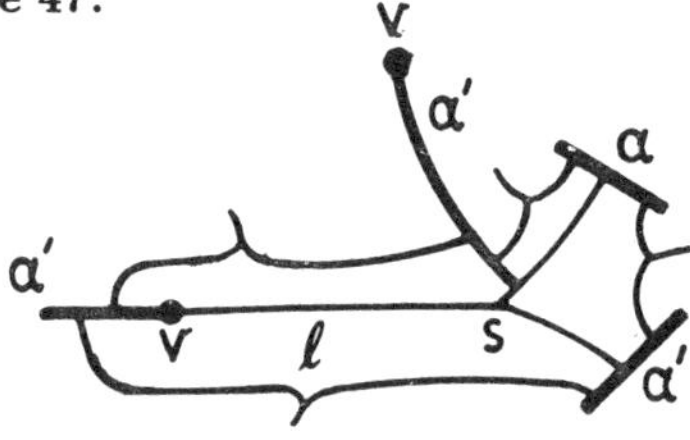

Figure 48.

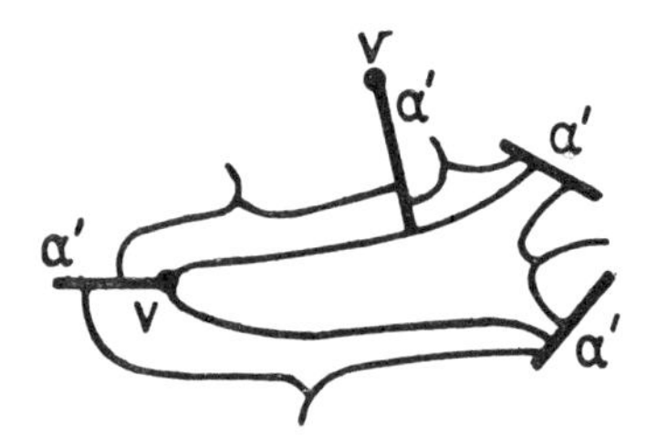

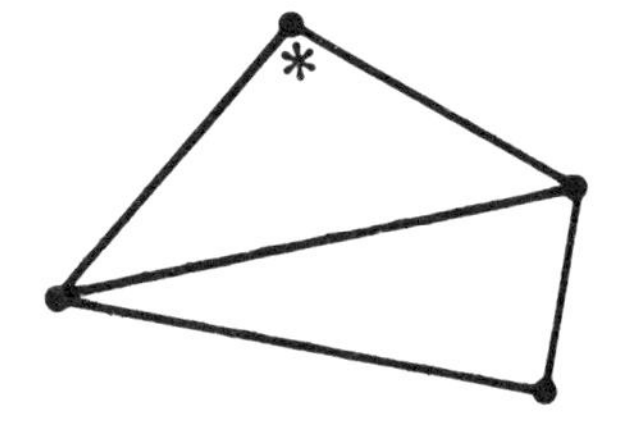

Figure 49.

Let Δ' be the resulting CDP. Note that $\Delta \rightarrow \Delta'$ is a splitting. Moreover, the splitting lemma shows that if $\mu' \in W(\Delta')$ is determined by the equation $\mu = M\mu'$, where M is a transition matrix of the splitting, then $L(\mu) = L(\mu')$.

To prove the "in this case" statement: choose the transversal α so that $\Delta(f, \alpha) = \Delta_0$. Then the string of CDPs $\Delta_0 = \Delta'_0 \rightarrow \Delta'_1 \rightarrow \ldots$ associated to f_μ and α satisfies the condition that, for each i, $f_\mu \in W(\Delta'_i)$. By uniqueness, $\Delta'_i = \Delta_i$.

We now prove the converse. Let $f = f_\mu$; choose a transversal α_0 to f such that $\Delta(f, \alpha_0) = \Delta_0$. By hypothesis, f never degenerates at any α_i, where α_i is defined inductively to be the first non-generic subtransversal of α_{i-1}.

Suppose there is some leaf segment **l** connecting a pair of singularities of f, neither of which is a 1-pronged singularity at v (call such an **l** a *connecting* leaf segment). f is so constructed that any connecting leaf segment intersects int(α). Moreover, there are only finitely many connecting leaf segments. It follows that there is a finite, non-empty set of points x in int(α) such that, if α' is the sub-transversal of α based at v with free end x, then there is a connecting leaf segment **l**, containing x, which does not intersect int(α'). Let α' be the largest such sub-transversal, i.e. the one whose free end x is closest to the free end of α.

Since f never degenerates at any α_i, it follows that $\bigcap_i \alpha_i$ contains α'; for if not, there is an N such that $\alpha' \subset \text{int}(\alpha_N)$, $\alpha' \not\subset \text{int}(\alpha_{N+1})$, and so f degenerates at α_N.

But the transversal measure of α' with respect to f is some positive number R, and since each α_i contains α',

$$\sup \mu_i(h) \geq R/A$$

where A is the number of arcs of Δ_i (the same for all i), and the sup is taken over all arcs $h \neq h^O$ of Δ_i. Since $R/A > 0$ is independent of i, this contradicts the hypothesis on μ_i.

Degeneracy Theorem 13

We now continue with the proof of Theorem 11. Let $\mathcal{L}$ be an R-L loop, Φ a mapping class whose conjugacy class is represented by $\mathcal{L}$, and $\Delta_0 \to \Delta_1 \to \cdots \to \Delta_M$ a representative sequence of splittings for Φ and $\mathcal{L}$. Choose a coordinate system for $W(\Delta_0)$. Recall that this is an indexing, by the index set $I = \{1, \ldots, K\}$, of all arcs of Δ_0 except h^0, where there are K such arcs. This induces a coordinate system on each $W(\Delta_i)$ in turn, indexed by the same set I. We call this the coordinate system on $W(\Delta_i)$ *induced by the sequence*.

Let A_m $(m=1, \ldots, M)$ be the transition matrix of the splitting $\Delta_m \to \Delta_{m+1}$, expressing in coordinates the inclusion $W(\Delta_M) \overset{\subset}{\to} W(\Delta_0)$. Then $A_1 \cdot \cdots \cdot A_M$ is the matrix expressing in coordinates the inclusion $W(\Delta_M) \overset{\subset}{\to} W(\Delta_O)$. Because $\Phi(\Delta_O) = \Delta_M$, then Φ: $W(\Delta_O) \to W(\Delta_M)$ is an isomorphism. However, the coordinate system of $W(\Delta_M)$ *inherited under* Φ may not coincide with the coordinate system on $W(\Delta_M)$ induced by the sequence. So if we let P be the permutation matrix which changes a column vector in $W(\Delta_M)$ coordinates inherited under Φ into a column vector in $W(\Delta_M)$ coordinates induced by the sequence, then $A_{\mathcal{L}}=A_1\cdot \cdots \cdot A_M\cdot P$ is the matrix which expresses in coordinates the linear self-inclusion $\Phi:L(W(\Delta_0)) \overset{\subset}{\rightarrow} L(W(\Delta_0))$. $A_{\mathcal{L}}$ has all non-negative, integer entries, and has determinant 1. Note that $A_{\mathcal{L}}$ is only defined up to conjugation by some non-negative, integral matrix of determinant 1. This shall not bother us.

Recall that a matrix of non-negative integers is *Perron-Frobenius* if some power of the matrix has all positive-entries. We shall need the following version of the Perron-Frobenius theorem (see[Fathi-Laudenbach-Poenaru, 1979 exposé 12, p. 227]).

Theorem. *If A is a Perron-Frobenius matrix, then A has a positive eigen-vector μ with eigenvalue $\lambda > 1$. Moreover, if π is the projectivization map on the collection $\mathbf{R}^n_+$ of positive vectors, then*

$$\bigcap_i \pi(A^i(\mathbf{R}^n_+)) = \pi(\mu)$$

To apply this theorem, we need a connection between the structure of the matrix $A_{\mathcal{L}}$ (which contains numerical information about the mapping class Φ) and the combinatorial properties of the loop $\mathcal{L}$.

Let $I' \subset I$ be the index set for the arcs of $[\Delta_0]$ contained in the maximal transparent set of $\mathcal{L}$. Given that there are N arcs in the maximal transparent set $(0 \leq N \leq K)$, we shall make the assumption that $I' = \{i \in I : i \leq N\}$. Of course, I' may well be empty.

Proposition 14. *Given a loop $\mathcal{L}$, the matrix $A_{\mathcal{L}}$ has the form*

$$\begin{array}{c c} & \begin{array}{cc} N & K-N \end{array} \\ \begin{array}{c} N \\ \\ K-N \end{array} & \left(\begin{array}{c|c} A' & \approx \\ \hline O & A'' \end{array}\right) \end{array}$$

where A' is an N by N permutation matrix, and A'' is a $K-N$ by $K-N$ Perron-Frobenius matrix.

Proof. In [Mosher, 1983] this was proven with the hands on analysis of the matrix. We shall prove it here using graph theoretic techniques. This may not improve on the overall grubbiness of the proof, but it at least provides a convenient bookkeeping tool.

Recall that $A_{\mathcal{L}} = A_1 \cdot \cdots \cdot A_M \cdot P$, where P is a permutation matrix. If $P = (p_{ij})$, then let ρ be the permutation on I defined by $p_{\rho(j), j} = 1$. Note that ρ leaves invariant the subset I'.

From the matrices $A_1, \ldots, A_M$, P we shall construct a directed graph $\mathfrak{D}$, and deduce the properties of $A_{\mathcal{L}}$ from the structure of $\mathfrak{D}$.

The vertices of $\mathfrak{D}$ consist of $M \cdot K$ points $x_{m,k}$ $(m = 0, \ldots, M\,;\ k = 1, \ldots, K)$ where $x_{0,k}$ is identified with $x_{M,\rho(k)}$.

Recall that each matrix A_m $(m=1, \ldots, M)$ is an elementary matrix $E_{i^m j^m}$. For each $m \geq 1$, we add to $\mathfrak{D}$ a directed edge $x_{m-1,k} \to x_{m,k}$; and in addition, for each $m \geq 1$ there is an edge $x_{m-1,i^m} \to x_{m,j^m}$.

It follows that the i,j entry of $A_{\mathcal{L}}$ is equal to the number of distinct paths

$$x_{0,i} \to x_{1,k^1} \to \cdots \to x_{M-1,k^{m-1}} \to x_{M,\rho(j)} = x_{0,j}$$

We now relate $\mathfrak{D}$ to the structure of $\mathcal{L}$. Let h be an arc of $[\Delta_M]$ in $\mathcal{L}$. h is indexed by an element $k \in \{1, \ldots, K\}$, so $x_{m,k}$ is the vertex of $\mathfrak{D}$ corresponding to h. Note that, if h' is the unique arc of $[\Delta_{m+1}]$ such that h, h' are invariant under the splitting $[\Delta_m]$ $[\Delta_{m+1}]$, then the vertex of $\mathfrak{D}$ corresponding to h' is $x_{m+1,k}$. We define a *straight path* to be a path of the form

$$x_{0,k} \to x_{1,k} \to \cdots \to x_{M,k} = x_{0,\rho^{-1}(k)}$$

A *straight cycle* is defined to be a simple cycle obtained by concatenating straight paths, i.e. a cycle of the form

$$x_{0,k} \to \cdots \to x_{0,\rho^{-1}(k)} \to \cdots \to \cdots \to x_{0,\rho^{-i}(k)} = x_{0,k}$$

where i is the lowest non-trivial power of ρ fixing k. Thus, to push an arc h around $\mathcal{L}$ by invariance, we need only go around the corresponding straight cycle of $\mathfrak{D}$.

Let MTS be the set of vertices of $\mathfrak{D}$ whose corresponding arcs in $\mathcal{L}$ are contained in the maximal transparent set of $\mathcal{L}$. Note that, if h, $h' \in [\Delta_{m-1}]$, $[\Delta_m]$ (resp.) are invariant, that h is in the maximal transparent set of L if and only if h' is. Therefore, a given straight cycle either consists entirely of vertices contained in MTS, or consists entirely of vertices not contained in MTS. Thus, $MTS = \{x_{m,k} : k \in I'\}$.

Recalling that $A_m = E_{i^m, j^m}$ $(m \geq 1)$, let us review the definition of i^m and j^m. Consider the splitting $[\Delta_{m-1}] \to [\Delta_m]$. i^m is the index of the arc of $[\Delta_{m-1}]$, adjacent to the distinguished prong, that the splitting is *not* performed on. j^m is the index of the arc of $[\Delta_{m-1}]$ that the splitting *is* performed on. Therefore, $x_{m-1, i^m} \notin MTS$. So a straight cycle is in MTS if and only if it does *not* contain a vertex of the form x_{m-1, i^m}. It follows, from the construction of $\mathfrak{D}$, that a straight cycle is in MTS if and only if it does not contain a vertex $x_{m-1,i}$ $(m \geq 1)$ for which there is a directed edge $x_{m-1,i} \to x_{m,j}$ where $i \neq j$.

Thus, if $k \in I'$ the *only* path going once around $\mathfrak{D}$ ending at $x_{0,k}$ is the straight path $x_{0,\rho(k)} \to \cdots \to x_{M,\rho(k)} = x_{0,k}$.

So if $k \in I'$, this gives the structure of the kth column of $A_{\mathcal{L}}$: given $i \in I$, the i, k entry of $A_{\mathcal{L}}$ is 1 if $i = \rho(k)$, and 0 otherwise. Since $\rho(k) \in I'$, this says that the upper left N by N block of $A_{\mathcal{L}}$ is the permutation matrix of the permutation $\rho|I'$, and the lower left $K-N$ by N block is all zeros.

We now know that $A_{\mathcal{L}}$ has the form

$$\left(\begin{array}{c|c} A' & \approx \\ \hline O & A'' \end{array}\right)$$

It remains to show that A'' is Perron-Frobenius. Recalling once again that $A_{\mathcal{L}} = A_1 \cdot \cdots \cdot A_M \cdot P$, we will first reduce to the case where P is the identity matrix.

Let $N = \text{order}(P)$. Notice that

$$A_{\mathcal{L}}^N = \left(\begin{array}{c|c} (A')^N & \approx \\ \hline O & (A'')^N \end{array}\right)$$

Notice also that

$$A_{\mathcal{L}}^N = [A_0 \ldots A_{M-1}] \; [(PA_0P^{-1}) \ldots \; (PA_{M-1}P^{-1})] \; [(P^2A_0P^{-2}) \ldots (P^2A_{M-1}P^{-2})] \ldots \; [(P^NA_0P^{-N}) \ldots \; (P^NA_{M-1}P^{-N})]$$

If we extend the sequence of splittings $\Delta_0 \to \cdots \to \Delta_M$ to an infinite sequence $\Delta_0 \to \cdots \to \Delta_M \to \cdots$ such that $\Phi(\Delta_i) = \Delta_{M+i}$, then the transition matrix of the splitting $\Delta_{kM+i} \to \Delta_{kM+i+1}$ is $P^k A_i P^{-k}$, where all coordinate systems are those induced by the sequence from the coordinate system on $W(\Delta_0)$.

Therefore, $A_{\mathcal{L}}^N$ is the matrix of the loop $\mathcal{L}^N$ going N times around $\mathcal{L}$, and the permutation part of the product expansion of $A_{\mathcal{L}}^N$ is the identity. Since A'' is Perron-Frobenius if $(A'')^N$ is, we have reduced to the case where P is the identity matrix.

Now we have that the permutation ρ on I is the identity. Therefore, every straight cycle in $\mathfrak{D}$ consists of a single straight path, so for each k, $x_{0,k} \to x_{1,k} \to \cdots \to x_{M,k} = x_{0,k}$ is a straight cycle which will be denoted c_k.

To show that A'' is Perron-Frobenius, it suffices to show that, for any pair of vertices of $\mathfrak{D}$ not in MTS, there is a directed path from the first vertex to the second. In fact, we shall construct a cycle d which intersects every straight cycle c_k not in MTS. Then if $x \in c_k$ and $x' \in c'_k$ are not in MTS, travel from x along c_k to d, then along d to c'_k, then along c'_k to x'. The construction of d follows.

Recall that $\mathcal{L}$ is $[\Delta_0] \to \cdots \to [\Delta_M] = [\Delta_0]$. Let $D_m \in \{\text{Left, Right}\}$ be the parity of the splitting $[\Delta_{m-1}] \to [\Delta_m]$ $(m=1, \ldots, M)$. We shall assume, without loss of generality, that the starting point $[\Delta_0]$ of $\mathcal{L}$ is chosen so that $D_M \neq D_1$ (recall that $\mathcal{L}$ does not have constant parity).

We shall define, by induction, a sub-sequence of $\{1, \ldots, M\}$: let $m(1) = 1$, and for each l, let $m(l+1)$ be the first integer after $m(l)$ such that $D_{m(l+1)} \neq D_{m(l+1)-1}$. Thus, $m(l)$ is defined for $l = 1, \ldots, L$, where L is the number of parity changes in the loop $\mathcal{L}$.

We can now define the cycle d, which starts at $x_{m(1)-1, i^{m(1)}}$ $(= x_{0, i^1})$, traverses an edge to $x_{m(1), i^{m(L)}}$, and then continues along $c_{i^{m(L)}}$ until $x_{m(L)-1, i^{m(L)}}$. In general, d is a concatenation of L paths, where for each $l \in \{1, \ldots, L\}$, the ith segment of d starts at $x_{m(l)-1, i^{m(l)}}$, traverses an edge to $x_{m(l), i^{m(l-1)}}$, then continues along $c_{i^{m(l-1)}}$ until $x_{m(l-1)-1, i^{m(l-1)}}$.

To verify this construction, we must show that for each $l \in \{1, \ldots, L\}$, $x_{m(l)-1, i^{m(l)}} \to x_{m(l), i^{m(l-1)}}$ is in the graph $\mathfrak{D}$.

Consider the sequence of CDPs and splittings

$$\Delta_{m(l-1)-1} \to \Delta_{m(l-1)} \to \cdots \to \Delta_{m(1)-1} \to \Delta_{m(l)}$$

Up until the last splitting, this sequence has constant parity. The final splitting $\Delta_{m(l)-1} \to \Delta_{m(l)}$ has opposite parity from what comes before. It follows that

$$i^{m(l-1)} = \ . \ . \ . \ = i^{m(l)-1} \qquad \ldots\ldots\ldots\ldots\ldots\ldots\ldots\ldots \quad (*)$$

This is true because, if $\Delta \to \Delta' \to \Delta''$ have the same parity, and if h, h' are the arcs of Δ, Δ' (respectively), adjacent to the (resp.) distinguished prongs, that the (resp.) splittings are not performed on, then h, h' are invariant under the splitting $\Delta \to \Delta'$ shown in Figure 50.

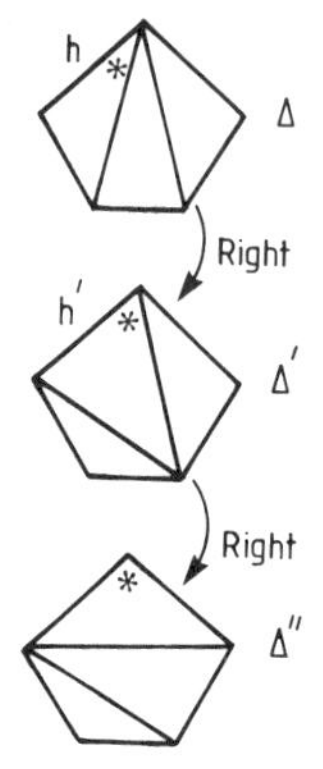

Figure 50.

Also,

$$i^{m^l-1} = j^{m^l} \qquad \ldots\ldots\ldots\ldots\ldots\ldots\ldots\ldots\ldots\ldots \quad (**)$$

This is true because, if $\Delta \to \Delta' \to \Delta''$ have opposite parity, if h is in the arc of Δ, adjacent to the distinguished prong, that the splitting $\Delta \to \Delta'$ is *not* performed on, and h' is the splitting of Δ' that the splitting $\Delta' \to \Delta''$ *is* performed on, then h, h' are invariant under $\Delta \to \Delta'$ (see Figure 52).

By definition of $\mathfrak{D}$, $x_{m(l)-1,\, i^{m(l)}} \to x_{m(l),\, j^{m(l)}}$ is an edge of $\mathfrak{D}$. But from (*) and (**), it follows that

$$j^{m(l)} = i^{m(l-1)}$$

We have verified that d is actually a cycle of $\mathfrak{D}$. it remains to show that d intersects every straight cycle c_k such that $k \notin I'$, i.e. $c_k \not\subset MTS$.

If $c_k \not\subset MTS$ then c_k contains a vertex of the form $x_{m-1,\, i^m}$, thus $k = i^m$. This vertex might not be in d; however, if we choose l so that $m(l)$ is the greatest element of the subsequence $m(1), \ldots, m(L)$ such that $m(l) \leq m$, then by (*),

$$i^{m(l)} = i^m = k$$

Moreover, d contains the vertex $x_{m(l)-1,\, i^{m(l)}} \in c_k$.

Proposition 14

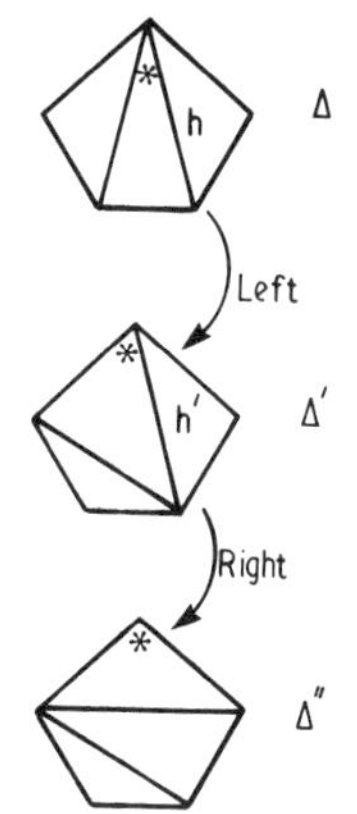

Figure 51.

Corollary 15. *The maximal transparent set of* $\mathcal{L}$ *is empty if and only if* $A_{\mathcal{L}}$ *is Perron-Frobenius.*

The graph $\mathfrak{D}$ constructed above is an important tool for further analysis of a pseudo-Anosov Φ. $\mathfrak{D}$ captures the symbolic dynamics of Φ; in particular, the nth order periodic points of Φ are (almost) in 1-1 correspondence with cycles travelling n times around $\mathfrak{D}$. These phenomena will be exploited in future works on the conjugacy problem for the mapping class group of a non-punctured surface.

The proof of Theorem 11 can now be completed.

Recall that $\Delta_0 \to \Delta_1 \to \cdots \to \Delta_M \to \Delta_{M+1} \to \cdots$ is an R-L string such that $\Phi(\Delta_i) = \Delta_{M+i}$ for all i, and the loop $\mathcal{L}$ is $[\Delta_0] \to [\Delta_1] \to \cdots \to [\Delta_M] = [\Delta_0]$.

Suppose that the maximal transparent set of **L** is empty. From corollary 15, it follows that the matrix $A_{\mathbf{L}}$, expressing in coordinates the linear self-inclusion $\Phi\colon L(W(\Delta_0)) \overset{\subset}{\to} L(W(\Delta_0))$, is Perron-Frobenius. By the Perron-Frobenius theorem, Φ has positive eigenvector μ_0 of eigenvalue $\lambda > 1$, and $L(\mu_0) \in L(W(\Delta_0))$.

Let $\mu_i \in W(\Delta_i)$ be the unique transverse weight in Δ_i such that $L(\mu_i) = L(\mu)$. We wish to show that the sup of the coordinates of μ_i goes to zero as i approaches ∞.

For multiples of M, $\mu_{kM} = \lambda^{-kM}\Phi^M(\mu_0)$ and since the $W(\Delta_{kM})$ coordinates of $\Phi^M(\alpha_0)$ are identical to the $W(\Delta_0)$ coordinates of μ_0, then the sup of the coordinates of μ_{kM} goes to zero as $k \to \infty$. Since the sup decreases monotonically, it goes to zero for all μ_i as $i \to \infty$.

By the Degeneracy Theorem, it follows that **L** is associated to a sector of Φ.

One direction of Theorem 11 is finished. To prove the other direction, suppose that Φ is pseudo-Anosov, **L** is a loop associated to a sector of Φ. $\Delta_0 \to \cdots \to \Delta_M \to \Delta_{M+1} \to \cdots$ is a string with $\Phi(\Delta_i) = \Delta_{M+i}$, and the loop **L** is $[\Delta_0] \to \cdots \to [\Delta_M] = [\Delta_0]$. If f is the stable foliation of Φ, by the construction of Proposition 9, $\{f\} \in L(W(\Delta_0))$; so take $\mu \in W(\Delta_0)$ such that $\{f\} \in L(\mu)$. It follows that μ is a positive eigenvector for the matrix $A_{\mathbf{L}}$.

If the maximal transparent set of **L** is *non-empty*, $A_{\mathbf{L}}$ has the form

$$\left(\begin{array}{c|c} A' & \approx \\ \hline O & A'' \end{array}\right)$$

where A' is a *non-trivial* permutation matrix, and A'' is Perron-Frobenius. Such a matrix cannot have a positive eigen-vector. Therefore, the maximal transparent set of **L** is empty.

Theorem 11

3.2. Loops with Non-trivial Transparent Set.

In order to complete the proof of the uniqueness half of Theorem 1, we need to carry through a combinatorial analysis of loops whose maximal transparent set is non-empty.

Let H be a subset of the arcs of a combinatorial CDP $[\Delta]$. We can also, of course, think of H as a subset of the arcs of any CDP Δ representing $[\Delta]$. We say that H *carries improper components* if, when the arcs in H and the arc h^0 are removed from Δ, then some complementary component is not a disc or a once-punctured disc.

There is evidently an effective procedure for checking whether H, as a set of arcs of $[\Delta]$, carries improper components. Basically, one checks for non-trivial cycles rel P such that the only arcs of Δ that they intersect are the arcs of $H \cup \{h^0\}$. This can be done by first looking for any such cycle connecting a pair of punctures; if none of these are found, one can then erase the arcs $H \cup \{h^0\}$ from the combinatorial picture of $[\Delta]$, giving a new combinatorial picture. Compute the genus of this new picture, and compare this to the genus of S. If they are the same, H does not carry improper components. Otherwise, H carries improper components.

Proposition 16. *Let $\{H_i\}$ be the maximal set of transparent arcs for a loop* **L**. *If any H_i carries improper components, then they all do, and the conjugacy class represented by* **L** *is reducible.*

Proof. What we show is the following: if $\Delta \to \Delta'$ is a splitting and H, H' are transparent under the splitting, then the boundary of a regular neighbourhood of arcs$(\Delta) - H - \{h^0\}$ is ambient isotopic rel P to the boundary of a regular neighbourhood of arcs$(\Delta') - H' - \{h^{0\prime}\}$. From this the proposition will follow, for these curves computed at Δ_0 will be ambient isotopic rel P to the curves computed at Δ_M, when $\Delta_0 \to \cdots \to \Delta_M$ is a representative sequence of splittings for **L** and Φ, and **L** represents the conjugacy class of Φ. But since $\Phi(\Delta_0) = \Delta_n$ and $\Phi(H_0) = H_n$, then Φ permutes this family of curves, so if any of them do not bound discs or once-punctured discs, Φ is reducible.

So suppose $\Delta \to \Delta'$ is a splitting, and H, H' are transparent under the splitting. There are two cases:

1) H does not contain the arc that the splitting is performed on. The Figure 52 shows an isotopy from the boundaries at Δ to the boundaries at Δ'.

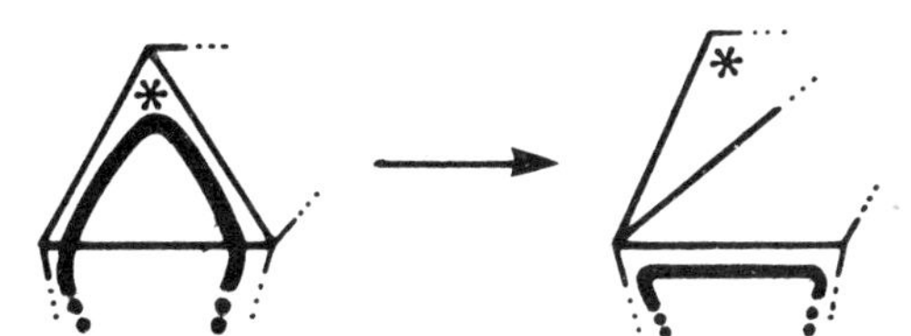

Figure 52.

2) H does contain the arc that the splitting is performed on. Then arcs$(\Delta) - H - \{h^O\}$ is the same set of arcs as arcs$(\Delta') - H' - \{h^{O\prime}\}$. Thus, no isotopy is even necessary. Figures 53 and 54 show some diagrams to illustrate this case.

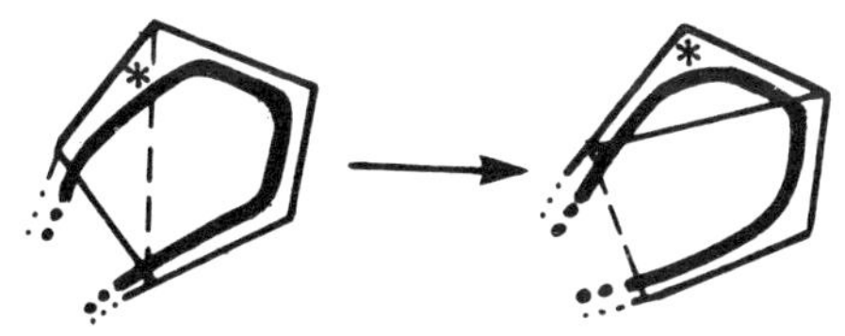

Figure 53.

Proposition 16

With a little more work, it can be shown that Φ is finite order on the improper components, and that the order is just the order of the

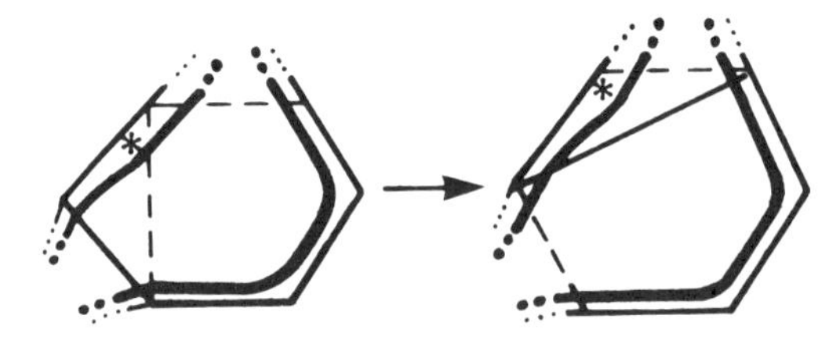

Figure 54.

permutation of the boundaries.

Now suppose **L** is a loop whose maximal transparent set $\{H_i\}$ contains no improper components. If $\{H_i\}$ is the trivial set, **L** is analysed by Theorem 11. Otherwise, we associate to **L** the R-L loop **L**$'$ obtained by "breaking the windows" of **L** to be defined below.

Suppose $\Delta \to \Delta' \to \Delta''$ is a sequence of splittings with opposite parity. Let h' be the arc of Δ' that the splitting is performed on, and let h, h', h'' be the invariant sequence of arcs containing h'. Then this sequence is not transparent under the splitting. This is so because h is necessarily the arc of Δ, adjacent to the distinguished prong of Δ, that the splitting is not performed on.

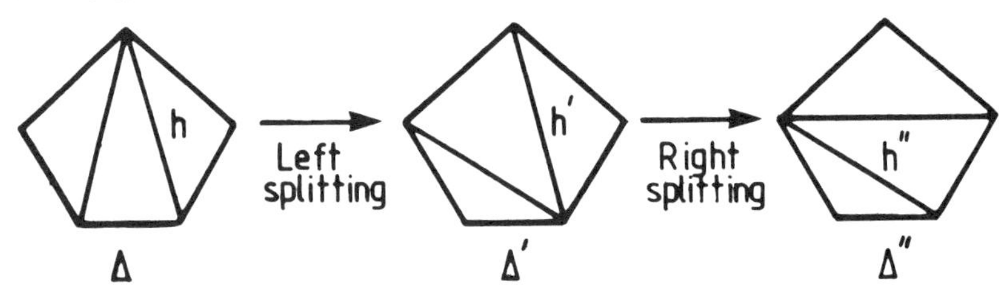

Figure 55.

Consider now the loop **L**, and assume, by changing the starting point for **L** if necessary, that $[\Delta_{M-1}] \to [\Delta_M] = [\Delta_0]$ and $[\Delta_0] \to [\Delta_1]$ have opposite parity. Thus, the arc h of $[\Delta_0]$ that the splitting $[\Delta_0] \to [\Delta_1]$ is performed on is not in the maximal transparent set of **L**.

We define inductively a subsequence of the $[\Delta_i]$ as follows: $[\Delta_{i_0}] = [\Delta_0]$; and let $[\Delta_{i_{j+1}}]$ be the first $[\Delta_i]$ after $[\Delta_{i_j}]$ such that the splitting $[\Delta_{i_{j+1}}] \to [\Delta_{i_{j+1}+1}]$ is not performed on an arc in $H_{i_{j+1}}$. Note that the splittings $[\Delta_{i_j}] \to \cdots \to [\Delta_{i_{j+1}}]$ all have the same parity. In general, if the parities of $[\Delta_{i-1}] \to [\Delta_i]$ and $[\Delta_i] \to [\Delta_{i+1}]$ are opposite, $[\Delta_i]$ is in the subsequence. The converse is not true, however.

Define $[\Delta'_j]$ to be the CDP obtained from $[\Delta_{i_j}]$ by deleting from $[\Delta_{i_j}]$ all the arcs in H_{i_j} ($[\Delta'_j]$ is guaranteed to be a CDP since H_{i_j} does not carry any improper components). Then $[\Delta'_j] \to [\Delta'_{j+1}]$ is a splitting of the same parity as the sequence of splittings $[\Delta_{ij}] \to \cdots \to [\Delta_{ij}+1]$. This construction is illustrated by Figure 56.

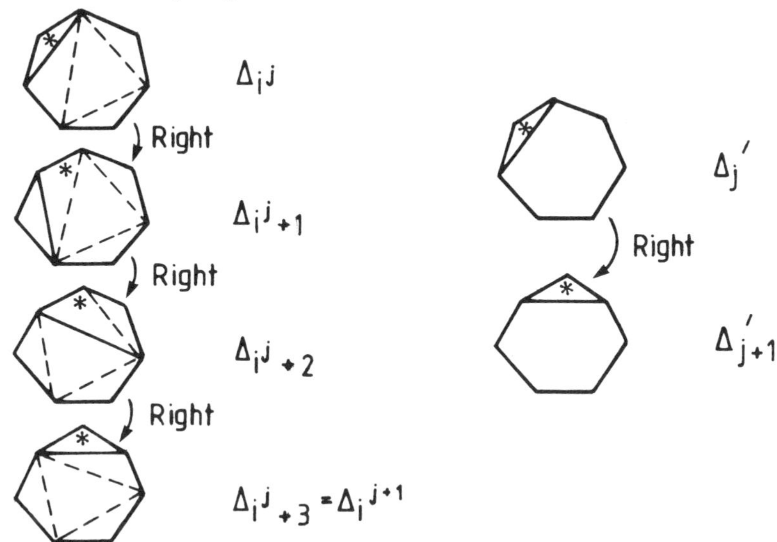

Figure 56.

If the loop **L** has an interior vertex, then the singularity prong of $[\Delta'_j]$ is defined to be the prong of $[\Delta'_j]$ "contained" in the singularity prong of $[\Delta_{ij}]$. It is easy to check that the rules for passing the singularity prong from $[\Delta'_j]$ to $[\Delta'_{j+1}]$ are satisfied.

Then $\mathbf{L}' = [\Delta'_0] \to \cdots \to [\Delta'_j] = [\Delta'_0]$ is a loop representing the same conjugacy class as **L**, and moreover, the maximal transparent set of **L** is trivial. We say that **L**′ is obtained from **L** by *breaking the windows*. By convention, if **L** already has only the trivial transparent set, then **L**′=**L**. Evidently, the maximal transparent set of **L**′ is empty.

We have proven the following:

Proposition 17. *If* **L** *is an R-L loop whose maximal transparent set of arcs carries on improper components, then R-L loop* **L**′ *obtained from* **L** *by breaking the windows represents the same conjugacy class as* **L**, *and has only trivial transparent set.*

3.3 Proof of Uniqueness.

We can now prove the uniqueness half of Theorem 1. It is a direct consequence of the following:

Proposition 18. *Let Φ be a pseudo-Anosov mapping class of singularity type $(r_i;s_j)$,* **L** *a loop representing the conjugacy class of Φ. Then* **L** *is associated to a sector of Φ if and only if the polygon type of* **L** *is given by equations 1), 2), or 3) of Theorem 1.*

Proof. One half of this is just Proposition 9.

So suppose that **L** is not associated to a sector of Φ. By Theorem 11, the maximal transparent set of **L** is non-empty, so let **L**$'$ be obtained by breaking the windows of **L**. By Proposition 17, the maximal transparent set of **L**$'$ is empty. By Theorem 11, **L**$'$ is associated to a sector of Φ. So by Proposition 9, the polygon type of **L**$'$ is given by equation 1), 2), or 3).

The idea is to show that the polygon types of **L** and **L**$'$ are related in such a way that it is impossible for **L** to also have its polygon type given by equation 1), 2), or 3).

We need a relation among polygon types which captures the relation of **L** to **L**$'$. Consider a convex N-gon **c** $(N \geq 1)$, and suppose that **c** is subdivided into a collection of smaller polygons by adding chords with ends on the vertices of **c**. Suppose this collection consists of a_i polygons with n_i sides, $(i=1, \ldots, k)$. Then

$$N = 2 + \sum a_i(n_i - 2) \tag{*}$$

One way to see this, in the case that $n_i \geq 3$ for each i, is to note that by an Euler-characteristic computation, if an n-gon is triangulated by chords connecting its vertices, there are $n-2$ triangles. The subdivision of **c** can be redefined to such a triangulation. Since the number of triangles in **c** is the sum of the number of triangles in each piece of the subdivision,

$$N - 2 = \sum a_i(n_i - 2)$$

In general, we wish also to allow for the possibility that one of the n_i is equal to 1 or 2. The formula can be easily extended to this case.

Suppose that the polygon types of **L** and **L**$'$ are $(p_m;q_n)$ and $(p'_m;q'_n)$ (we ignore the s part of the polygon type), and suppose at first that all of the transparent arcs of **L** are contained in a single **L**$'$ N-gon (this should actually be done on the level of the CDPs in **L** and **L**$'$; we leave the details to the reader). Then for some a_i, n_i, where N, a_i, n_i are related by equation (*), and $n_1 < n_2 < \cdots < N$, if the N-gon is non-punctured,

$$q'_N = q_N + 1,\ q'_{n_i} = q_{n_i} - a_i$$

and if the N-gon is punctured, then there is an $I \in \{1, \ldots, k\}$ such that

$$p'_N = p_N + 1,\ p'_{n_I} = p_{n_I} - 1,\ q'_{n_i} = \begin{cases} q_{n_i} - a_i & \text{if } i \neq I \\ q_{n_I} - a_I + 1 & \text{otherwise} \end{cases}$$

If $(p_m;q_n)$ and $(p'_m;q'_n)$ are two polygon types for which the above conditions hold, we say that $(p'_m;q'_n)$ *degenerates simply* to $(p_m;q_n)$. If there is a sequence of simple degenerations from $(p'_m;q'_n)$ to $(p_m;q_n)$, then we say that $(p'_m;q'_n)$ *degenerates* to $(p_m;q_n)$. In general, since $\mathbf{L}'$ is obtained by breaking the windows of $\mathbf{L}$, the polygon type of $\mathbf{L}'$ degenerates to $\mathbf{L}$ (as defined, this is a *strict* partial order on polygon types; so the above statement holds only when the maximal transparent set of $\mathbf{L}$ is non-empty).

For a polygon type $(p_m;q_n)$, define the *number of sides* of $(p_m;q_n)$ to be

$$\sum p_m \cdot m + \sum q_n \cdot n$$

From equation (*) it follows that if $(p'_m;q'_n)$ degenerates to $(p_m;q_n)$, then the number of sides of $(p'_m;q'_n)$ is strictly greater than that of $(p_m;q_n)$, by at least 2 for each simple degeneration. This follows from equation (*), since

$$\sum a_i n_i - N = 2 \cdot \sum a_i - 2$$

and $\sum a_i \geq 2$, which follows because the degeneration is strict.

Recall that the singularity type of the pseudo-Anosov Φ is $(r_i,\ s_j)$. Then the possible polygon types for $\mathbf{L}'$ are given by equations 1), 2) and 3) of Theorem 1, which are repeated here for reference (we drop the primes on p_i and q_j for convenience).

$$p_1 = r_1 - 1 \quad \dots\dots\dots\dots \quad 1)$$

$$p_i = r_i \qquad i \geq 2$$

$$q_j = s_j \qquad j \geq 3$$

or, for some $m \geq 2$,

$$p_m = r_m - 1 \qquad \ldots\ldots\ldots \; 2)$$

$$p_i = r_i \qquad i \neq m$$

$$q_{m+1} = s_{m+1} + 1$$

$$q_j = s_j \qquad j \neq m+1$$

or, for some $n \geq 3$,

$$p_i = r_i \qquad i \geq 1 \qquad \ldots\ldots\ldots\ldots\ldots\ldots\ldots\ldots \; 3)$$

$$q_n = s_n - 1$$

$$q_{n+1} = s_{n+1} + 1$$

$$q_j = s_j \qquad j \neq n, n+1$$

By computing the number of sides of any of these polygon types, it is easy to see that they are equal, except that polygon type 1) has 2 fewer sides than polygon types 2) or 3). so the question reduces to: can polygon types 2) or 3) degenerate to polygon type 1) ? Clearly this can only happen by a simple degeneration, where $N = 1$, $a_1 = 1$, $n_1 = 1$, $a_2 = 1$, $n_2 = 2$.

Since q_i is only defined for $i \geq 3$, this simple degeneration must change p_1 and p_2. However this is clearly not allowed by the rules for a simple degeneration.

In plain language, the maximal transparent set of **L** would have to be an arc separating a punctured 1-gon from a punctured 2-gon, which implies that Φ is reducible.

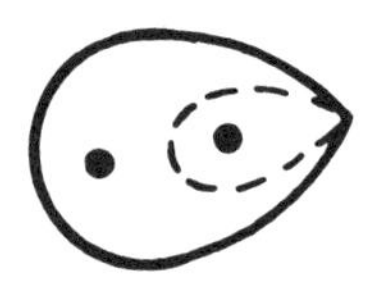

Figure 57.

Proposition 18

3.4 Counting Multiplicities.

In order to complete the proof of Theorem 1, we need to count multiplicities. In order to do this, we need to fix a few ideas about "sectors".

Let f be a canonical model for an arational measured foliation, and let $\Psi \in M(S,P)$ be such that $\Psi(\pi\{f\}) = \pi\{f\}$. Thus, there exists a unique $\lambda > 0$ such that $\Psi\{f\} = \{\lambda \cdot f\}$. Then the action of Ψ on the sectors of f is well-defined, in the following sense: Suppose that ψ, $\psi' \in \mathrm{Homeo}^+(S,P)$ are both in the mapping class Ψ, and suppose that $\psi(f) = \psi'(f) = \lambda \cdot f$.

Proposition 19. *Given a singularity s of f, $\psi(s) = \psi'(s)$; and, given a transversal α to f based at s, $\psi(\alpha)$ and $\psi'(\alpha)$ are in the same sector f at s.*

The proof of this proposition is relegated to the appendix. We use it to make sense out of the following:

Proposition 20. *Let Φ be a pseudo-Anosov, f the stable foliation, and suppose that Φ fixes all sectors of f. Let α, α' be non-generic transversals to f in different sectors, and let $\mathbf{L}$, $\mathbf{L}'$ be the loops associated to the sectors containing α and α'. Then $\mathbf{L} = \mathbf{L}'$ if and only if there exists $\Psi \in M(S,P)$ such that $\Psi(\pi\{f\}) = \{f\}$, and Ψ takes the sector containing α to the sector containing α'.*

Proof. If there exists such a Ψ, then by Proposition 19 and the naturality of Proposition 9, $\mathbf{L} = \mathbf{L}'$.

Conversely, suppose $\mathbf{L} = \mathbf{L}'$. Thus, we can chose sub-transversals of α, α' (denoted still by α, α') such that

1) The strings associated to f, α and f, α' are $\Delta_0 \to \Delta_1 \to \cdots$ and $\Delta'_0 \to \Delta'_1 \to \cdots$

2) There exists $M > 0$ such that, for all $i \geq 0$, $\Phi(\Delta_i) = \Delta_{i+M}$ and $\Phi(\Delta'_i) = \Delta'_{i+M}$. The loops $[\Delta_0] \to [\Delta_1] \to \cdots \to [\Delta_M] = [\Delta_0]$ and $[\Delta'_0] \to [\Delta'_1] \to \cdots \to [\Delta'_M] = [\Delta'_0]$ are equal.

Therefore, by choosing a different starting point for one of the loops (which amounts to taking another subtransversal of α or α') we can assume that $[\Delta_i] = [\Delta'_i]$ for all $i = 0, \ldots, M$. It follows that there exists $\Psi \in M(S,P)$ such that $\Psi(\Delta_i) = \Delta'_i$ for all $i \geq 0$.

Consider the two strings in 1) above. Since the maximal transparent set of $\mathbf{L}$ is empty (by Theorem 11), it follows from Corollary 15 and the Perron-Frobenius theorem that

$$\pi(\bigcap_k L(W(\Delta_{k \cdot M}))) = \pi(\bigcap_k L(W(\Delta'_{k \cdot M}))) = \pi\{f\}$$

Since $\Psi(\Delta_{k \cdot M}) = \Delta'_{k \cdot M}$ for all k, it follows that $\Psi(\pi\{f\}) = \pi\{f\}$.

Suppose that ψ is in the mapping class Ψ, and $\psi(f) = \lambda\cdot f$. Then $\Delta(f,\psi(\alpha)) = \Delta(\lambda\cdot f,\psi(\alpha)) = \Delta(\psi(f),\psi(\alpha)) = \Psi(\Delta(f,\alpha)) = \Psi(\Delta_0) = \Delta_0' = \Delta(f,\alpha')$

Therefore, by Proposition 5 we can replace α' by α'' in the same sector, and we can find $\theta \in \mathrm{Homeo}^+(S,P)$, whose mapping class is the identity, such that $\theta(f) = f$ and $\theta(\psi(\alpha)) = \alpha''$. But the identity map fixes sectors of f, so by Proposition 19, $\psi(\alpha)$ and α'' must be in the same sector of f. Therefore $\psi(\alpha)$ and α' are in the same sector of f.

Proposition 20

From this, the multiplicity count of Theorem 1 follows directly, for if N is the subgroup of $M(S,P)$ fixing $\pi\{f\}$, the action of N on sectors of f is well-defined; so if $\mathbf{L}$, $\mathbf{L}'$ are loops associated to different sectors of f, then by the above proposition, $\mathbf{L} = \mathbf{L}'$ if and only if the two sectors are in the same orbit under the action of N.

4. Proof of Theorem 2

Given pseudo-Anosovs Ψ, $\Psi \in M(S,P)$ such that $n(\Phi) = n(\Psi) = 1$, it is evident that the pseudo-Anosov invariants of Φ and Ψ are conjugacy invariants.

Suppose conversely that Φ, Ψ have identical pseudo-Anosov invariants. To prove that Φ and Ψ are conjugate, we first reduce to the case where Φ and Ψ have the same stable foliation. Let f, g be the respective stable foliations of Φ, Ψ. By the Stability Lemma, and using the fact that the invariants of Φ and Ψ are identical, there exist sectors of f and g and transversals in these sectors such that the associated R-L strings are combinatorially equivalent. So there exists $\Theta \in M(S,P)$ taking the string of f to the string of g. Using the Perron-Frobenius Theorem, it follows that $\Theta(f)=g$. Therefore Ψ and $\Theta^{-1}\circ\Phi\circ\Theta$ have the same stable foliation g; and they evidently both fix the singular prongs of g.

Assuming now that Φ and Ψ have the same stable foliation g, we show that $\Phi = \Psi$. Φ and Ψ have the same expansion factor λ, for they have the same matrix, since the pseudo-Anosov invariants are identical. So $\Psi\circ\Psi^{-1}$ fixes f (not just up to a factor), and fixes the sectors of f (since Φ and Ψ both do). By Proposition B of the appendix, $\Psi\circ\Phi^{-1}$ is the identity.

Now suppose that $\Phi,\Psi \in M(S,P)$ are arbitrary pseudo-Anosovs.

If Φ^m, Ψ^m are conjugate for some $m \geq 1$, note that $n(\Phi^m) = n(\Psi^m)$, since $n(\Theta)$ is a conjugacy invariant for a pseudo-Anosov $\Theta \in M(S,P)$; let n be the common value. It follows that Φ^{m-n} and Ψ^{m-n} fix all their singular prongs. Moreover, Φ^{m-n} and Ψ^{m-n} are conjugate, and so have identical pseudo-Anosov invariants.

Conversely, suppose that for some $n \geq 1$, Φ^n and Ψ^n both fix all their singular prongs and have identical pseudo-Anosov invariants. By the first part of Theorem 2, proven above, it follows that Φ^n and Ψ^n are conjugate.

Theorem 2

One might hope to obtain a complete classification for pseudo-Anosovs by keeping track of more information: for instance, $n(\Phi)$ is evidently a conjugacy invariant of a pseudo-Anosov Φ, as is the collection of pseudo-Anosov invariants of $\Phi^{n(\Phi)}$. This is not enough though, for two pseudo-Anosovs with identical invariants thus far may not act on their singular prongs by conjugate permutations. So one might hope to augment the invariants by adding on information about how the singular prongs are permuted, plus information about the loop associated to each cycle in this permutation, *etc.* This tack does not prove completely successful; the obstruction seems to be that for an arbitrary measured foliation f, the subgroup N_f may have outer automorphisms which are not induced by elements of $M(S,P)$; this is in contrast to the entire group $M(S,P)$, whose only non-trivial outer automorphism, in general, is induced by an orientation reversing mapping class of (S,P) (see [McCarthy, 1984]).

A more successful approach, suggested by the approach of [Fathi-Laudenbach-Poenaru, 1979], might be to study *families* of transversals to a measured foliation f, look at how such a family cuts (S,P) up into rectangles, and study how the combinatorial structure of this rectangle decomposition is altered when you change the family by isotopy. This would lead to a higher dimensional analogue of a loop: a cell-complex whose cells are labelled by combinatorial types of certain kinds of decompositions of (S,P); in this picture, the loops of Theorem 1 would appear as a 1-dimensional subcomplex.

Appendix

We prove here some facts about the action of a mapping class on the sectors of a measured foliation. The first result is repeated from Section 3.4.

Proposition A. *Let f be a canonical arational measured foliation, and let $\Psi \in M(S,P)$ be such that $\Psi(\pi\{f\}) = \pi\{f\}$. Thus, there is a unique $\lambda > 0$ such that $\Psi\{f\} = \{\lambda \cdot f\}$. Suppose that $\psi, \psi' \in \mathrm{Homeo}^+(S,P)$ are both in the mapping class Ψ, and suppose that $\psi(f) = \psi'(f) = \lambda \cdot f$. Then ψ and ψ' induce the same action on the singularities and sectors of f. More formally, given a singularity s of f, $\psi(s) = \psi'(s)$; and, given a transversal α to f based at s, $\psi(\alpha)$ and $\psi'(\alpha)$ are in the same sector of f at $\psi(s)$.*

Proposition B. *Let f be as above, and let $\phi \in \mathrm{Homeo}^+(S,P)$ be such that $\phi(f) = f$. Then ϕ is in the mapping class of the identity if and only if ϕ fixes all singularities of f, and all sectors at each singularity of f*

Proposition A follows directly from Proposition B, for $\psi'\psi^{-1}$ is in the mapping class of the identity, and $\psi'\psi^{-1}(f) = f$.

The idea of proving the "only if" part of Proposition B is to use the fact that a measured foliation on (S,P) is "half" of a conformal structure on (S,P). If $\phi \in \mathrm{Homeo}^+(S,P)$ is isotopic to the identity on (S,P), it is well known that if ϕ fixes some conformal structure on (S,P), then ϕ is the identity. While the fact that ϕ fixes f is not quite sufficient to guarantee that ϕ is the identity, it is sufficient to guarantee that ϕ maps every leaf of f to itself (by some homeomorphism in the leaf topology). The "only if" part follows from such considerations; the "if" part will fall out of the proof almost for free.

We shall not give all the details of the proof. In particular, we shall quote the following result from the theory of measured foliations and train tracks.

Fact. *There exists a measure foliation g transverse to f.*

This means that near each regular point, f and g are as in Figure 58;

near a singularity they are as in Figure 59.

Now we shall isotope ϕ, along leaves of f, until $\phi(g) = g$. This isotopy will not affect the action of ϕ on the singularities and sectors of f.

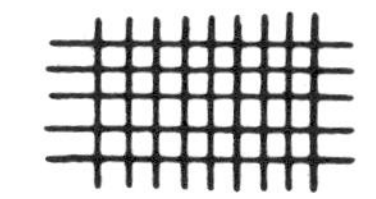

Figure 58.

or

Figure 59.

Let s be a singularity of f. Out of each sector s emerges a singular leaf in g. Choose a transversal to f based at s, lying on this singular leaf. Do this for each singularity and each sector of f. Let $\{\alpha_i\}$ be the family of transversals so chosen.

Figure 60.

The collection $\{\alpha_i\}$ can be chosen so that each transversal has the same f-length. Furthermore, by choosing this length short enough, we can guarantee that for each α_i, if α'_i is the chosen transversal in the sector containing $\phi(\alpha_i)$, then $\phi(\alpha_i)$ and α'_i are isotopic along leaves, without moving the common fixed endpoint. (This is just the stability property for measured foliations, used just before the statement of the Stability lemma). Note that if ϕ fixes sectors, $i = i'$.

Now alter ϕ, by isotoping along leaves of f, so that the resulting homeomorphism ϕ' actually permutes the family of transversals $\{\alpha_i\}$ by homeomorphisms, not just up to isotopy. Note that ϕ' acts as an f-isometry on each α_i.

The family of transversals $\{\alpha_i\}$, together with the measure foliation f, now determine a unique family of rectangles $\{H_j\}$. (We use the definition of rectangle appropriate to punctured surfaces, where the vertical side of a rectangle is allowed to fold around a 1-pronged singularity at the base of a transversal α_i).

ϕ' permutes by homeomorphisms the collection $\{\beta_k\}$ of horizontal sides of the rectangles $\{H_j\}$. Note that each β_k is transverse to g. Since some power of ϕ' takes each horizontal side β_k to itself by a homeomorphism, ϕ' must preserve the total g-length of each β_k.

Now alter ϕ', again isotoping along leaves of f, so that the resulting ϕ'' acts as a g-isometry on each β_k; this isotopy can be chosen to be stationary on the collection of vertical sides $\{\alpha_i\}$.

ϕ'' acts as a f-isometry on the collection of vertical sides $\{\alpha_i\}$, and as a g-isometry on the collection of horizontal sides $\{\beta_k\}$. Moreover, ϕ'' preserves leaves of f. Therefore, ϕ'' can be isotoped, along leaves of f, so that the resulting ϕ''' is a g-isometry on each interior f-leaf of a rectangle. This isotopy can be chosen to leave stationary the boundaries of the rectangles.

To complete the proof of the "if" part, suppose ϕ preserves sectors. It follows that ϕ' is the identity on each α_i, and maps each β_k to itself and each H_j to itself. Then ϕ'' is the identity on each α_i and β_k, and maps each H_j to itself. ϕ''' is the identity on each H_j, and so ϕ''' is the identity on (S, P).

We now continue the proof of the "only if" part. It is easy to see that ϕ''' preserves leaves of g in each rectangle. Therefore, $\phi'''(g) = g$. Also, since all isotopies were performed along leaves of f, $\phi'''(f) = f$.

f and g together determine a conformal structure on (S, P), i.e. a conformal structure on S, with P as a collection of distinguished points. To see this at a regular point: use f-measure as a y-coordinate, use g-measure as an x-coordinate, and take the underlying conformal structure of the resulting local Euclidean structure. At an n-pronged singularity $(n \geq 1)$, the f and g measures are given by the imaginary and real parts of a complex variable w, where $w^n = z^2$, and z is some local complex coordinate for the surface S at the singularity.

Since ϕ''' preserves f and g, ϕ''' is isotopic rel P to ϕ, and therefore to the identity. It is well-known that this implies that ϕ''' is the identity on (S, P). One way to see this is with the Uniformisation Theorem, together with Theorem 18 of exposé 3 of [Fathi-Laudenbach-Poenaru, 1979]. It follows that the action of ϕ''' on the singularities and sectors of f is the identity. Since the action of ϕ and the action of ϕ''' are identical, we have finished the proof of Proposition B.

Proposition B

References

Casson, 1983.
A. Casson, *Automorphisms of Surfaces after Nielsen and Thurston,* University of Texas at Austin, 1983. Lecture Notes by S.Bleiler

Fathi-Laudenbach-Poenaru, 1979.
A. Fathi, F. Laudenbach, and V. Poenaru, "Travaux de Thurston sur les Surfaces," *Société Mathématique de France*, **66-67**, 1979.

Gilman, 1981.
J. Gilman, "On the Nielsen type and the classification for the mapping class group," *Advances in Mathematics*, **40**, pp. 68-96, 1981.

Harer-Penner, 1986.
J. Harer and R. Penner, "Combinatorics of Train Tracks," *Annals of Math. Studies*, Princeton University Press, 1986. (with an Appendix by N. Kuhn) (to appear)

Hemion, 1979.
G. Hemion, "On the classification of homeomorphisms of 2-manifolds and the classification of 3-manifolds," *Acta. Math.*, **142**, pp. 123-155, 1979.

McCarthy, 1984.
J. McCarthy, *The Outer Automorphism Group of the Mapping Class Group of a Surface*, 1984. preprint

Miller, 1982.
R. Miller, "Geodesic Laminations from Nielsen's Viewpoint," *Advances in Mathematics*, **45**, pp. 189-212, 1982.

Mosher, 1983.
L. Mosher, *Pseudo-Anosovs on Punctured Surfaces*, 1983. Princeton Dissertation

Mosher, 1986.
L. Mosher, *Conjugacy Invariants of Mapping Class Groups of Surfaces*, 1986. In preparation

Nielsen, 1944.
J. Nielsen, "Surface Transformation classes of algebraically finite type," *Danske Vid. Selsk. Mat.-Fys. Medd.*, **XXI**, pp. 1-89, 1944.

Thurston, 1979.
W. P. Thurston, *The Geometry and Topology of 3-Manifolds,* Princeton University Mathematics Department, 1979. Part of this material – plus additional material – will be published in book form by Princeton University Press

L. Mosher,
Harvard University,
Mass. 02138,
U.S.A.

An Introduction to Train Tracks

Robert C. Penner

We begin by posing a problem which will motivate everything we do. Let F be a smooth, oriented, compact surface of negative Euler characteristic. We define the *mapping class group* $MC(F)$ of F to be the group of orientation-preserving diffeomorphisms of F modulo isotopy. We furthermore define the collection of *multiple curves* $S'(F)$ of F to be the set of isotopy classes of (non-oriented) closed one-manifolds embedded in the interior of F with each component essential and non-boundary parallel in F.

The action of diffeomorphisms of F on curves in F descends to a well-defined action of $MC(F)$ on $S'(F)$, and our motivating problem can be stated as

Motivating Problem MP. *Describe the action of $MC(F)$ on $S'(F)$. In particular, when does $\phi \in MC(F)$ fix some $\xi \in S'(F)$?*

A first idea towards approaching this problem is to consider the action of a fixed $\phi \in MC(F)$ on *subsets* of $S'(F)$. That is, we seek a geometric object so that the action of a representative of ϕ on this object determines the action of ϕ on a corresponding subset of $S'(F)$.

A *branched one-submanifold* of F is a CW complex T embedded in the interior of F so that

a) Each vertex of T is at least tri-valent.

b) Each edge of T is C^1, and for each vertex v of T, there is a smooth arc in T through v.

c) If e and e' are edges of T incident on the vertex v, then the one-sided tangents to e and e' at v either agree or differ by rotation by π.

An example of a branched one-submanifold in the two-holed torus is given in Figure 1a. (The numbers in the figure will be explained presently.)

We consider a close up of a vertex of a branched one-submanifold in Figure 2. The set of edges incident on the vertex partition into two disjoint

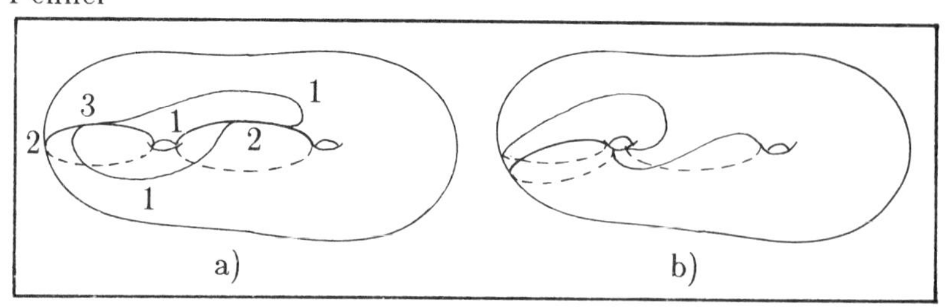

Figure 1.

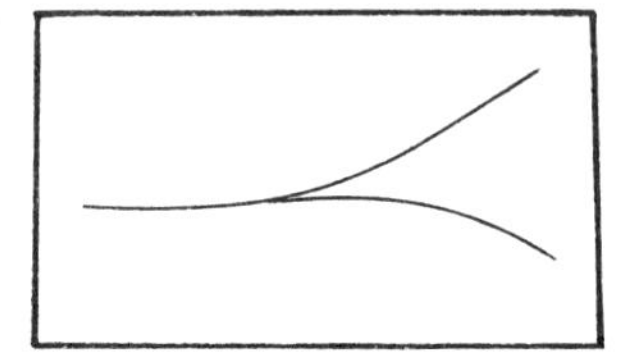

Figure 2.

sets depending on the direction of the one-sided tangent at the vertex. We arbitrarily choose to call one set of this partition *outgoing* and the other *incoming*.

Given a branched one-submanifold T of F, we will specify certain families in $S'(F)$ by assigning non-negative integers to the 1-cells of T. A one-submanifold is constructed from this set of integers by arranging the corresponding number of disjoint strands of string parallel to the branch and gluing the strands of string together at the switches. Figure 1b depicts the one-submanifold determined by the integers in Figure 1a. Clearly, to guarantee consistency of the gluings, we require a condition on the assignment of integers: for each vertex of T, the sum of assignments of integers to incoming edges must agree with the sum of assignments of integers to outgoing edges. We call this collection of conditions the *switch conditions* for T. We call the assignment of non-negative integers which satisfy the switch conditions a *weight* on T and let $Z(T)$ denote the set of all weights supported on T.

Having described subsets of $S'(F)$ by considering integral weights on branched one-submanifolds, we return to the problem MP and consider the action of $\phi \in MC(F)$ on multiple curves described by elements of $Z(T)$ for some fixed T. The key observation is that if $\mathbf{x}$ in $Z(T)$ describes $\xi \in S'(F)$ and f represents ϕ, then the weight in $Z(fT)$ corresponding to $\mathbf{x}$ describes $\phi\xi$; that is, f describes a one-to-one correspondence between the edges of T and the edges of fT, and we associate the weights on corresponding edges. Thus, the action of f on T describes the action of ϕ on the collection of multiple curves represented by weights in $Z(T)$. We illustrate this by example.

Example 1 If c is a simple closed curve in F, we let τ_c^{+1} (τ_c^{-1}) denote the right (left) Dehn twist along c. (Recall that a Dehn twist along a curve is a diffeomorphism supported on a neighbourhood of the curve, which is obtained by cutting the surface along c, performing a full twist to the right or left, and regluing.) We will read words in Dehn twists from right to left.

Let m and ℓ denote the meridian and longitude of the torus-minus-a-disc as depicted in Figure 3a, and consider the homeomorphism $f = \tau_m \tau_\ell^{-1} \tau_m^{-1}$. Let T be the branched one-submanifold depicted in Figure 3b, and consider the sequence of Dehn twists and isotopies depicted in Figure 3c-3f. (We write the letters a, b, c next to the edges of T in Figure 3 in order to keep track of the image of edges.)

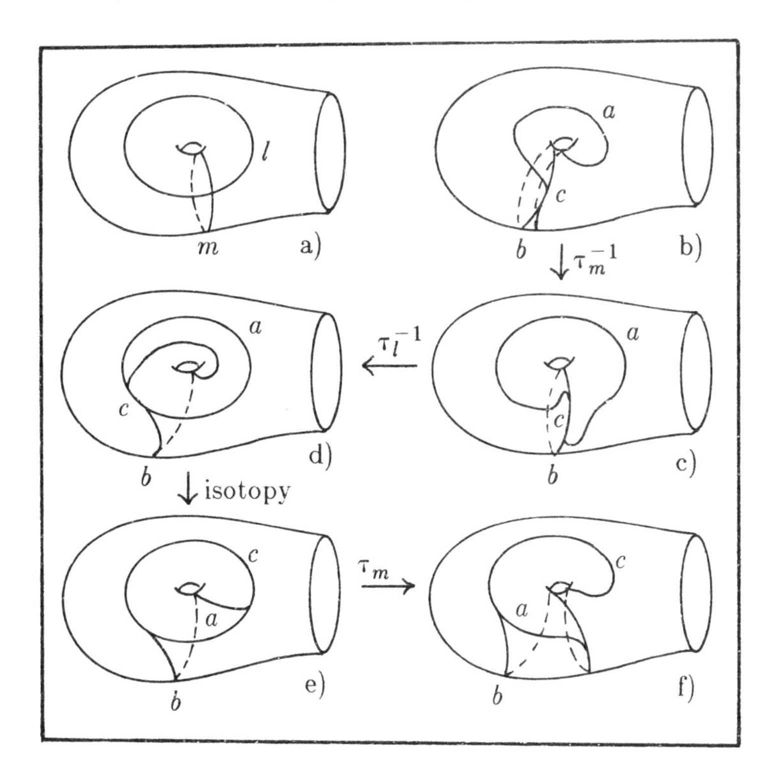

Figure 3.

Since f is the conjugate of the Dehn twist τ_ℓ^{-1} by τ_m, one sees that the isotopy class of f must fix the isotopy class of $\tau_m \ell$. The multiple curve corresponding to $\tau_m \ell$ is given by assigning the weights (1, 0, 1) to (a, b, c). Assigning these weights to the image track depicted in figure 3f, we again describe the isotopy class of $\tau_m \ell$, as expected.

The problem with considering arbitrary branched one-submanifolds is that a weight may not describe a multiple curve. Furthermore, distinct weights in $Z(T)$ can describe the same points of $S'(F)$. It is left to the reader to construct examples of such weights on the branched one-submanifold depicted in Figure 4. Thus, we will put an additional restriction on the branched manifolds we consider and arrive at the definition of train tracks.

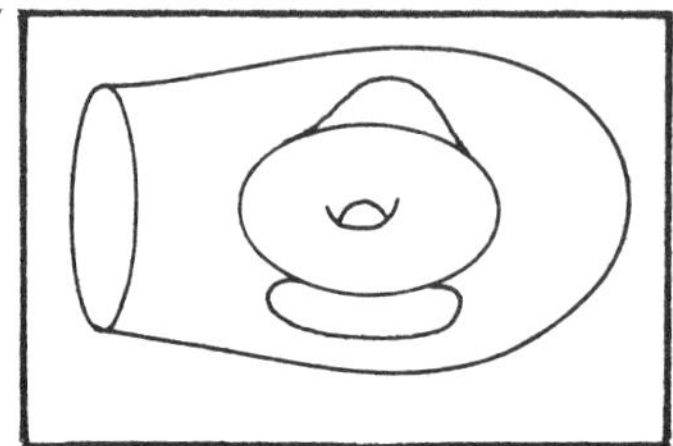

Figure 4.

A *train track* τ in F is a collection of smooth simple closed curves and branched one-submanifolds disjointly embedded in the interior of F so that no component of $F - \tau$ is one of the regions depicted in Figure 5. Some examples of train tracks are given in Figure 6.

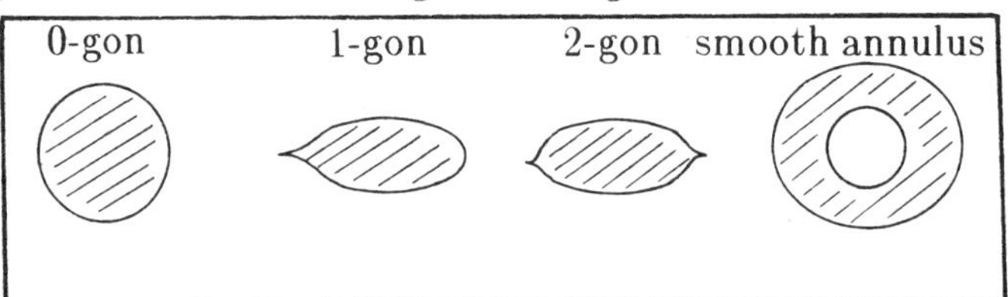

Figure 5.

Remark (exercise) A surface of non-negative Euler characteristic contains no train tracks.

We define the *branches* of the train track τ to be the open 1-cells of the branched manifold together with the simple closed curve components. The *switches* of τ are the vertices of the branched manifold components. We will sometimes regard an integral weight on a track τ as a function

$$\mu: \{\text{branches of } \tau\} \to \mathbb{Z}^+ \cup \{0\}$$

and sometimes as a vector of integers. The reader should consider examples of integral weights on the train tracks illustrated in Figure 6. Not every track τ supports a weight $\mu > 0$, and we say a track is *recurrent* if it supports such a weight. An example of a non-recurrent track is the "bifocals" depicted in Figure 7 (and embedded in an appropriate surface so that it is a train track). If a multiple curve ζ is represented by an integral weight in $Z(\tau)$, then we say that τ *carries* ζ.

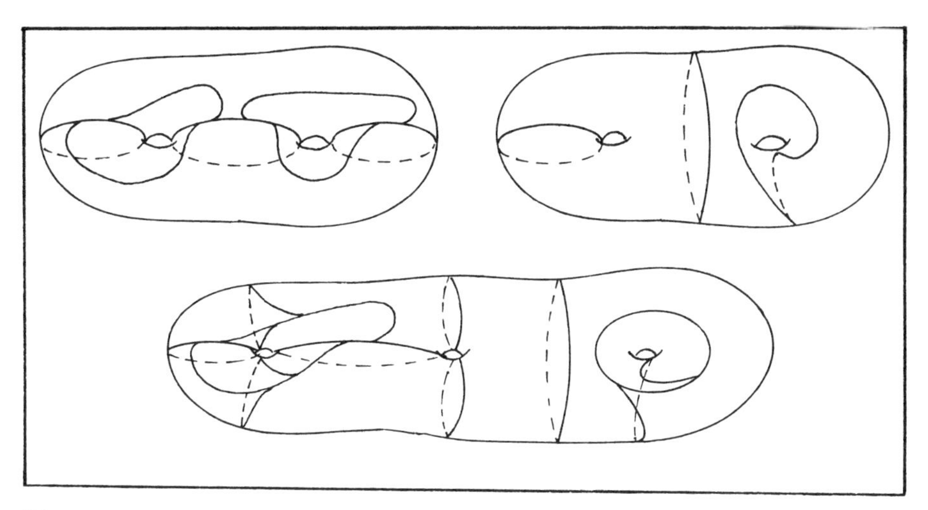

Figure 6.

Figure 7.

Proposition.

1) *If* $\mathbf{x}$ *is in* $Z(\tau)$, *then* $\mathbf{x}$ *describes a multiple curve in* F.

2) *If* $\mathbf{x} \not\doteq \mathbf{y}$ *in* $Z(\tau)$, *then* $\mathbf{x}$ *and* $\mathbf{y}$ *describe distinct elements of* $S'(F)$.

Proof. (exercise): Hint: it suffices to consider weights which describe *connected* one manifolds. Use the right hand rule for getting out of mazes. ∎

Thus, train tracks are pleasant from the point of view of carrying multiple curves. However, distinct tracks can carry the same curve. For instance, we may isotope a weighted train track about in the surface without altering the multiple curve defined. Figure 8 depicts two further moves, called *shifting* and *splitting* which we can perform on weighted tracks without altering the multiple curve defined. If two weighted tracks are related by isotopy, shifting and splitting, we will call them *equivalent*. Note that we may always shift a measured track so that all the switches are tri-valent, and we assume that this is the case from now on.

Note that unlike isotopy, the topological type of the track is changed by shifting and splitting. Furthermore, note that we may shift a track with no reference to the weights, while splitting is an operation which we can

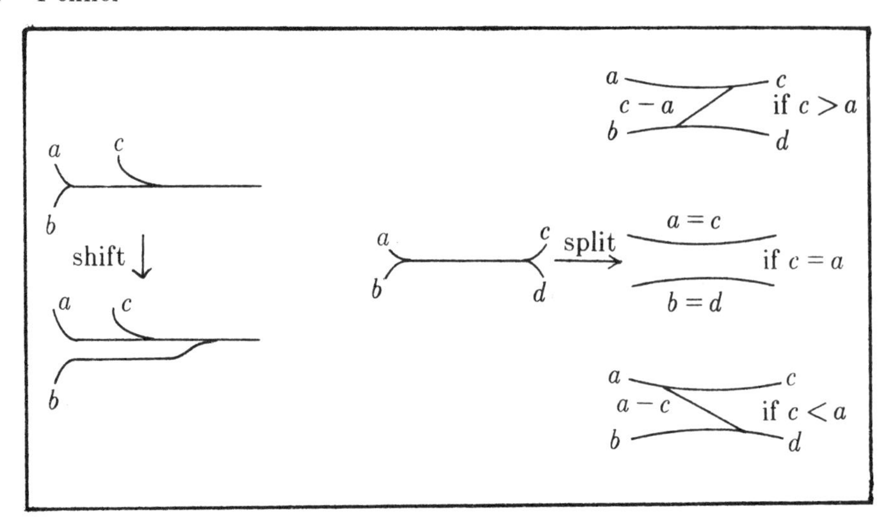

Figure 8.

perform only on weighted tracks.

We call the inverse of a split a *collapse*. If τ_1 is a track which collapses to τ, a weight $\mu_1 \in Z(\tau_1)$ gives rise to a weight $\mu \in Z(\tau)$ so that (τ_1, μ_1) and (τ, μ) describe the same elements of $S'(F)$. For example, in the collapse depicted in Figure 9, the weights are related by the following linear formula

$$\begin{pmatrix} a' \\ b' \\ c' \\ d' \\ e' \end{pmatrix} = \begin{pmatrix} 1 & 0 & 0 & 0 & 0 \\ 0 & 1 & 0 & 0 & 0 \\ 0 & 0 & 1 & 0 & 0 \\ 0 & 0 & 0 & 1 & 0 \\ 1 & 0 & 0 & 1 & 1 \end{pmatrix} \begin{pmatrix} a \\ b \\ c \\ d \\ e \end{pmatrix}$$

Thus, a collapse acts as an invertible integral linear map $Z(\tau_1) \to Z(\tau)$ in the natural coordinates.

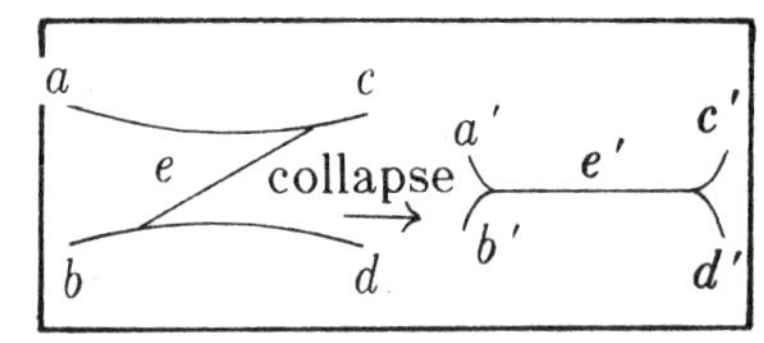

Figure 9.

Returning to our motivating problem, suppose that $\phi \in MC(F)$ and that τ is a train track with n branches and $\phi(\tau)$ collapses to τ. The action of ϕ on $Z(\tau)$ is given by a non-negative $n \times n$ integral matrix called the *incidence matrix* of ϕ for τ. Spectral properties of the incidence matrix translate into information about our motivating problem. For instance, in the previous example, the track depicted in Figure 3f collapses to the original track T of Figure 3b. The incidence matrix is given by $\begin{pmatrix} 0 & 0 & 1 \\ 0 & 1 & 0 \\ 1 & 2 & 0 \end{pmatrix}$ which has eigen-values 1 and -1 with corresponding eigen-vectors $(1, 0, 1)$ and $(1, 0, -1)$, respectively. We see our invariant multiple curve arising as an eigen-vector of the incidence matrix.

This example suggests a general procedure for our problem MP: given $\phi \in MC(F)$, find a track τ so that $\phi(\tau)$ collapses to τ and spectrally analyse the incidence matrix in search of invariant elements of $S'(F)$. Let's apply this reasoning to another example.

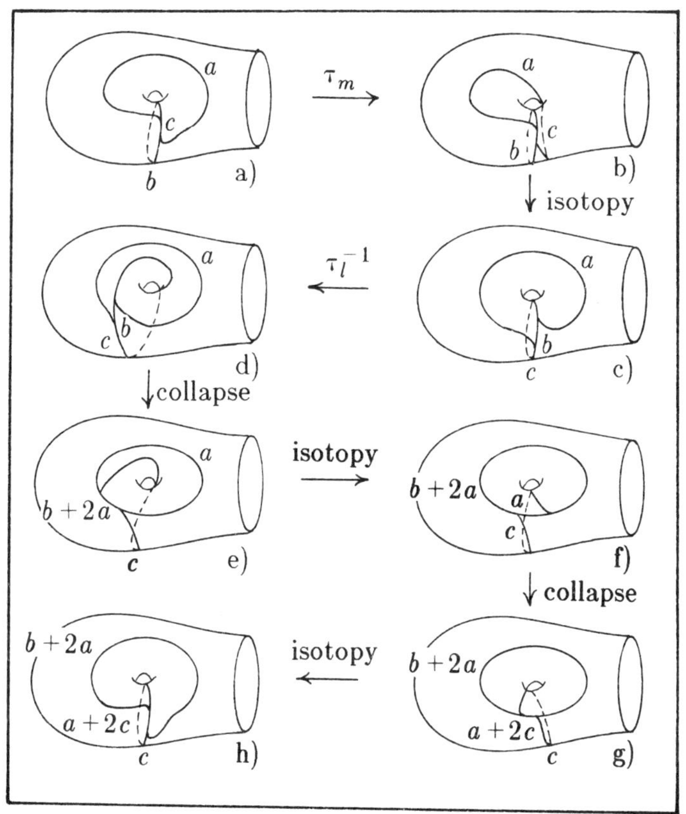

Figure 10.

Example 2. Again letting m and ℓ denote the meridian and longitude of the torus-minus-disc, we consider the action of homeomorphism $f = \tau_m \tau_\ell^{-1}$ on the train track τ depicted in Figure 10a. The sequence of Dehn twists, isotopies and collapses illustrated in Figure 10 allows us to compute that the incidence matrix is $\begin{pmatrix} 2 & 1 & 0 \\ 0 & 0 & 1 \\ 1 & 0 & 2 \end{pmatrix}$ which has eigen-values $(3+\sqrt{5})/2$,

$(3-\sqrt{5})/2$, and 1 with corresponding eigen-vectors $(1, \frac{\sqrt{5}-1}{2}, \frac{\sqrt{5}+1}{2})$, $(1, \frac{-1-\sqrt{5}}{2}, \frac{1-\sqrt{5}}{2})$, and $(-1, 1, 1)$, respectively. None of these eigen-vectors is non-negative integral, so we conclude that there is no multiple curve carried by τ which is invariant under the isotopy class of f. However, the unique non-negative eigen-vector satisfies the switch conditions for the track τ.

We are forced to define $V(\tau)$ to be the set of all functions

$$\mu : \{\text{branches of } \tau\} \to \mathbb{R}^+ \cup \{0\}$$

which satisfy the switch conditions. Elements of $V(\tau)$ are called *measures* on τ. Before we try to give a geometrical interpretation to elements of $V(\tau)$, we will investigate what $V(\tau)$ looks like. Suppose that τ has n branches. The switch conditions cut out a linear subspace of $(\mathbb{R}^+ \cup \{0\})^n$ (which is non-empty by recurrence). Since the switch conditions are projectively invariant, $V(\tau)$ is a finite-sided polyhedron. The faces of $U(\tau)$ are given by setting certain of the measures equal to zero. Thus, faces of $U(\tau)$ are given by $U(\sigma)$ for $\sigma \subset \tau$ a sub-track. The operation of successive splitting sub-divides $\mu(\tau)$ into smaller and smaller polyhedra, as illustrated:

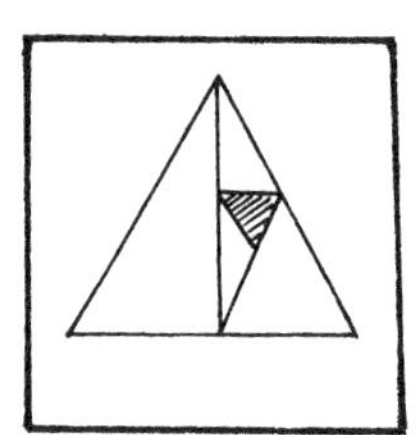

Figure 11.

What is the geometrical significance of an element of $V(\tau)$? Let's follow the intuition of "strands of string" which brought us to the notion of the train track in the first place. Choose a metric ρ on F and consider rectangular neighbourhoods R_i of each branch b_i of τ of width $\epsilon\mu(b_i)$ for some small $\epsilon > 0$. Relative to ρ, each R_i has a natural foliation by arcs parallel to b_i.

We may glue the R_i together at switches (by the switch conditions) to get a foliated neighbourhood $(\mathcal{F}, N(\tau))$ of τ, as illustrated in Figure 12. we define the *singular leaves* of $\mathcal{F}$ to be the leaves which begin at vertices of $\partial N(\tau)$. See Figure 12. Notice that $\mathcal{F}$ consists of various simple closed and/or bi-infinite paths in F together with the singular leaves of the foliation.

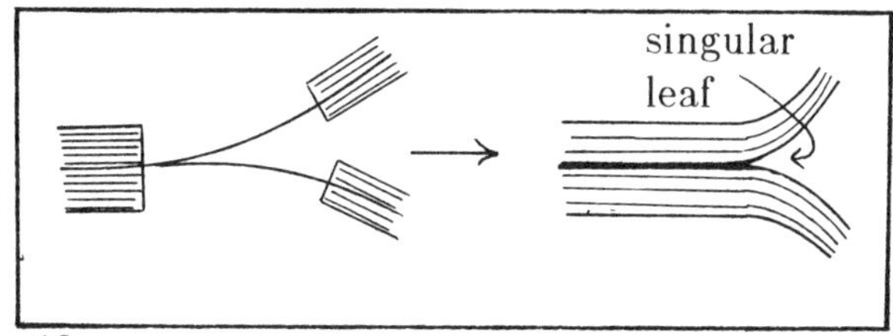

Figure 12.

We define a *lamination* to be a foliation of a closed subset of the interior of F. We say a lamination is a *geodesic lamination* for a hyperbolic metric ρ on F if the leaves of the lamination are geodesic. A *transverse measure* for a geodesic lamination G is a function Λ: $\{$arcs transverse to the leaves of G with endpoints in $F - G\} \to \mathbb{R}^+ \cup \{0\}$ so that the following conditions hold: Λ is invariant in the sense that

a) $\Lambda\alpha = \Lambda\beta$ whenever α is homotopic to β through arcs transverse to G with endpoints in $F - G$.

b) Λ is σ-additive in the sense that $\Lambda(\alpha) = \Sigma\Lambda(\alpha_i)$ whenever $\alpha = \cup\, \alpha_i$ for $\alpha_i \cap \alpha_{i+1} \subset \partial\alpha_i \cap \partial\alpha_{i+1}$ with each α_i transverse to G with endpoints in $F - G$.

c) Λ has full support in the sense that if α is transverse to G with endpoints in $F - G$ and $\alpha \cap G \neq \emptyset$, then $\Lambda(\alpha) \neq 0$.

We will usually denote a measured geodesic lamination (*mgl*) by G, suppressing the transverse measure from the notation.

The simplest example of an *mgl* is a collection of disjoint simple geodesics C_i in F together with a weighting $m_i \subset \mathbb{R}^+$, where we define $\Lambda\alpha = \sum_i m_i \,\#\, (\alpha \cap C_i)$.

Typically, however, the topology of how an *mgl* sits inside F is rather complicated. Since geodesic, no two leaves are parallel. $F - G$ has only finitely many components and the Lebesgue measure of G in F is zero. A typical transverse cross section to G in F is a Cantor set together with a finite union of isolated points (corresponding to closed leaves of G).

We have defined an *mgl* with respect to a fixed hyperbolic metric on F. However, the notion does not depend so closely on the metric as it seems at first. Suppose that G is an *mgl* for a metric ρ on F. If ρ' is another hyperbolic metric on F, then it turns out that there is an isotopy of G to an *mgl* G' for ρ' and the isotopy tells us how to give G' a transverse measure. The isotopy thus "straightens the leaves of G to geodesics for ρ'", as we see that the salient part of an *mgl* is the set of "homotopy classes" of leaves (and the transverse measure). For instance, consider the case described above, where the leaves of G are all simple closed curves.

Theorem. *There is a natural way to associate an mgl G to a measured train track (τ, μ). (Naturality means that if (τ, μ) determines the mgl G and $\phi \in MC(F)$ with f representing ϕ, then the measure on $f\tau$ arising from μ describes the mgl ϕG.)*

The idea of the proof is simple enough to describe, though the technical details are quite involved. It turns out that the non-singular leaves of $\mathcal{F}$ do determine a lamination on F (a foliation of a *closed* subset). We straighten the leaves to geodesics (throwing away multiple copies of the same "homotopy class" of leaf), and these steps require another technical condition on the track τ. We say τ is *transversely recurrent* if for any branch b of τ there is a simple closed essential non boundary parallel curve C in F so that C intersects b and no complementary region of $F - (C \cup \tau)$ is a bi-gon. An example of a recurrent non transversely recurrent track is:

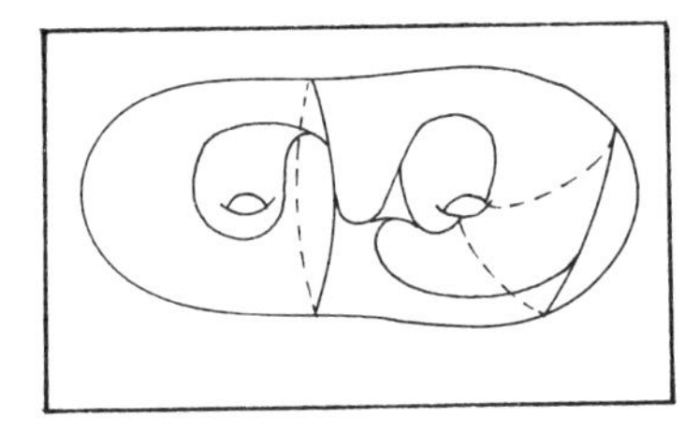

Figure 13.

We need this condition on the track τ to guarantee that the straightening process is possible with the result a lamination. We say a track is *bi-recurrent* if it is both recurrent and transversely recurrent. It turns out that we may always split a measured track to get a transversely recurrent one, so we do not require transverse recurrence as a hypothesis of the previous theorem. The transverse measure on the *mgl* G arises from considerations of how arcs intersect the neighbourhood $N(\tau)$ of τ.

Corollary. *If f is a diffeomorphism of F representing $\phi \in MC(F)$ and τ is a train track in F with $f\tau$ collapsing to τ, then there is an mgl G in F so that $\phi G = \lambda G$ for some $\lambda \in \mathbb{R}^+$. (This means that for any hyperbolic metric ρ on F, we may choose a representative homeomorphism f for ϕ which fixes the class of the mgl G for ρ multiplying measures by λ).*

Proof. Apply the Brouwer fixed point theorem on $U(\tau)$. Moreover, one can compute the *mgl* by considering spectral properties of the incidence matrix, taking care that the computed eigen-vector satisfies the switch conditions. **Exercise** If the incidence matrix, has a positive iterate, then the eigen-vector corresponding to the eigen-value of maximum modulus is positive real and satisfies the switch conditions.

□

We give the collection of all *mgl*'s on the surface F a topology as follows. If (G, Λ) is an *mgl*, then we define the *variation function*

$$V_{(G,\Lambda)} : S'(F) \to \mathbb{R}^+ \cup \{0\}$$

$$V_{(G,\Lambda)}\, \xi = \inf_{\gamma \in \xi} \Lambda(\gamma),\ \gamma$$

with γ transverse to G for $\xi \in S'(F)$. Let me remark that the infimum is always achieved, and the variation of a weighted system of disjoint simple closed geodesics is simply the weighted intersection number of an arc with these geodesics. Furthermore, distinct *mgl*'s have distinct variation functions.

We give the collection of all *mgl*'s the weak-star topology for the variation functions, so that the two *mgl*'s are close together if their variation functions are close together in a pointwise sense. The (geodesic) *measured lamination space* is denoted by $ML(F)$. Multiplying measures by a positive constant gives a $\mathbb{R}^+$-action on $ML(F)$; the quotient $PL(F) = [ML(F) - \{0\}]/\mathbb{R}^+$ is the *projective lamination space*. $MC(F)$ acts on both $ML(F)$ and $PL(F)$ in the natural way. Some answers to what the spaces $ML(F)$ and $PL(F)$ look like are provided in the following

Theorem.

0) *Any mgl arises from a measure on some bi-recurrent track (and we say the track carries the mgl).*

1) $V(\tau) \to ML(F)$ *is a continuous injection for any bi-recurrent track* τ. *This map is open if* τ *is not a subtrack of any train track in* F.

2) (τ_1, μ_1) *and* (τ_2, μ_2) *give rise to the same mgl's if and only if there is a sequence of isotopies and splits (and collapses) relating them.*

Consequences.

1) The variation topology agrees with the topology of train tracks.

2) $ML(F)$ (and $PL(F)$) are manifolds with coordinate charts given by $V(\tau)$ (and $U(\tau)$) for τ a bi-recurrent track.

3) $ML(F)$ has a natural (with respect to the action of $MC(F)$) *PIL* (piecewise integral linear) structure: the transition functions from

one track to another are given by invertible *integral* linear maps. (This structure is natural because the image of a track under a diffeomorphism is itself a track.) Similarly, $PL(F)$ has a natural projective PIL structure.

4) As with $S'(F)$, the action of $MC(F)$ on $ML(F)$ is described by the action on train tracks. In particular, $MC(F)$ acts continuously on $ML(F)$ and $PL(F)$.

5) $PL(F)$ is the completion of $S'(F)$ in the following sense. A multiple curve is itself an *mgl*, and the collection of projective variation functions of multiple curves in dense in $PL(F)$.

So $PL(F)$ is in particular a PL manifold with PL charts given by train tracks. One wonders what $PL(F)$ is topologically. A very concrete approach to this question is via the following:

Theorem. *There is a finite collection* $\{\tau_i\}_1^{N(F)}$ *of bi-recurrent train tracks so that* $\{U(\tau_i)\}_1^{N(F)}$ *describes a cell-decomposition of* $PL(F)$*; that is 1)* $U(\tau_i)$ *and* $U(\tau_j)$ *have disjoint interiors and meet along a common (possibly empty) face and 2) any measured track* (τ, μ) *is equivalent to a measure on some track* τ_i *in the collection.*

By studying these "standard" tracks, one can build $PL(F)$ a cell at a time to get an explicit description of $PL(F)$ as a cell complex and analyse the topology. For instance, on the torus-minus-a-disc, there are four standard tracks which are depicted next to the edges of the square in Figure 14. Each standard track has $V(\tau)$ a triangle so $U(\tau)$ is a line segment. These four line segments meet at pairs of endpoints, so $PL(F)$ is a circle. The multiple curves which correspond to the endpoints of these segments are also depicted in Figure 14.

Corollary. *If* F *has genus* g *and* r *boundary components, then* $PL(F)$ *is a* $(6g-7+2r)$*-sphere.*

There is another approach to seeing what $PL(F)$ is which explains the appearance of the famous number $6g-6+2r$ and will return us to our motivating problem. Recall the Teichmüller space $T(F)$ of F is the space of all complete metrics on a fixed topological model for F of constant curvature -1 and finite area, modulo push-forward by diffeomorphisms isotopic to the identity. $T(F)$ is an open $6g-7+2r$ dimensional ball (by Teichmüller theory). We will compare $T(F)$ and $ML(F)$ by considering them as subsets of

$$(\mathbb{R}^+ \cup \{0\})^{S'(F)} = \{\text{ functions from } S'(F) \text{ to } \mathbb{R}^+ \cup \{0\}\}.$$

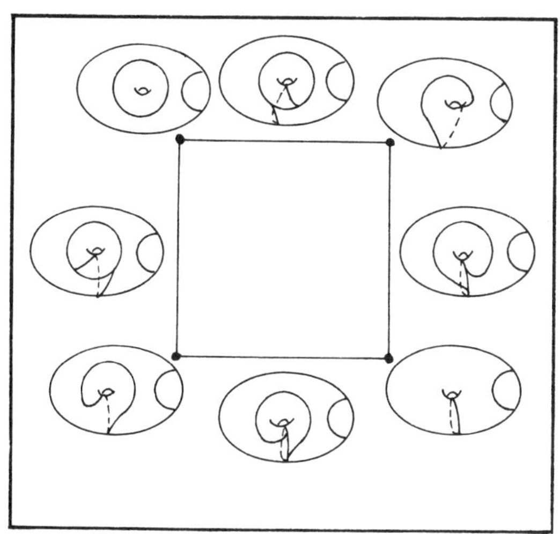

Figure 14.

We regard $T(F)$ as a subset of $(\mathbb{R}^+ \cup \{0\})^{S'(F)}$ by assigning to $\rho \in T(S)$ the geodesic length function l_ρ of ρ, and we regard $ML(F)$ as a subset of $(\mathbb{R}^+ \cup \{0\})^{S'(F)}$ via the variation function V_G as before. There is a formal isomorphism between $T(F)$ and $ML(F)$ given by an isomorphism between parametrisations of these spaces (the Fenchel-Nielsen parametrisation of $T(F)$ and the Dehn-Thurston parametrisation of $ML(F)$.)

Theorem. *$l_\rho \sqcup V_G : T(F) \sqcup ML(F) \to (\mathbb{R}^+ \cup \{0\})^{S'(F)}$ is an embedding. Moreover, a sequence $\rho_i \in T(F)$ converges projectively if and only if the corresponding sequence $G_i \in ML(F)$ does, and these sequences converge to the same point.*

Corollary. *$PL(F)$ forms a boundary for a compactification of $T(F)$ so that $T(F) \cup PL(F) = B^{6g-6+2r}$. Furthermore, the natural action of $MC(F)$ on $T(F)$ extends continuously to the natural action on $PL(F)$.*

Corollary. *If $\phi \in MC(F)$ is not finite order, then ϕ fixes some projective geodesic lamination.*

Proof. Apply the Brouwer fixed point theorem on $B^{6g-6+2r}$ together with the fact that any isometry of a hyperbolic surface is of finite order. $\square$

In fact, somewhat more is true.

Technical Lemma. *Suppose that* $\phi \in MC(F)$ *and* $G \in ML(F)$ *with* $\phi(G) = \lambda G$. *If 0 is not in the image of* V_G, *then there is an mgl* H *transverse to* G *with* $\phi H = \frac{1}{\lambda} H$ *and 0 not in the image of* V_H.

We say that $\phi \in MC(F)$ is *periodic* if ϕ is of finite order in $MC(F)$. ϕ is *reducible* if there is a point of $S'(F)$ which is invariant under ϕ. ϕ is *pseudo-Anosov* if there exist transverse *mgl*'s G_s and G_u (s for stable and u for unstable) so that 0 is not in the image of either V_{G_s} or V_{G_u} with $\phi G_s = \frac{1}{\lambda} G_s$ and $\phi G_u = \lambda G_u$ for some $\lambda > 1$.

Theorem. *If* $\phi \in MC(F)$, *then* ϕ *is either pseudo-Anosov, periodic, or reducible. The only overlap in this classification is between periodic and reducible mapping classes.*

In particular, we obtain a solution to our problem MP.

Corollary. *Given* $\phi \in MC(F)$, *either* ϕ *is pseudo-Anosov, or there is an*
Laudenbach-Poenaru, 1979] [Harer-Penner, 1986]

References

Casson, 1983.
A. Casson, *Automorphisms of Surfaces after Nielsen and Thurston,* University of Texas at Austin, 1983. Lecture Notes by S.Bleiler

Fathi-Laudenbach-Poenaru, 1979.
A. Fathi, F. Laudenbach, and V. Poenaru, "Travaux de Thurston sur les Surfaces," *Société Mathématique de France*, **66-67**, 1979.

Harer-Penner, 1986.
J. Harer and R. Penner, "Combinatorics of Train Tracks," *Annals of Math. Studies*, Princeton University Press, 1986. (with an Appendix by N. Kuhn) (to appear)

R.C. Penner,
University of Southern California,
Los Angeles,
Calif 90007,
USA.

Earthquakes in two-dimensional hyperbolic geometry

William P. Thurston

1. Introduction

A hyperbolic structure on a surface is related to any other by a left earthquake (to be defined presently). I proved this theorem in the case of closed surfaces several years ago, although I did not publish it at that time.

Stephen Kerckhoff made use of the earthquake theorem in his proof of the celebrated Nielsen Realization Conjecture [Kerckhoff, 1983], and he presented a proof in the appendix to his article.

The original proof in some ways is quite nice, but it has shortcomings. It is not elementary, in that it makes use of the understanding and classification of measured laminations on a surface as well as the classification of hyperbolic structures on the surface. It also uses some basic but non-elementary topology of $\mathbb{R}^n$. Given this background, the proof is fairly simple, but developed from the ground up it is complicated and indirect.

In this paper, I will give a more elementary and more constructive proof of the earthquake theorem. The new proof is inspired by the construction and analysis of the convex hull of a set in space. The new proof also has the advantage that it works in a very general context where the old proof would run into probably unsurmountable difficulties involving infinite-dimensional Teichmüller spaces.

The first part of this paper will deal with the hyperbolic plane, rather than with a general hyperbolic surface. This will save a lot of fussing over extra definitions and cases. The general theory will be derived easily from this, in view of the fact that the universal cover of any complete hyperbolic surface is the hyperbolic plane.

The standard definition for a hyperbolic structure on a manifold is an equivalence class of hyperbolic metrics, where two metrics are equivalent when there is a diffeomorphism isotopic to the identity which acts as an isometry from one metric to the other. However, this definition is not appropriate for studying hyperbolic structures on manifolds via their universal covers, since by this definition there is only one possible structure on the universal cover.

The solution to this difficulty is to use a smaller equivalence relation which keeps track of more information — information which describes the behavior of the metric at infinity.

1.1 Definition. A *relative hyperbolic structure* on $(\mathbb{H}^2, S^1_\infty)$ is a complete hyperbolic metric on the hyperbolic plane, up to the relation that two metrics are equivalent if there is an isometry ϕ between them which

a). is isotopic to the identity, and

b). which extends continuously to be the identity at infinity.

(The condition on isotopy is redundant here, but it will be needed for the more general case).

There are relative hyperbolic structures of varying quality. Let *adjective* be a quality which describe maps of the circle to itself: for instance, continuous, Lipschitz, quasi-symmetric, etc. We suppose that an *adjective* map composed on the left and right with any element of $PSL(2,\mathbb{R})$ is still *adjective*. Then a relative hyperbolic structure is *adjective* at infinity if there is an isometry to the standard hyperbolic structure which extends to continuously to a map which is *adjective* at infinity.

It is well-known that a homeomorphism between two closed hyperbolic surfaces lifts to a map between their universal covers which extends uniquely to a continuous map on $\mathbb{H}^2 \cup S^1_\infty$. Therefore, a hyperbolic structure on a surface determines a continuous relative hyperbolic structure on $(\mathbb{H}^2, S^1_\infty)$.

One nice thing about studying continuous relative hyperbolic structures on $(\mathbb{H}^2, S^1_\infty)$ is that there is an easy classification of them. For any continuous relative hyperbolic structure h, there is a homeomorphism f_h of the disk which takes h isometrically to the standard hyperbolic structure on the hyperbolic plane. Clearly, $f_h | S^1_\infty$ is determined by the relative hyperbolic structure h up to post-composition (that is, composition on the left) by homeomorphisms of the circle which come from isometries of the hyperbolic plane, that is, elements of $PSL(2,\mathbb{R})$.

1.2. Proposition: homeomorphisms classify hyperbolic structures. *There is a one-to-one correspondence between the set of right cosets*

$$PSL(2,\mathbb{R})\backslash homeomorphisms(\mathbb{H}^2)$$

and the set of continuous relative hyperbolic structures on $(\mathbb{H}^2, S^1_\infty)$.

Proof. homeomorphisms classify hyperbolic structures: The only remark needed to complete the proof is that every homeomorphism of the circle extends to a homeomorphism of the disk. Any such homeomorphism pulls back a continuous relative hyperbolic structure. Every relative hyperbolic structure arises in this way. Two relative hyperbolic structures are equivalent iff they are in the same right coset of $PSL(2,\mathbb{R})$.

homeomorphisms classify hyperbolic structures

2. What are hyperbolic earthquakes?

A left earthquake is intuitively a change in hyperbolic structure which is obtained by shearing toward the left along a certain set of geodesics, the *faults* ; the union of these geodesics is called the *fault zone* .

The simplest example of a left earthquake on the hyperbolic plane is obtained by cutting along a geodesic g and gluing the two half-planes A and B back together again by attaching the boundary of B to the boundary of A by an isometry which moves to the left a distance d along the boundary of A, as viewed from A. (In other words, the displacement is in the positive sense with respect to the orientation of $g = \partial A$ induced from A.) Such an earthquake is an *elementary* left earthquake. The description is exactly the same if the roles of A and B are interchanged: the attaching map is replaced by its inverse, and the induced orientation of g is reversed, so the attaching map still shifts to the left by a distance d.

For this definition to make sense, it is necessary that our hyperbolic plane be equipped with an orientation. This orientation will be preserved throughout; whenever we speak of isometries of the hyperbolic plane, they will be understood to preserve the orientation.

The cutting and regluing procedure, strictly speaking, does not define a new hyperbolic structure on the hyperbolic plane, since the map — called an *earthquake map*— between the hyperbolic plane and the new hyperbolic manifold is discontinuous. This difficulty can be resolved by approximating the earthquake map by continuous maps. However, since we are

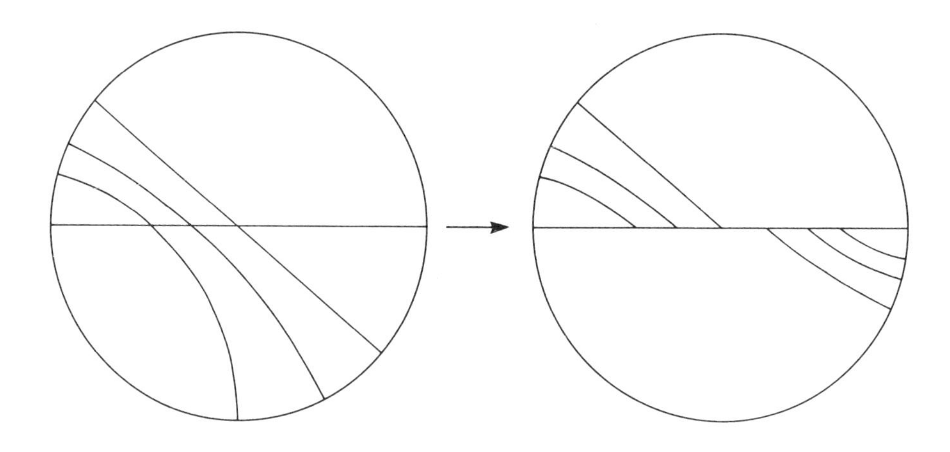

2.1 Figure. *The effect of an elementary left earthquake.*

working with the hyperbolic plane, an easier way to think of it is in terms of the circle at infinity, where the limit of the earthquake map is continuous. To identify the two hyperbolic planes which differ by a given elementary earthquake, we can use any homeomorphism between the two hyperbolic planes which extends the continuous map at infinity. In any event, the earthquake determines a relative hyperbolic structure on $(\mathbb{H}^2, S^1_\infty)$.

In a similar way, we can construct left earthquakes whose faults are finite sets of disjoint geodesics. We could also construct left earthquakes whose faults are countable sets of geodesics , although if the countable set has accumulation points we would have to worry about considerations of summability for the set of displacements. What we really need, however, is a definition for left earthquakes whose faults consist of an arbitrary closed set of disjoint geodesics.

2.2 Definition. A *geodesic lamination* λ on a Riemannian surface is a closed subset L (the locus of λ) of the surface together with a foliation of the set by geodesics. In other words, for any point p in the locus of λ, there is a neighborhood U of p on the surface such that $L \cap U$ is given the structure of a product $X \times (0,1)$ where each $x \times (0,1)$ is a piece of a geodesic; the geodesics in the product structures assigned to different neighborhoods are required to match.

The geodesics in these product structures can be extended uniquely and indefinitely in both directions, assuming the Riemannian metric is complete, although the extension will repeat periodically, if the geodesic is closed.

These geodesics are called the *leaves* of λ. The locus of a lamination is the disjoint union of its leaves.

The components of the complement of a lamination are its *gaps*. Sometimes it is useful to refer to the metric completions of the gaps, with respect to their Riemannian metrics (so that they become complete Riemannian surfaces with geodesic boundary). We will call these the *complete gaps*.

The gaps together with the leaves of a lamination, are its *strata*.

2.3 Definition. If λ is a geodesic lamination on a the hyperbolic plane, S, a λ-*left earthquake map* E is a (possibly discontinuous) injective and surjective map E from the hyperbolic plane to the hyperbolic plane which is an isometry on each stratum of λ. E must satisfy the condition that for any two strata $A \neq B$ of λ, the *comparison isometry*

$$cmp(A, B) = (E|A)^{o-1} \circ (E|B) : \mathbb{H}^2 \to \mathbb{H}^2$$

is a hyperbolic transformation whose axis weakly separates A and B and which translates to the left, as viewed from A. Here, $(E|A)$ and $(E|B)$ refer to the isometries of the entire hyperbolic plane which agree with the restrictions of E to the given strata. A line l weakly separates two sets A and B if any path connecting a point $a \in A$ to a point $b \in B$ intersects l. Shifting to the left is defined as before: the direction of translation along l must agree with the orientation induced from the component of A in $\mathbb{H}^2 - l$. One exceptional case is allowed: if one of the two strata is a line contained in the closure of the other, then the comparison isometry is permitted to be trivial.

A *left earthquake* on the hyperbolic plane is a structure consisting of two copies of the hyperbolic plane (the source and target for the earthquake), laminations λ_s and λ_t on the source and target hyperbolic planes, and a λ_s-earthquake map between them which sends the strata of λ_s to the strata of λ_t. A *right earthquake* is defined similarly. The inverse of a left earthquake is a right earthquake. It is obtained by interchanging source and target hyperbolic planes, source and target laminations, and using the inverse of the left earthquake map.

Why not allow earthquakes to shift both to the right and to the left? There is no particular reason, except that the theory becomes very different. A comparison can be made between the properties of earthquakes and the properties of functions $\mathbb{R} \to \mathbb{R}$. Left and right earthquakes are like not necessarily continuous monotone functions (either non-increasing or non-decreasing). A general earthquake would be like a completely general function. The theory of monotone functions is neat and useful, and closely related to the theory of measures on the line. The theory of general functions (without additional conditions and structure) is terrible.

This definition of an earthquake is a bit different from the definitions that have been used previously in the case of hyperbolic surfaces of finite area, where less care was needed because geodesic laminations on such surfaces automatically have measure 0. To motivate and justify the present definition, we prove

2.4. Proposition: simple left earthquakes are dense. *If λ is a finite lamination, a map $E: \mathbb{H}^2 \to \mathbb{H}^2$ which is an isometry on each stratum of λ is a left earthquake map iff the comparison maps for adjacent strata are hyperbolic transformations which translate the separating leaf of λ to the left.*

Left earthquake maps with finite laminations are dense in the set of all left earthquake maps, in the topology of uniform convergence on compact sets.

This says that left earthquakes could have been alternatively defined as the closure of the set of left earthquakes which have finite laminations. Such earthquakes we will call *simple* left earthquakes.

Proof. simple left earthquakes are dense:

First we need to verify that if s and t are hyperbolic transformations with disjoint axes such that they translate in the "same" direction, then the axis of $s \circ t$ separates the axes of s from t. (To translate in the "same" direction means that the region between the axes of s and t induces an orientation on the axes so that which agrees with one of the translation directions and disagrees with the other.) One way to see this is by looking at the circle at infinity: in the interval cut off by the axis of s (or by the axis of t), both s and t are moving points in directions which agree. This forces $s \circ t$ to have a fixed point in each of the two intervals between the axes of s and t, since the composition maps one of these intervals into itself, and the inverse of the composition maps the other of these intervals into itself.

By induction on the number of strata separating two strata of the finite lamination λ of the proposition, it follows that the comparison isometry of an arbitrary pair of strata shifts to the left if all the comparison isometries for adjacent strata do.

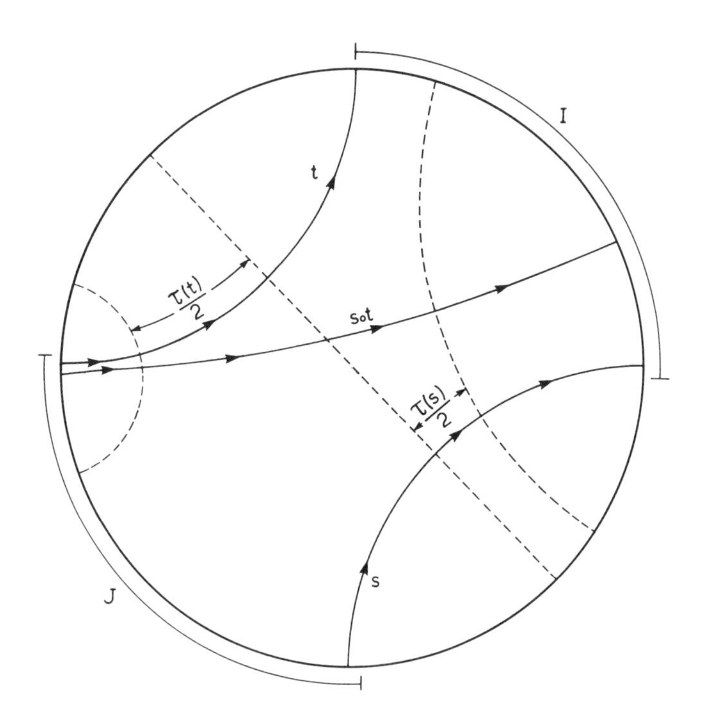

2.5 Figure. *The composition of two hyperbolic isometries which translate in the "same" direction is a hyperbolic isometry whose axis separates the axes of the original two isometries which also translates in the same direction. The interval I is mapped into itself by the composition, while the interval J is mapped over itself. The fixed points of the composition within these two intervals are the endpoints of its axis. Any translation is the product of two reflections.*

The proof of the density of simple left earthquakes among all left earthquakes is quite easy.

First, observe that the image of a compact set by a left earthquake map has a bounded diameter, hence a compact closure. The proof of this is in analogy to the theorem that the image of compact interval by a monotone function is bounded and is left to the reader.

Given any left λ-earthquake E, then for any compact subset K of the hyperbolic plane, there is a finite subset of strata which intersect K such that the union of the images of this finite set of strata in the graph of $E \subset \mathbb{H}^2 \times \mathbb{H}^2$ is ϵ-dense in the graph of E restricted to K.

Define a new left earthquake F by the condition that the isometries obtained by restricting F to its strata are exactly the isometries obtained by restricting E to the chosen finite set of strata. The lines separating adjacent strata of F are axes of certain comparison isometries for E. These are disjoint, as any two are separated by at least one of the strata, so they form a lamination which is respected by F. The simple earthquake map F approximates E in the set K.

simple left earthquakes are dense

Even though earthquake maps are not (in general) continuous, they have an extension which is continuous at infinity:

2.6. Proposition: quake at infinity. *If E is a λ-earthquake map from $\mathbb{H}^2$ to $\mathbb{H}^2$, then there is a unique map*

$$E_\infty: S^1_\infty \to S^1_\infty$$

such that E together with E_∞ form a map which is continuous at each point $x \in S^1_\infty$.

Proof. quake at infinity: First we will make sure that E_∞ is well-defined.

If the closures of A and of B at infinity intersect, then $E|A$ and $E|B$ agree in this intersection (which consists of 1 or 2 points), because the axis of the comparison isometry $cmp(A, B)$ must hit these points at infinity.

If a point $x \in S^1_\infty$ is not in the closure of any stratum of λ, then it has a neighborhood basis (in D^2) consisting of neighborhoods bounded by leaves l_i of λ. The diameters of the image leaves $E(l_i)$ must go to zero, for otherwise a limit point in $\mathbb{H}^2$ of their images would have to be on the image of some stratum which would violate the fact that E preserves separation properties. This determines a unique point which is defined to be $E_\infty(x)$.

This defines an injective and surjective map of the circle to itself which preserves the circular order. Any such map is continuous.

quake at infinity

Just as in the case of an elementary earthquake, an arbitrary left earthquake determines a relative hyperbolic structure on $(\mathbb{H}^2, S^1_\infty)$. In fact, the earthquake map up to the equivalence relation of post-composition (composition on the left) with isometries of $\mathbb{H}^2$ determines a relative hyperbolic structure. This equivalence relation on earthquake maps is a nice one: the equivalence classes are described by the lamination associated with a left earthquake, together with the collection of comparison isometries. Later, we shall see that the set of comparison isometries is determined by a much simpler, well-understood structure: a transverse measure for the lamination.

3. Associating earthquakes to maps of the circle

Here is the main theorem:

3.1. Theorem: geology is transitive. *Any two continuous relative hyperbolic structures on* $(\mathbb{H}^2, S^1_\infty)$ *differ by a left earthquake. The earthquake is unique, except that there is a range of choices for the earthquake map on any leaf where it has a discontinuity. The choices are maps which all have the same image, but differ by translations ranging between the limiting values for the two sides of the leaf.*

Equivalently, every continuous orientation-preserving map of S^1_∞ *to itself arises as the limiting value* E_∞ *of a left earthquake map* E, *unique up to the same set of ambiguities.*

3.2. Corollary: rightward geology. *Any two continuous relative hyperbolic structures on* $(\mathbb{H}^2, S^1_\infty)$ *differ by a right earthquake, unique except on leaves where the earthquake has a discontinuity.*

The corollary immediately follows by reversing orientation. It is rather curious that the effect of a left earthquake can also be obtained by a right earthquake. The reader may enjoy puzzling out the right earthquake map which corresponds to an elementary left earthquake. In the next section, after we have proven the earthquake theorem, we will show how this example and similar ones can be worked.

Proof. geology is transitive:

Let $f: S^1 \to S^1$ be any orientation preserving homeomorphism. We will prove the theorem by looking at the left coset of the group of isometries of hyperbolic space in the group of homeomorphisms of the circle which passes through f, that is, $PSL(2,\mathbb{R})\cdot f$. This coset is a three-manifold, homeomorphic to $PSL(2,\mathbb{R})$ which we name C. Choose a reference point $x_0 \in \mathbb{H}^2$. Then $PSL(2,\mathbb{R})$ has a fibration over $\mathbb{H}^2$, with projection map $\gamma \mapsto \gamma(x_0)$. The fibers are circles. This fibration carries over to a fibration of $p: C \to \mathbb{H}^2$ via the homeomorphism $(\gamma \leftrightarrow \gamma \circ f)$.

Some elements of C have fixed points as they act on the circle, others have none. If a homeomorphism h of the circle has at least one fixed point, then there is a unique lift of the h to a homeomorphism $\tilde{h}$ of the universal cover of the circle (namely $\mathbb{R}$) which also has fixed points. We will call h an *extreme left* homeomorphism if it has at least one fixed point, and if $\tilde{h}$ satisfies $\tilde{h}(x) \geq x$ for all $x \in \mathbb{R}$. Less formally, this says that h moves points counterclockwise on S^1, except for those points that it fixes. Let $XL \subset C$ be all extreme left homeomorphisms in C.

... 3.3. Lemma: XL is a plane. *The fibration p maps XL homeomorphically to the hyperbolic plane.*

Proof. XL is a plane: Let $F \subset C$ be any fiber of p, and let $f_0 \in F$ be any element of the fiber. The other elements of F are obtained by post-composing f_0 with a isometry which fixes $p(F)$. Using appropriate coordinates we can take the isometries which fix $p(F)$ to act as rotations of the circle.

The proof is easiest to complete in the universal cover. Let $\tilde{f}_0$ be any lift of f_0 to $\mathbb{R}$. The set of all lifts of all elements of F are the homeomorphisms $\tilde{f}_0 + T$, for constants T. Obviously, exactly one of these has the property that the periodic function $\tilde{f}_0(x) + T$ has its minimum value 0.

We have demonstrated that $p|XL$ gives a 1-1 correspondence between XL and $\mathbb{H}^2$. Construction of a continuous inverse reduces to a choice of T depending continuously on f_0. This follows from the fact that the infimum of a function depends continuously on the function, using the uniform topology on functions.

XL is a plane

With any element $g \in XL$ is associated its fixed point set fix$(g) \subset S^1$. Let $H(g) \subset D^2 = \mathbb{H}^2 \cup S^1_\infty$ be the convex hull of fix(g). The convex hull is taken in the hyperbolic sense; this agrees with the Euclidean convex hull if we use the projective model for $\mathbb{H}^2$, so that the compactified hyperbolic plane is the ordinary disk in the Euclidean plane, and hyperbolic straight lines are Euclidean straight lines.

The sets $H(g)$ often consist of a single point on the S^1_∞. In fact, for any $g \in XL$ and any $x \in$ fix(g), if g is postcomposed with a parabolic transformation which fixes x and moves all other points counterclockwise, the resulting homeomorphism is also in XL and has x as its sole fixed point.

Other possibilities are that *fix*(g) consists of two points, so that $H(g)$ is a compactified hyperbolic line, or that *fix*(g) is a finite set of points, so that $H(g)$ is a compactified ideal polygon, or that fix(g) is infinite.

... **3.4. Lemma: convex hulls do not cross.** *The sets $H(g)$ do not cross each other. Formally, if g_1 and g_2 are elements of XL such that the $H(g_i)$ are distinct sets which are not points, then $H(g_2)$ is contained in the closure of a single component of $D^2 - H(g_1)$.*

Proof. convex hulls do not cross:

Consider any two elements g_1 and g_2 of XL. Suppose that l_i $[i=1,2]$ is a line connecting two fixed points of g_i. Suppose the two lines intersect other than at their endpoints. The (hyperbolic) angle of intersection of their image under an element $g \in C$ does not depend on the choice of g.

It follows from the fact that g_1 is an extreme left homeomorphism that a certain one of the two angles formed by l_1 and l_2 cannot decrease. Since g_2 is also an extreme left homeomorphism, the complementary angle cannot decrease. This means that all four endpoints of the two lines must be fixed by g_1 and by g_2. Since an element of $PSL(2, reals)$ which fixes three or more points is the identity, it follows that $g_1 = g_2$.

The proposition readily follows by applying this reasoning to lines joining various pairs of points in $H(g_i) \cap S^1_\infty$.

convex hulls do not cross

As you probably suspect by now, the sets $H(f)$ for $f \in XL$ will turn out to be the closed strata for an earthquake map. The remaining difficulty in constructing an earthquake map is to show that these strata cover all of D^2. The reason they cover the whole disk is related to Brouwer's theorem that any map of the disk to itself which is the identity on the boundary is surjective. The correspondence $f \mapsto H(f)$ which maps XL to subsets of the disk is, in fact, much like a continuous map from the disk to the disk. To picture this similarity, consider the set $G \subset \mathbb{H}^2 \times \mathbb{H}^2$ which describes the graph of this correspondence. In other words, $G = \{(p(f), x) | x \in H(f)\}$. It should seem plausible that G is topologically a disk, and that it can be approximated by a disk which is the graph of a continuous function. What we will prove is related but slightly different.

Here is a continuity property for H:

... **3.5. Proposition: continuity of H.** *For any $f \in XL$ and any neighborhood U of $H(f) \subset D^2$, there is a neighborhood V of f in XL such that*

$$g \in V \Rightarrow H(g) \subset U.$$

Intuitively, this says that as f moves a little bit, what $H(f)$ can do is move a little bit and decrease to a subset of its former size.

Proof. continuity of H: We can assume that the neighborhood U is convex. In that case, it suffices to take V small enough so that the fixed points of homeomorphisms $g \in V$ are contained in $U \cap S^1_\infty$. This can easily be accomplished in view of the fact that there is a lower bound to the distance that f moves points in the complement of $U \cap S^1_\infty$.

continuity of H

The set XL is homeomorphic to an open disk; it has a compactification $\overline{XL}$ to a closed disk, copied from the hyperbolic plane. Extend the map H to the circle at infinity for XL (which we also call S^1_∞) by the rule that

$$H(x) = \{x\} \quad [x \in S^1_\infty].$$

... **3.6. Proposition: continuity of H bar.** *The preceding statement remains true for the extension of H to $\overline{XL}$.*

Proof. continuity of H bar: What we need to show is that for any $x \in S^1_\infty$ and for any neighborhood U of x in S^1, there is a neighborhood V of x in XL consisting of homeomorphisms whose fixed point sets are all contained in U.

Any $\gamma \in PSL(2,\mathbb{R})$ can be written in the form $\gamma = \tau \circ \rho$, where τ is the hyperbolic element whose axis passes through x_0 and takes x_0 to $\gamma(x_0)$, and ρ is an elliptic element, fixing x_0. Let us take x_0 to be the center of the disk.

If $\gamma(x_0)$ is near $x \in S^1_\infty$, then τ translates a great distance. If $\gamma \circ f_0$ has any fixed point y which is not close to x, then $\tau^{-1}(y)$ is close to the point diametrically opposite hx. Consider what $\gamma \circ f_0$ does to points in a neighborhood of y. A point y_1 close to y in the clockwise direction is first sent to a point clockwise from $f_0(y)$ (how far is estimated by constants for uniform continuity of f_0). Then ρ sends y and y_1 to a pair of points spaced exactly the same amount; since $\rho \circ f_0(y) = \tau^{-1}(y)$ is near the point antipodal to x, y_1 is also near this antipodal point. Now τ expands the interval between the two points by a tremendous factor. If τ is chosen to overcome whatever shrinking f_0 caused on the interval, then y_1 moves clockwise, so $\gamma \circ f_0$ was not an extreme left homeomorphism.

The only possibility is for an extreme left homeomorphism near x is that its fixed points are near x.

continuity of H bar

To prove that every point in $\mathbb{H}^2$ is in some $H(f)$, we will replace H with a continuous map by an averaging technique. We can do the averaging using the Euclidean metric on the disk, considered as the projective model of $\mathbb{H}^2$. This disk is identified with both the compactified hyperbolic plane and $\overline{XL}$. There is an obvious measure defined on the set $H(f)$: either 2-dimensional, 1-dimensional or 0-dimensional Lebesgue measure. Let $h(f)$ be the center of mass of $H(f)$, taken with respect to this measure.

Choose a bump function β around the origin with small support, that is, a positive continuous function with integral 1 having support contained in an ϵ-neighborhood of the origin. The convolution $\beta{\star}h$ of β with h is a new function from XL the plane. ($\beta{\star}h$ is the average of the values of h over a small neighborhood in XL, with weighting governed by β). $\beta{\star}h$ is continuous, since the difference of its values at nearby points can be expressed as the integral of the bounded function h times a small function (the difference of copies of β with origin shifted to the two nearby points).

For any $f \in XL$, $\beta{\star}h(x)$ is contained in the convex hull of the union of the $H(g)$ where g ranges over the ϵ-neighborhood of f. By proposition 3.6 *(continuity of H bar)*, this is contained in a small neighborhood of $H(f)$.

The map $\beta{\star}h$ is close to the identity near S^1_∞. Therefore (by consideration of degree), it is surjective onto all except possibly a small neighborhood of S^1_∞ whose size depends on ϵ. For any $x \in \mathbb{H}^2$, let f be an accumulation point of elements f_i such that $\beta_i{\star}h(f_i) = x$, as the diameter of the support of β_i goes to zero. It follows that $x \in H(f)$.

The union of lines of $H(XL)$ with boundary components of two-dimensional regions of $H(XL)$ forms a lamination λ.

To complete the proof of existence of an earthquake, we must construct a λ-earthquake map which agrees with f_0 at infinity. For each stratum A, choose an $f_A \in XL$ such that $H(f) \supset A$. There is only one choice possible except for some cases when A is a line. Define E on the stratum A to be γ_A, where γ_A is determined by the equation

$$f_0 = \gamma_A \circ f_A$$

It is clear that E is a 1-1 correspondence from $\mathbb{H}^2$ to itself, and that it agrees with f_0 at infinity. What remains to be shown is that any comparison isometry

$$cmp(A, B) = \gamma_A^{-1} \circ \gamma_B = f_A \circ f_B^{-1}.$$

are hyperbolic, that its axis weakly separates A and B, and that it translates to the left as viewed from A.

Let I and J be the two intervals of S^1_∞ which separate A from B, named so that I is to the left of J as viewed from A. (We allow the possibility that one or both of them is a degenerate interval, that is, a point.) The two homeomorphisms f_A and f_B^{-1} both map I into itself. The $cmp(A, B)$ maps I into itself, so by the Brouwer fixed point theorem it must have a fixed point somewhere in that interval. Similarly, $cmp(B, A) = (cmp(B, A))^{-1}$ must have a fixed point in J.

Therefore, $cmp(A, B)$ is hyperbolic, and its axis separates A from B. The direction of translation along the axis is from the interval J to the interval I, since I the isometry maps I inside itself. It translates to the left as viewed from A, so the map E that we have constructed is indeed a left earthquake map.

The uniqueness part of the main theorem is easy. Suppose that E' is any earthquake having the same limiting values on S^1_∞ as the earthquake map E we have constructed. If A is any stratum of E', then the composition h of E' with an isometry which makes h the identity on A acts as an extreme left homeomorphism on S^1_∞, with two or more fixed points. Therefore A is a stratum of E, and the two left earthquakes agree on A unless it is a leaf where they are discontinuous.

geology is transitive

4. Examples

Given a map of the circle to itself, the main theorem gives a reasonably constructive procedure to find a left earthquake with the given limiting values. (How constructive the procedure really is depends on the form in which the map of the circle is given, and whether certain inequalities involving its values can be answered constructively. Technically, the theorem is not constructive except under very restrictive hypotheses. The natural stratification of the boundary of the convex hull of a curve in $\mathbb{R}^3$ into zero-dimensional, one-dimensional, and two-dimensional flat pieces is a quite similar problem, and it is not technically constructive, either. For most practical purposes, both procedures are reasonably constructive.)

To illustrate, consider an elementary left earthquake map L which shifts by a distance d along a fault l, and let us find the corresponding right earthquake. Let p be a point on l; choose coordinates so that p is at the center of the disk. There is an extreme left homeomorphism h in the coset of L which acts as a hyperbolic transformation which moves a distance of $d/2$ on the two sides of l, in opposite directions. If h is composed with a suitable clockwise rotation, an extreme right homeomorphism g will be obtained. The fixed points of g cannot be at the ends of l, so g is differentiable at its fixed points. The qualitative picture immediately shows that the derivative of an extreme left homeomorphism (when it exists) can only be 1.

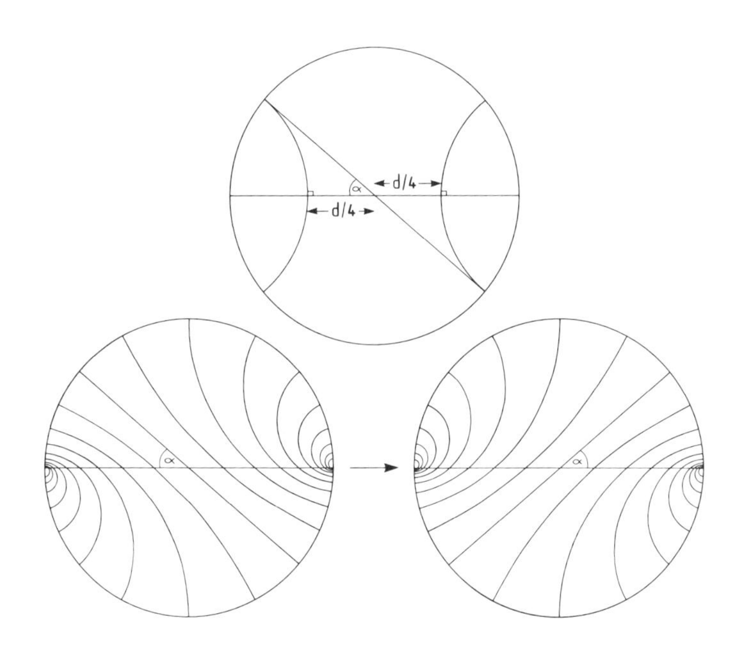

4.1 Figure. *The faults of the right earthquake which corresponds to an elementary left earthquake consists of lines making a constant angle* α *to the fault of the left earthquake. The angle* α *is such that after the endpoints are transformed by the left earthquake, the lines which now join them meet the left earthquake fault with the complementary angle* $\pi - \alpha$.

A little reflection shows that the stratum containing p is a leaf. It can be constructed by drawing the two lines perpendicular to l at a distance of $d/4$ on either side of p; the leaf is one of the diagonal lines m connecting their endpoints on S^1_∞. The right earthquake map R

corresponding to L has as its lamination the set of all lines meeting l at the same angle as m. R acts as the identity on l, and takes each leaf of its lamination to a leaf meeting l in the complementary angle at the same point.

There are other maps $f_{\alpha,\beta}$ which can be described in a similar way, acting as the identity on l and sending the foliation by lines which meet l with angle α to the foliation by lines which meet l with angle β. These are not earthquake maps except when α and β are complementary angles. The intuitive reason is that the spacing between the leaves of the foliation is not preserved except when α and β are complementary. The part of the definition that is violated is that the axes of the comparison isometries cannot meet the strata being compared.

A similar procedure works to find the right earthquake corresponding to any simple left earthquake. The leaves of the lamination for the right earthquake fall into a finite number of groups. The leaves in any group make a constant angle to a line which is the axis of a comparison isometry between two of the strata for the left earthquake. Whether or not there is a group of leaves for a given comparison isometry, and if so, how big it is, depends on the amount of shearing along the faults of the left earthquake and the spacing of the leaves.

5. Earthquakes on hyperbolic surfaces

As indicated in the introductory section, the earthquake theorem carries over without difficulty to general complete hyperbolic surfaces, even surfaces of infinite genus or infinitely many ends. To make this generalization, we have to extend the definitions.

5.1 Definition. A *left earthquake map* between two surfaces is a 1-1, surjective map between the surfaces, together with a lifting to a map between the universal covers which is a left earthquake. A right earthquake map is defined similarly.

If there is a closed leaf in the earthquake lamination on the surface, then the map between surfaces does not determine the lifting: there are an infinite number of different liftings, differing by twists along the geodesic, which would give different earthquakes, even though the physical map is the same. It is because of these examples, which it is useful to distinguish (so that an earthquake map determines a well-defined homotopy class of continuous maps) that the lifting is included as part of the data.

5.2 Definition. If M^2 is a complete hyperbolic surface, the *Fuchsian enlargement* of M^2 is defined to be

$$\overline{M}^2 = (\mathbb{H}^2 \cup \Omega)/\pi_1(M^2)$$

where Ω is the domain of discontinuity of $\pi_1(M^2)$ as it acts on the circle at infinity. $\overline{M}^2$ is a surface with boundary.

Relative hyperbolic structures are defined on $(M^2, \partial\overline{M}^2)$ just as for $(\mathbb{H}^2, S^1_\infty)$. A relative hyperbolic structure is a complete hyperbolic metric on M^2, up to homeomorphisms which extend to homeomorphisms on $\overline{M}^2$ which are isotopic to the identity while remaining constant on ∂M^2. A relative hyperbolic structure M^2_1 is *adjective* if there is a representative metric in the equivalence class so that the identity map extends to a homeomorphism which is *adjective* from $\partial\overline{M}^2$ to $\partial\overline{M}^1$.

Note that cusps need not correspond in arbitrary relative hyperbolic structures. For example, if τ is a hyperbolic transformation and $M^2 = \mathbb{H}^2/\tau$ is the quotient cylinder, there are relative hyperbolic structures on $(M^2, \partial M^2)$ having either end a cusp. The same is true if τ is parabolic. For continuous relative hyperbolic structures, an end cannot change from having a cusp to having a boundary component of $\overline{M}^2$, or vice versa.

The significance of the Fuchsian enlargement of a complete hyperbolic surface comes about through its close relation to the compactification of the hyperbolic plane:

5.3. Proposition: continuity at infinity. *If $h: M_1 \to M_2$ is a homeomorphism between hyperbolic surfaces and $\tilde{h}: \mathbb{H}^2 \to \mathbb{H}^2$ is a lift of h to the universal covers, then h extends to a continuous map $\bar{h}: \overline{M}_1 \to \overline{M}_2$ iff $\tilde{h}$ extends to a continuous map $\bar{\tilde{h}}: \overline{\mathbb{H}} \to \overline{\mathbb{H}}$.*

Proof. continuity at infinity:

If $\tilde{h}$ extends continuously to $\bar{\tilde{h}}: \overline{\mathbb{H}}^2 \to \overline{\mathbb{H}}^2$, a continuous extension of h is directly constructed as $\bar{\tilde{h}} | \mathbb{H}^2 \cup \Omega$ modulo $\pi_1(M_1)$.

Conversely, suppose that $\bar{h}$ is a continuous extension of h. A neighborhood basis of the circle at infinity for $\tilde{M}_2$ can be formed using two kinds of sets: half-planes bounded by axes of hyperbolic elements, and half-planes bounded by lines which have both endpoints in Ω. (Actually, the first kind of set suffices if $\Omega = \emptyset$, and the second kind of set suffices otherwise.) Since a monotone function can fail to be injective only on a countable number of intervals, all but a countable set of the arcs with endpoints in Ω have preimages which converge at both ends, and these will suffice to form a neighborhood basis.

For any point p on the circle at infinity for $\tilde{M}_1$, there is a unique point q on the circle at infinity for $\tilde{M}_2$ so that a neighborhood basis for p maps to a family of sets cofinal in a neighborhood basis for q. This means there is a unique continuous extension $\bar{\bar{h}}$.

continuity at infinity

We remark that the analogous proposition in three dimensions is false, which may explain why this proof sounds a bit strained. In order for there to be a homeomorphism between universal covers of hyperbolic three-manifolds, it is necessary that there be a homeomorphism between their Kleinian enlargements, but in addition it is necessary (at least) that ending laminations for any geometrically infinite ends be the same.

5.4. Corollary: earthquakes on surfaces. *If M^2 is any complete hyperbolic surface, and if N^2 is any other continuous relative hyperbolic structure on $(M^2, \partial\overline{M}^2)$, then there is a left earthquake between $(M^1, \partial M^1)$ and $(N^1, \partial M^1)$.*

Proof. earthquakes on surfaces: Since there is a homeomorphism between the Fuchsian enlargements, by the preceding proposition there is a homeomorphism between the compactified universal covers. By theorem 3.1 *(geology is transitive)* there is an earthquake between the universal covers. The earthquake can be taken to be equivariant with respect to the action of the fundamental group of M, if choices are made consistently at its discontinuities. This gives an earthquake map between the surfaces.

earthquakes on surfaces

6. The measure and cause of earthquakes

Part of the data comprising an earthquake is the source lamination λ_s, but there is additional information, namely a transverse measure for λ_s, which is determined by the map. The transverse measure measures the amount of shearing along the faults. Mathematically, the shearing can be determined from the translation distances of comparison isometries. Given any transverse arc and any partition of it into subintervals, associated with each partition element is a comparison isometry between the strata containing its endpoints. The composition of these isometries is the comparison isometry for the strata of the two endpoints of the entire interval. To specify the transverse shearing measure, we require that the sum of the translation lengths of the comparison isometries for the partition should approximate the total measure of the arc.

We need to check that these approximations converge. The translation distance of a hyperbolic isometry is the log of its derivative at the repelling fixed point on the circle at infinity. If coordinates are chosen so that the axis of the isometry is a diameter of the circle, the norm of the derivative at a nearby point p on the circle differs by $O(T.d^2)$, where d is the distance of p from the fixed point and T is the translation distance. More intrinsically, it follows that the translation distance of the composition of two hyperbolic transformations with disjoint nearby axes pointing the same direction differs from the sum of translation distances by $O(T.d^2)$, where d now represents the distance between the axes and T is the smaller of the two translation distance of the composition.

Applying this iteratively for a comparison isometries of a partition, we conclude that the sum of translation distances for a partion differs from the translation distance T of the comparison isometry for the endpoints of a transverse arc by at most $O(T.l^2)$, where l is the length of the arc. This implies the sums for partitions must converge, as the partition is refined.

6.1. Proposition: metering earthquakes. *Associated with any earthquake is a transverse lamination on its source lamination, which approximates the translation distances of comparison isometries for nearby strata.*

For any two earthquakes with the same source lamination and the same shearing measure, there is an isometry between the targets which conjugates the two earthquake maps.

Proof. metering earthquakes: We have already established assertion of the first paragraph, that there is a well-defined measure associated with an earthquake.

It remains to show that different earthquakes are distinguished by their measures.

As a special case, we need to make sure that a non-trivial earthquake has a non-trivial measure. Actually, this case is quite clear: for any transverse arc with total measure 0, there are arbitrarily fine partitions such that the total translation distance is arbitrarily close to 0. By the preceding discussion, it follows that the translation distance of the composition is arbitrarily close to 0. Another way to visualize this is to keep track of how far each of the comparison isometries for neighboring points moves the center of the disk. It never gets very far, so the comparison isometry is the identity.

Given any two earthquake maps which have the same source lamination and the same shearing measure, consider now the inverse of one earthquake map composed with the other. We do not immediately know that this is an earthquake. However, we can still consider the comparison isometries. For any two nearby strata, the comparison isometry is the composition of two hyperbolic isometries whose axes are geometrically restricted so that they nearly coincide, and whose translation distances are equal but moving in opposite directions. The composition is close to the identity -- its distance from the identity is $O(T.d)$. As the partition is refined, the total size of the comparison isometries goes to 0.

metering earthquakes

The question of existence of an earthquake, given a lamination and a transverse invariant measure, is more delicate. There is always an earthquake of sorts, but without some restrictions on the measure, it can easily happen that the target surface is not complete. For instance, the three-punctured sphere can be divided into two ideal triangles. There is a sort of earthquake which shears to the left by a distance of 1 along each of the edges; the target surface is the interior of a pair of pants whose boundary components all have length 1. (This and related examples are discussed in [Thurston]).

6.2 Definition. An earthquake is *uniformly bounded* if for every constant $a \geq 0$ there is a constant C such that for any two strata A and B whose distance does not exceed a, the comparison isometry $cmp(A, B)$

A measured lamination is *uniformly bounded* if for every constant $a \geq 0$ there is a constant C such that any transverse arc of length less than or equal to a has total measure less than C.

To show that an earthquake or a measured lamination is uniformly bounded, it clearly suffices to produce a constant C for one value of a greater than 0.

6.3. Proposition: Making bounded earthquakes. *For any uniformly bounded measured lamination* μ, *there is a uniformly bounded earthquake* E_μ *having* μ *as shearing measure.*

7. Quasisymmetries and quasi-isometries

A *quasi-isometry* of a metric space is a homeomorphism h for which there is a constant K such that for all x, y,

$$d(h(x), h(y)) \leq K d(x, y) \quad \text{and} \quad d(x, y) \leq K d(h(x), h(y)).$$

For a Riemannian manifold, this is the same as saying that h is bi-Lipschitz with constant K.

There is a certain sense in which uniformly bounded earthquakes can be approximated by quasi-isometries. This is not uniform approximation as maps: the uniform limit of continuous maps is always continuous. Instead, we consider again the graph of the earthquake map as a relation. At the discontinuities, where there is ambiguity in the earthquake map, adjoin all possible choices of the earthquake map to the graph, thus obtaining a canonical earthquake relation.

7.1. Proposition: Approximating by quasi-isometries. *The graph of a uniformly bounded earthquake can be uniformly approximated by a quasi-isometric diffeomorphism. The approximation can be done in a natural way.*

There are two methods which work well, both using smoothing. Method (a) is to integrate a vector field, flowing through the one-parameter family of earthquakes. Method (b) is to construct a foliation extending the lamination to a neighborhood, together with a transverse foliation, and to use these as local coordinates for a leaf-preserving diffeomorphism. The second method is more difficult to do in a natural way, so that it does not depend on ad hoc choices, so we will describe the first.

Proof. Approximating by quasi-isometries:

Sketch. Construct a three-manifold fibered by hyperbolic 2-manifolds resulting from earthquakes with the same faults, but linearly increasing measure. Construct a transverse vector field by averaging the local isometries coming from infinitesimal earthquake maps between slices.

Now integrate.

Approximating by quasi-isometries

There is a famous characterization by Ahlfors of which maps of the circle can extend to quasi-conformal maps of the disk. This ties in quite nicely with the theory of earthquakes.

7.2 Definition. A homeomorphism $f: S^1 \to S^1$ is *quasisymmetric* if there are constants K and $\epsilon > 0$ such that, for $0 < t < \epsilon$ and $x \in S^1$,

$$\frac{1}{K} \leq \frac{|f(x+t) - f(x)|}{|f(x-t) - f(x)|} \leq K.$$

7.3. Theorem: Extending quasisymmetry. *A quasisymmetric homeomorphism of the circle extends in a natural way to a quasi-isometry of the hyperbolic plane.*

Proof. Extending quasisymmetry: Sketch. It is easy to check that an earthquake is uniformly bounded if the homeomorphism of the circle at infinity is quasisymmetric. Apply the preceding result.

Extending quasisymmetry

7.4. Corollary: quasisymmetry on surfaces. *Let M and N be complete hyperbolic surfaces, and suppose that $\phi: M \to N$ is a quasi-isometry between them. Then for any quasi-symmetric homeomorphism at infinity from ∂M to ∂N which is isotopic to $\partial\phi$, there is a uniformly bounded earthquake map from M to N, and a quasi-isometry from M to N, with given boundary values at infinity.*

References

Kerckhoff, 1983.

S. P. Kerckhoff, "The Nielsen realization problem," *Annals of Math.*, **117**, pp. 235-265, March, 1983.

Thurston.

W. P. Thurston, "Three-dimensional geometry and topology," *Princeton University Press.* (to appear)

W.P. Thurston,
Fine Hall,
Princeton,
N.J. 08544,
U.S.A.

Part B:
Knots and 3-manifolds

Augmented Alternating Link Complements Are Hyperbolic

Colin C. Adams

1. Introduction.

W. Menasco proved in [Menasco, 1984] that if L is a nonsplittable prime alternating link which is not a $(2,q)$ torus link, then $S^3 - L$ is a complete hyperbolic 3-manifold. In this paper we extend this result to a class of links obtained by adding particular trivial components to links of the type considered by Menasco. We call these links *augmented alternating links.*

There are several reasons for our interest in these links. First, each of the added trivial components bounds an incompressible twice-punctured disk. Hence the results of [Adams, 1985] can be applied to find numerous examples of distinct hyperbolic link complements sharing the same volume. In addition, these link complements furnish us with the building blocks from which more complicated hyperbolic link complements can be constructed as in Section 4 of [Adams, 1985].

Secondly, sequences of alternating links with hyperbolic complements can be obtained by performing $(p,1)$ Dehn surgeries on the added trivial components in an augmented alternating link Q. Results from [Thurston, 1979] in conjunction with our result imply that the sequence of hyperbolic complements of these links is converging to the hyperbolic structure on the complement of Q.

Finally, utilising results from [Adams, 1985], [Bonahon-Siebenmann], [Menasco, 1984], [Oertel, 1984] and the results from this paper, we can determine which links of more than one component in the tables at the end of [Rolfsen, 1976] do and do not have hyperbolic complements.

After the preliminaries, we begin by proving a theorem which provides sufficient conditions under which the complement of a knot in a finite volume noncompact hyperbolic 3-manifold is still hyperbolic. This theorem is then utilized to prove augmented alternating link complements are hyperbolic. Applications of these results are then considered.

2. Preliminaries.

We work throughout in the *PL* category. We refer to [Jaco, 1977] for the definitions of *incompressible, ∂-incompressible* and *∂-parallel surfaces,* and *irreducible* and *∂-irreducible* 3-manifolds. A *knot* K in a 3-manifold M is a simple closed curve in the interior of M. We let $N(K)$ denote a closed regular neighbourhood of K. A *knot exterior* will be the complement of the interior of a regular neighbourhood of a knot in S^3. A (p,q) curve on a torus with specified longitude l and meridian m where $(p,q) = 1$ is a simple closed curve representing $[pl + qm]$.

An *essential torus* in a 3-manifold M is an incompressible torus embedded in M which is not ∂-parallel. An *essential annulus* in M is an incompressible, ∂-incompressible annulus properly embedded in M.

A *link* in a 3-manifold M is the union of disjoint knots in M. We refer to [Rolfsen, 1976] for the definitions of *nonsplit, prime* and *alternating* as applied to links in S^3 and for the definition of a *regular projection* of a link in S^3. Given a regular projection of a link L in S^3, a *region* of the projection plane is a disk in the projection plane bounded by strands of L and containing no strands of L in its interior.

A *hyperbolic* 3-manifold will be a 3-manifold with a complete Riemannian metric of constant sectional curvature -1. A compact 3-manifold with hyperbolic interior of finite hyperbolic volume will have only toroidal ∂-components. We refer to [Thurston, 1979] for details. All hyperbolic 3-manifolds considered in this paper are the interiors of 3-manifolds of this type.

If M is a compact orientable 3-manifold with boundary, then Thurston has proved that M is a finite volume hyperbolic 3-manifold if and only if M is irreducible, ∂-irreducible, contains no essential annuli or tori and has only toroidal ∂-components. See [Thurston, 1982] for instance.

3. Main Theorem.

We prove the following.

Theorem 3.1. *Let M be a compact orientable 3-manifold with non-empty boundary such that $\mathring{M}$ is a finite volume hyperbolic 3-manifold. Let K be a knot in $\mathring{M}$ such that*

(i) *There does not exist $S^1 \times S^1 \times I \subset M$ such that $K \subset S^1 \times S^1 \times I$ where $S^1 \times S^1 \times \{0\} \subset \partial M$.*

(ii) *There exists an incompressible thrice-punctured sphere E properly embedded in $M - \mathring{N}(K)$ such that exactly one component μ of ∂E is contained in $\partial N(K)$.*

(iii) *K is not a torus knot in a solid torus $V \subset \mathring{M}$ such that ∂V is incompressible in $M - \mathring{V}$.*

(iv) *There does not exist an essential torus in $M' = M - \mathring{N}(K)$ which bounds a knot exterior in M'.*

Then $M' = M - \mathring{N}(K)$ has a hyperbolic interior.

Proof. According to Thurston's theorem, it is enough to show M' is irreducible, ∂-irreducible and contains no essential annuli or tori.

M' is irreducible, as if not, K would be contained in a 3-cell B contained in M. We could then construct an $S^1 \times S^1 \times I$ which would contradict condition (i). It follows from irreducibility that M' is also ∂-irreducible.

In order to prove that M' contains no essential tori, we need the following lemma:

... **Lemma 3.2.** *Let T be an essential torus in $\mathring{M}'$ in general position with respect to E. Then there is an isotopy which eliminates all intersections of E with T except possibly for those intersection curves which are both isotopic on E to μ and nontrivial on T.*

Proof. Suppose there is a trivial innermost intersection curve α on T (or E). Then it must be trivial on E (or T) by incompressibility. By the irreducibility of M', there is an isotopy eliminating α.

Now suppose there is a nontrivial intersection curve β on T which is isotopic to one component γ of $\partial E \cap \partial M$ on E and is the innermost such. Compressing T to ∂M along the annulus bounded by β and γ on E yields an incompressible annulus A, which is essential in M' but ∂-compressible in M. The irreducibility of M implies K is contained in a solid torus V bounded by A and an annulus in ∂M such that ∂A-curves each intersect a meridian of ∂V once. Then if Q is the corresponding boundary torus of M, $V \cup \partial N(Q)$ is an $S^1 \times S^1 \times I$ which contradicts condition (i).

Lemma 3.2

Let T be an essential torus in M'. If T were ∂-parallel in M, we would contradict condition (i). Hence T compresses in M. Let D be a compressing disk. Then $\partial N(D \cup T)$ has a 2-sphere component which must bound a 3-cell B. If $D \subset B$, then $T \subset B$. However, since T is incompressible in M', T would then bound a knot exterior in M', contradicting condition (iv). Hence $D \cap B = \emptyset$ implying that T bounds a solid torus V in M containing K such that K intersects every meridional disk in V. By lemma 3.2, we can remove all intersections of T and E except for at least one intersection curve which is nontrivial on T and isotopic to μ on E. Hence there exists an annulus A contained in E which is properly embedded in $V - \mathring{N}(K)$ with one ∂-component nontrivial on $\partial N(K)$ and the other nontrivial on ∂V.

Suppose $\partial A \cap \partial V$ is not a meridional curve on ∂V. Let $A' = \partial N(A \cup N(K))$ in V. Then A' splits V into two solid tori V_1 and V_2, such that $\partial A'$-curves are $(1,p)$ curves on ∂V_1 and (r,s) curves on ∂V_2. If $V_1 = N(A \cup N(K))$, then ∂A-curves are $(1,p)$ curves on $\partial N(K)$. Hence K is isotopic to a simple closed curve in ∂V. By condition (iii), K must be a core curve of V, contradicting the fact T is not ∂-parallel in M'. If $V_2 = N(A \cup N(K))$, then K is a core curve of V_2 and hence also of V, again making T ∂-parallel in M'.

Suppose now $\partial A \cap \partial V$ is meridional on V. Replacing annuli on E with those on T and pushing off slightly, we can form a new incompressible thrice-punctured sphere E' such that $E' \cap T$ contains only one intersection curve which is meridional on T.

Let $A'' = (E' - (E' \cap V)) \cup D$ where D is a disk in V with $\partial D = E' \cap T$. Since A'' is an incompressible annulus in M, A'' must ∂-compress in M. This implies E ∂-compresses in M. Let D' be the compressing disk. Since incompressibility of E in M' implies ∂-incompressibility of E in M', K must intersect D'. Since $E \cap \mathring{D}' = \emptyset$, μ must be meridional on $\partial N(K)$. However in that case, let A''' be the incompressible annulus in M obtained by taking the union of E with a meridional disk $D'' \subset N(K)$ where $\partial D'' = \mu$. Since A''' must ∂-compress in M, A''' and an annulus in ∂M together bound a solid torus V' in M. However, this yields a contradiction as K intersects ∂V transversely only once.

Finally, we must show that M' contains no essential annuli. Suppose first that A is an essential annulus in M' with both ∂-components on ∂M. Both ∂-components of A must lie on the same component Q of ∂M, as otherwise A is essential in M. Then A must ∂-compress in M, hence by irreducibility of M, A is ∂-parallel in M. We can then construct an $S^1 \times S^1 \times I$ which contradicts condition (i).

Suppose now that A is an essential annulus with either one or both ∂-components on $\partial N(K)$ and non-meridional. By Lemma 3.7 of [Hatcher], M' Seifert fibers with fibers along $S^1 \times \{i\}$ where $A = S^1 \times I$. This Seifert fibration can be extended to M, contradicting the fact that M is hyperbolic.

In the case A is an essential annulus with one ∂-component meridional on $\partial N(K)$ and the other ∂-component on ∂M, then after filling in $N(K)$, ∂M compresses in M, contradicting hyperbolicity of $\mathring{M}$.

Finally, suppose A is an essential annulus in M' with both of its ∂-components meridional on $\partial N(K)$. Put E and A in general position. Let α be a trivial innermost simple closed intersection curve in $E \cap A$ on A (or E). Then by incompressibility of E (or A), α is trivial on E (or A). By irreducibility, we can isotope to remove α.

Now let δ be an arc in $E \cap A$ such that there is disk $D \subset A$ bounded by δ and an arc in ∂A which contains no other intersection arcs. By ∂-incompressibility of E in M', irreducibility and incompressibility of $\partial N(K)$, we can isotope to remove δ.

There remains at least one intersection arc in $E \cap A$, as otherwise, μ would be meridional on $\partial N(K)$, a case which we have already eliminated. Since ∂E is a (p,q) curve on $\partial N(K)$ where $p \neq 0$, we can assume ∂E intersects each ∂-component of A $|p|$ times alternating between them as we traverse $\partial E \cap \partial N(K)$.

Suppose every intersection arc separates the two components of $\partial E \cap \partial M$ on E. There must be an even number of such arcs. Let α_1 be an intersection arc which is nearest on E to one of the two components of $\partial E \cap \partial M$. Let α_2 be the intersection arc on E nearest to α_1. Let D_1 be the strip on E bounded by α_1 and α_2 and let D_2 be the strip on A bounded by α_1 and α_2. Then $D_1 \cup D_2$ is a Möbius band with boundary either trivial or meridional on $\partial N(K)$. In either case, we have a projective plane embedded in a 3-cell when we fill $N(K)$ in.

Thus there must exist an intersection arc which does not separate the components of $\partial E \cap \partial M$ on E. This implies A ∂-compresses in M'.

Theorem 3.1

4. Augmented Alternating Links.

Let L be a nontrivial nonsplittable prime alternating link that is not a $(2,q)$ torus link. Suppose L is in a particular regular alternating projection without any unnecessary crossings.

Let P be the projection plane. Let $J_1, \ldots, J_n$ be n nonisotopic embedded 1-spheres in $S^3 - \mathring{N}(L)$ such that each J_i intersects P in exactly two points, where each of the two intersections occurs in a different region of the projection plane and such that each J_i bounds a vertical disk E_i in $\mathbb{R}^3$ where

(i) $E_i \cap E_j = \varnothing$ for $i \neq j$

(ii) Each E_i intersects L in exactly two points.

Then we call the link $L \cup (\bigcup_{i=1}^{n} J_i)$ an augmented alternating link. Figure 1 denotes an augmented figure eight knot.

Figure 1.

Theorem 4.1. *Let Q be an augmented alternating link. Then $S^3 - Q$ has a complete hyperbolic structure.*

Proof. Let $R = L \cup \bigcup_{i=1}^{n-1} J_i$ and let $J = J_n, E = E_n$. We assume the result is true for R and then prove it for $R \cup J$. We will check that each of the four hypotheses of Theorem 3.1 are satisfied where $M = S^3 - \mathring{N}(R)$ and K corresponds to J.

We first demonstrate that E is incompressible. Since the punctures on E correspond to nontrivial curves on $\partial(S^3 - N(L))$, they cannot compress. Hence if E compresses, there exists a disk $D \subset S^3 - N(R \cup J)^o$ such that $D \cap E = \partial D$ where ∂D is isotopic on E to J. Letting E' be the twice-punctured disk on E bounded by ∂D, we can form $S = E' \cup D$ a twice punctured sphere in $S^3 - N(R \cup J)^o$. Moving E' and hence S slightly off E, we can replace J with a link component K as in Figure 2 in such a way that $L \cup K$ is alternating.

Figure 2.

We utilize the notation and results from [Menasco, 1984]. By theorem 1(a) of [Menasco, 1984], $L \cup K$ is nonsplit.

By the proof of lemma 1 of, [Menasco, 1984] either S can be isotoped in $S^3 - N(L \cup K)^o$ to satisfy the conclusions of lemma 1 without moving E', or S contains a meridional loop γ which does not intersect E'. The second case cannot occur as any meridional loop on S must separate the punctures.

Lemma 2 of [Menasco, 1984] then implies S intersects the projection plane in a single circle which intersects L transversely twice. By theorem 1(b) of [Menasco, 1984], L is a trivial arc to one side of the circle. However, this implies that both of the intersections of J with the projection plane occur in the same region, contradicting our definition of an augmented alternating link.

We next prove $S^3 - (R \cup J)$ is irreducible. If not, then J is contained in a 3-cell in $S^3 - \mathring{N}(R)$ and hence bounds a disk D in $S^3 - \mathring{N}(R)$. We can isotope D into general position with respect to E such that $D \cap E$ consists only of simple closed curves. Choose an intersection curve α in $D \cap E$ that is innermost on D. Since E is incompressible, α must be trivial on E and hence by the irreducibility of $S^3 - \mathring{N}(R)$, D can be isotoped to remove the intersection. We repeat this until $D \cap E = \emptyset$. However, since both D and E are bounded by J, we can use D to compress E, contradicting the incompressibility of E.

We next show $R \cup J$ is a prime link. Suppose not. Then there exists a twice-punctured sphere S in $S^3 - (R \cup J)$ such that R is a trivial arc to the side of S containing J. Placing E and S in general position, we examine the simple closed curves making up $E \cap S$. An innermost trivial intersection curve on E must bound a disk on S by ∂-irreducibility of $S^3 - \mathring{N}(L)$, hence this type of intersection can be removed by isotoping S, using the irreducibility of $S^3 - N(R \cup J)^o$. An innermost intersection curve α on E looping one puncture on E must separate the punctures on S. Let E' be the once-punctured disk α bounds on E and let D_1 and D_2 be the once-punctured disks α bounds on S. Let B_1 and B_2 be the 3-cells bounded by S in S^3 such that B_1 contains J and the trivial arc of L.

In the case $E' \subset B_1$, J lies to one side of E' in B_1 and hence we can isotope one of D_1 or D_2 to E' through the other side as $R \cup J$ is just a trivial arc of R to that side. In the case $E' \subset B_2$, E' divides B_2 into two 3-cells C_1 and C_2 where $\partial C_1 = E' \cup D_1$ and $\partial C_2 = E' \cup D_2$. Then since $E' \cup D_1$ is a twice-punctured sphere in $S^3 - R$, either C_1 or $B_1 \cup C_2$ contains only a trivial arc of R. If C_1 contains the trivial arc, isotope D_1 to E' through C_1. If $B_1 \cup C_2$ contains the trivial arc, isotope D_2 to E' through C_2. Thus we can remove all intersections of this type.

The only possible remaining intersection curves loop both punctures on E. Since R must puncture any sphere an even number of times, these intersection curves must all be trivial on S. Replacing the twice-punctured disk α bounds on E with the disk α bounds on S yields a disk bounded by J in $S^3 - R$, contradicting irreducibility of $S^3 - (R \cup J)$.

Thus, we now have $E \cap S = \emptyset$. Then exactly as we did previously, we replace J with K as in Figure 2 and then isotope S in $S^3 - (L \cup K)$ so that it intersects the projection plane in a single circle C which intersects L transversely twice. We know L is trivial to the side of C containing K. Since theorem 1(b) of [Menasco, 1984] implies $L \cup K$ is prime, L must be trivial to the other side of C as well. This contradicts the hypothesis that L is nontrivial.

We now show there does not exist $S^1 \times S^1 \times I \subset S^3 - \mathring{N}(R)$ such that $J \subset S^1 \times S^1 \times I$ and $S^1 \times S^1 \times \{0\} \subset \partial N(R)$. Supposing such an $S^1 \times S^1 \times I$ does exist, let T be the incompressible torus in $S^3 - \mathring{N}(R)$ corresponding to $S^1 \times S^1 \times \{1\}$ and let K_1 be the link component of R such that $\partial N(K_1) = S^1 \times S^1 \times \{0\}$. Putting E and T in general position, we first note $E \cap T \neq \emptyset$ as otherwise, after filling in $N(K_1)$, E would become a disk in a solid torus. Then there would exist a meridional disk which avoids E. Hence T could be compressed along this once-punctured meridional disk to form an incompressible ∂-incompressible annulus in $S^3 - N(R \cup J)$, contradicting primeness of $R \cup J$.

An innermost trivial intersection curve α on E (or T) must be trivial on T (or E) by incompressibility of T (or E) and hence can be removed by irreducibility. Let β be an innermost intersection curve looping one puncture on E and nontrivial on T. Again, the existence of the annulus obtained by compressing T along the once-punctured disk on E bounded by β contradicts primeness of $R \cup J$.

Let γ be an innermost intersection curve looping both punctures on E and therefore nontrivial on T. Let E' be the twice-punctured disk on E bounded by γ. Then $\partial E'$ is a (p,q)-curve on T bounding a disk with interiors in $S^3 - T$, implying (p,q) is either $(0,1)$ or $(1,0)$. Thus J is isotopic on E to either a $(0,1)$ or a $(1,0)$ curve on T. The first possibility cannot occur by primeness of $R \cup J$, hence J is isotopic to K_1. By our hypotheses, this forces K_1 to be a component of our original alternating link L.

Taking the union of E' and an annulus A with boundary components γ and K_1, we can form a twice-punctured disk E'' in $S^3 - L$ with $\partial E'' = K_1$. We again utilize the notation as in [Menasco, 1984].

After an isotopy, $E'' \cap S^2_{\pm}$ each consists of disjoint simple closed curves, some of which may consist in part of arcs in $\partial E''$. As E'' is obtained in part from E', we can isotope E'' so that there exists a simple closed curve μ in $E'' \cap S^2_+$ such that $w_+(\mu) = P^2 S^j$ for some $j \geq 0$. Note that if μ lies on K_1 as it passes over an overcrossing, we count it as a bubble contact of μ.

We will show such a twice-punctured disk cannot exist in an alternating link complement. Since E'' is incompressible and ∂-incompressible, there is an isotopy of E'' as in [Menasco, 1983] such that if α is an intersection loop in $E'' \cap S^2$ then:

(i) α is not contained totally in any region of the projection plane or on a bubble.

(ii) α bounds a disk hugging the plane on the side it is viewed from.

(iii) the number of saddle contacts α has on each side of a bubble is either 0 or 1.

(iv) α does not run along a segment of the link adjacent to the same side of a bubble on which α has a saddle contact.

Note that we can perform the isotopes necessary to put E'' in this type of embedding without changing the fact that $w_+(\mu) = P^2S^j$ for some $j \geq 0$.

Suppose some simple closed curve β in $E'' \cap S_+^2$ intersected both sides of a bubble. Then, as in the proof of lemma 1 of [Menasco, 1984], there exists a simple closed curve in E'' constructed from an arc on β and an arc on the corresponding saddle disk which bounds a once-punctured disk D in $S^3 - L$ such that $D \cap E'' = \partial D$. Then ∂D must bound a once-punctured disk on E''. However, since if we remove from μ the two points corresponding to the punctures, one of the remaining two arcs will not intersect any bubbles, ∂D must intersect μ. This contradicts the fact that ∂D was constructed from an arc in β and an arc in a saddle disk.

Hence any given simple closed curve in $E'' \cap S_+^2$ intersects a particular bubble at most once. Suppose some curve in $E'' \cap S_+^2$ does have a bubble intersection. Let γ be such a curve such that γ bounds a disk F in S_+^2 which does not contain any other curves in $E'' \cap S_+^2$ which have bubble intersections. Then γ must intersect a bubble such that the other side of the bubble is inside F. This contradicts the fact that a saddle disk must intersect the top half of a bubble along arcs contained in intersection curves in $E'' \cap S_+^2$.

Thus $w_+(\mu) = P^2$. The primeness of L then implies L is a trivial arc to one side of μ in the projection plane. This contradicts the ∂-incompressibility of E''.

We next demonstrate that J is not a torus knot in a solid torus V with incompressible boundary. If such was the case, then since J is unknotted in S^3, V must be unknotted and J must be a $(p,1)$ torus knot in V. Exactly as before, we can eliminate intersections of E and ∂V until only those intersection curves looping both punctures on E and nontrivial on ∂V remain. At least one such intersection curve exists. Let E' be the twice-punctured disk on E bounded by an innermost such intersection curve. Then $\partial E'$ must be a $(1,0)$ curve on ∂V. This forces J to be isotopic in V to a $(1,0)$ curve on ∂V, contradicting the fact that J is a torus knot in V.

Finally, we show that there does not exist an essential torus T in $S^3 - N(R \cup J)^o$ which bounds a knot exterior in $S^3 - N(R \cup J)^o$. Suppose such a torus T exists. Then T bounds a solid torus V in S^3 which contains $R \cup J$. We can again isotope to remove all intersections of T with E except for those simple closed curves which loop both punctures on E and are nontrivial on T. Let α be an innermost such intersection curve on E and let E' be the twice-punctured disk on E bounded by α. Then E' is a meridional disk of V.

Let A be the annulus on E bounded by α and the next concentric intersection curve on E. Then ∂A consists of two meridional loops on ∂V, cutting ∂V into two annuli, denoted B_1 and B_2. If A is isotopic in $S^3 - \mathring{V}$ to either B_1 or B_2, then we isotope A to lower the number of intersection curves in $E \cap T$. Otherwise, the core of V is not a prime knot. In this case, take a new torus $T' = A \cup B_1$ which, after pushing off E slightly, has two less intersections with E. Then T' still bounds a solid torus V' containing $R \cup J$ to one side and a knot exterior to the other side such that T' is essential in $S^3 - N(R \cup J)^o$. We continue this process until the resulting torus does not intersect E. After repeating this procedure for each of the twice-punctured disks bounded by $J_1, \ldots, J_{n-1}$, the resulting torus, which we again denote T, satisfies $T \cap (\bigcup_{i=1}^{n} E_i) = \emptyset$.

Again utilising the notation of [Menasco, 1984], we have a set of intersection curves in $T \cap S_+$. Since T is incompressible and avoids $\bigcup_{i=1}^{n} E_i$, at least one of the intersection curves must have a bubble contact. The proof of Theorem 2 of [Menasco, 1984] then implies that there is a disk D' once-punctured by L in $S^3 - N(R \cup J)^o$ that intersects T in its boundary. The annulus A' obtained by compressing T to L along D' is essential, contradicting the fact $R \cup J$ is a prime link.

Theorem 4.1

5. Applications.

In [Thurston, 1979], Thurston proved that if one does Dehn surgery on some subset of the cusps of a finite volume complete hyperbolic 3-manifold M, then if for each pair of surgery coefficients (p_i, q_i), $p_i^2 + q_i^2$ is large enough, the resulting 3-manifold is also hyperbolic. In addition, as $p_i^2 + q_i^2$ approaches ∞ for all i, the hyperbolic structures on the resulting manifolds approach the hyperbolic structure on M and in particular the hyperbolic volumes approach the hyperbolic volume of M from below.

Let L be a nonsplittable prime alternating link in a particular alternating projection where L is not a $(2,q)$ torus link. Choosing any two strands l_1 and l_2 of L bounding the same region of the projection plane as in Figure 3a, we can twist the two strands about one another, introducing an even number of additional crossings such that the resulting link L' is still alternating as in Figure 3b. Then L' is obtainable from a $(p,1)$ surgery on the trivial component of the augmented alternating link L'' as in Figure 3c where p is half the number of new crossings introduced.

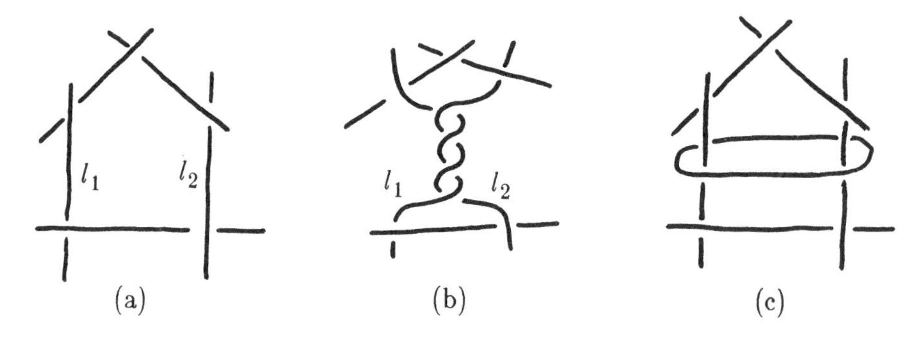

Figure 3.

Hence, the sequence of alternating link complements obtained by twisting in this manner as the number of crossings introduced increases will have hyperbolic structures converging to the hyperbolic structure on the complement of L''. Note, in addition, that Corollary 5.1 of [Adams, 1985] implies the sequence of alternating link complements obtained by adding an odd number of new crossings has hyperbolic volumes converging to the volume of the complement of L''.

An additional application of Theorem 4.1 is an alternative proof that every closed orientable 3-manifold can be obtained from Dehn surgery on some hyperbolic link complement, a fact Thurston proves in [Thurston, 1984]. Since every closed orientable 3-manifold can be obtained from Dehn surgery on some link in S^3 as in [Lickorish, 1962] or [Wallace, 1960], it suffices to show that any link L in S^3 comes from Dehn surgery on a link in S^3 with hyperbolic complement. We can add components to L to make it nonsplit and prime. Then about each crossing in L, add a trivial component to L except where such an addition would make two such trivial components isotopic. Then by twisting along the twice-punctured disks bounded by these trivial components, we can see this link complement is homeomorphic to the complement of an augmented alternating link and hence hyperbolic.

Finally, utilizing the results of [Adams, 1985], [Bonahon-Siebenmann], [Menasco, 1984], [Oertel, 1984] and this paper, we determine which links of more than one component in the tables at the end of [Rolfsen, 1976] do or do not have hyperbolic complements. All such links which do not have hyperbolic complements appear in Table 1. Of those which do have hyperbolic complements, only 9^2_{50}, 9^2_{51}, 9^2_{52}, 9^2_{54}, 9^2_{59}, 9^2_{60} and 9^3_{17} are not either hyperbolic alternating links, augmented alternating links or obtainable from augmented alternated links via half or full twists along twice-punctured disks. Of these exceptions, the first four are star links and hence can be seen to have hyperbolic complements by the results of [Bonahon-Siebenmann] and [Oertel, 1984]. The last three exceptions can be obtained from one another by twisting along twice-punctured disks, hence the following corollary to Theorem 3.1 suffices to show that all three have hyperbolic complements.

Table 1: Links with nonhyperbolic complements.

0^2_1	9^2_{49}
2^2_1	9^2_{53}
4^2_1	9^2_{61}
6^2_1	6^3_3
7^2_7	8^3_7
8^2_1	8^3_{10}
9^2_{43}	9^3_{21}
8^4_3	

Corollary 5.1. *The complement of 9^2_{60} is hyperbolic.*

Proof. We will apply Theorem 3.1, where $\mathring{M}$ will correspond to $S^3 - L$, L being the knotted complement of 9^2_{60} and K will correspond to the unknotted complement of 9^2_{60}. Then $\mathring{M}$ is the 5_2 knot complement and hence hyperbolic by. [Menasco, 1984]

We check each of the hypotheses of Theorem 3.1.

First suppose there exists an $S^1 \times S^1 \times \{\mathbf{0}\} \subset \partial M$. Denote $S^1 \times S^1 \times \{1\}$ by T. Then there must exist a disk D once-punctured by L in $S^3 - (K \cup L)$ and intersecting T in its boundary, as otherwise K would be knotted in S^3. Compressing T to L along D yields an essential annulus contradicting the primeness of 9^2_{60}. Thus condition (i) of Theorem 3.1 is satisfied.

The fact that condition (ii) is satisfied follows immediately from the nonsplittability of 9^2_{60}. The same argument as in the proof Theorem 4.1 applies to show condition (iii) is satisfied.

Finally, suppose there exists an essential torus in $S^3 - (L \cup K)$ which bounds a knot exterior in $S^3 - (L \cup K)$. Then, exactly as in the proof of Theorem 4.1, there exists such a torus T which does not intersect the incompressible twice-punctured disk E bounded by K. Let V be the solid torus in S^3 bounded by T L. We replace K by a component J as in Figure 4. Let E' be the twice-punctured disk bounded by J. Note that $L \cup J$ is an augmented alternating link.

Figure 4.

Then T remains incompressible in $S^3 - \mathring{V}$ as V is still knotted. If T compressed in $V - (L \cup J)$, then the compressing disk D' must intersect E' as otherwise, T would have compressed in $V - (L \cup K)$. However, the incompressibility of E' implies that all intersections of E' and D' can be removed. Hence T is an essential torus in the augmented alternating link complement, contradicting Theorem 4.1.

Corollary 5.1

References

Adams, 1985.
C. Adams, "Thrice-punctured spheres in hyperbolic 3-manifolds," *T.A.M.S.*, **287**, 2, pp. 645-56, 1985.

Bonahon-Siebenmann.
F. Bonahon and L. Siebenmann, *Seifert 3-orbifolds and their role as natural crystalline parts of arbitrary compact irreducible 3-orbifolds,* London Mathematical Society Lecture Notes, C.U.P..

Hatcher.
A. Hatcher, *Torus Decompositions.* xeroxed notes

Jaco, 1977.
W. Jaco, "Lectures on 3-manifold topology," *CBMS regional conference series in mathematics*, **43**, 1977.

Lickorish, 1962.
W.B.R. Lickorish, "A representation of orientable combinatorial 3-manifolds," *Annals*, **76**, pp. 531-538, 1962.

Menasco, 1983.
W. Menasco, *Incompressible surfaces in the complement of alternating knots and links*, 1983. Unpublished

Menasco, 1984.
W. Menasco, "Closed incompressible surfaces in alternating knot and link complements," *Topology*, pp. 37-44, 1984.

Oertel, 1984.
U. Oertel, "Closed incompressible surfaces in complements of star links," *Pac.J.Math.*, **111**, pp. 209-230, 1984.

Rolfsen, 1976.
Rolfsen, D., *Knots and links,* Publish or Perish, Berkeley, 1976.

Thurston, 1979.
W. P. Thurston, *The Geometry and Topology of 3-Manifolds,* Princeton University Mathematics Department, 1979. Part of this material – plus additional material – will be published in book form by Princeton University Press

Thurston, 1982.
W. P. Thurston, *Hyperbolic Geometry and 3-Manifolds,* London Mathematical Society Lecture Notes, 48, pp. 9-26, C.U.P., 1982. Written by Scott from lectures of Thurston.

Thurston, 1984.
W. P. Thurston, *Universal Links*, 1984. Preprint

Wallace, 1960.
A. O. Wallace, "Modifications and Cobounding Manifolds," *Can. J. Math.*, **12**, pp. 503-528, 1960.

C.C. Adams,
Williams College,
Mass 01267
U.S.A.

Incompressible Surfaces in 3-Manifolds: The Space of Boundary Curves

*William J. Floyd**

If M is a Haken 3-manifold with incompressible boundary and S is an incompressible surface in M, then there is a natural way to view ∂S as a point $[\partial S]$ in the projection lamination space $PL(\partial M)$. The set $\{ [\partial S] : S$ is an incompressible surface in $M \}$ is dense in $PL(\partial M)$, since one can choose S to be a union of boundary parallel annuli. The situation is much different, however, if one requires also that S be ∂-incompressible. Our main result is the following:

Theorem. *Let M be a Haken 3-manifold with incompressible boundary. Then $\{ [\partial S] : S$ is a two-sided, incompressible, ∂-incompressible surface in $M \}$ is a dense subset of a finite union of closed cells in $PL(\partial M)$, each of dimension less than one half the dimension of $PL(\partial M)$.*

This theorem answers a question raised by Hatcher in [Hatcher, 1982], where he proved the special case when ∂M is a union of tori. Our proof, like Hatcher's, is based on the result of [Floyd-Oertel, 1984] that there are finitely many incompressible branched surfaces in M which carry all the 2-sided, incompressible, ∂-incompressible surfaces in M. The new ingredient here is the use of Thurston's construction of a local symplectic structure on the measured lamination space $ML(F)$ of a surface F. If B is an incompressible branched surface in M, we show that $\{ r[\partial S] : S$ is a surface carried by M and $r \in \mathbb{R}_+ \}$ lies in a finite union of Lagrangian subspaces, in this local symplectic structure, of $ML(\partial M)$.

I would like to thank Allen Hatcher for raising this question, William Goldman for first pointing out to me the local symplectic structure on $ML(\partial M)$, and William Thurston for giving me details about this structure.

* Supported in part by NSF grant MCS 81-02469

1. Train tracks and the projective lamination space

In this section we give an exposition of some results of Thurston on the projective lamination space $PL(F)$ of a surface F and on the use of train tracks in studying $PL(F)$. For more information, see [Thurston, 1979].

Let F be a closed, oriented surface such that each component F_i has genus $g_i > 1$ and is equipped with a Riemannian metric of constant sectional curvature -1. A *lamination* γ on F is a closed subset $X \subset F$ together with a finite covering of a neighbourhood of X by open sets $\{ U_i \}$, $i \in \{ 1, \ldots, n \}$, and maps $f_i : U_i \to \mathbb{R} \times \mathbb{R}$ such that for each i there is a closed set $N_i \subset \mathbb{R}$ with $f_i(X \cap U_i) = \mathbb{R} \times N_i$, and so that the transition functions $f_i \circ f_j^{-1}$, restricted to $f_j(X)$, preserve horizontal lines. A *geodesic lamination* on F is a lamination γ on F such that for each $x \in N_i$, $f_i^{-1}(\mathbb{R} \times \{ x \})$ is a geodesic arc. Thus γ is a union of geodesics, which are called *leaves* of the lamination. A *measured lamination* (γ, υ) on F is a geodesic lamination γ on F together with a measure υ_i on each N_i so that the transition functions $f_i \circ f_j^{-1}$ are measure preserving. Given an essential, simple closed surface α in F, let $g(\alpha)$ be the geodesic in F homotopic to α. We will denote the measured lamination $(g(\alpha), 1)$ by α, where 1 is counting measure.

Let $ML(F)$ be the space of measured laminations on F, with a topology coming from integrating finitely many continuous functions on finitely many arcs in F which are transverse to γ. Let $PL(F) = (ML(F) - \{0\})/\mathbb{R}_+$, where $\mathbb{R}_+$ acts on measures by scalar multiplication and $\{0\}$ is any geodesic lamination with measure 0. Then $ML(F_i) \cong \mathbb{R}^{6g-6}$ and $PL(F_i) \cong S^{6g-7}$. $ML(F)$ and $PL(F)$ were introduced in [Thurston, 1977], under the related theory of measured foliations, in Thurston's study of surface diffeomorphisms (see [Fathi-Laudenbach-Poenaru, 1979] for a detailed exposition). If F is connected, then simple closed geodesics are dense in $PL(F)$ and $PL(F)$ can also be defined in terms of functions on isotopy classes of simple closed curves. A crucial result in this study is that $PL(F)$ forms a natural boundary for the Teichmuller space $T(F)$.

A *train track* τ in F is a subspace of F such that

i) each point in τ has a neighbourhood diffeomorphic to an open set in the space Y shown in Figure 1a, and

ii) for each component C of F split along τ, the double of C along the smooth arcs of ∂C has negative Euler characteristic.

If instead of (ii) we only require that no component C of F split along τ is a disk bounded by a simple closed curve which is smooth or smooth except at one point, we will call τ a *digon track*. If C is a component of $F - \tau$, then the points at which the boundary of C is not smooth will be called *cusps*. In each case, the points at which τ is not a manifold are called *vertices* or *switches*, and the components of $\tau - \{\text{vertices}\}$ are called *edges* or *branches*. We say τ is *complete* if each component of F split along τ is a disk with exactly 3 cusps. If V_j is a vertex of a digon track τ, then locally there are 3 edges coming into V_j (globally, they may not be distinct edges). Following the model in Figure 1a, we will label the left edge $e_{j,i}$ and the two right edges $e_{j,o}$ and $e_{j,o}'$. A *transverse measure* on τ is a function $\mu : \{\text{edges}\} \to \mathbb{R}_{\geq 0}$ (where $\mathbb{R}_{\geq 0} = \{x \in \mathbb{R} : x \geq 0\}$) such that for each vertex v_j,

$$\mu_{j,i} = \mu(e_{j,o}) + \mu(e_{j,o}').$$

If e is an edge, then $\mu(e)$ is called the *weight* on e. The above condition is that the weight on the incoming edge is the sum of the weights on the outgoing edges. If one thinks of the weights as recording the path of a train, then the condition says that each time the train enters a switch it leaves it again. Let $V(\tau)$ denote the set of transverse measures on τ and let $V_+(\tau) = \{\mu \in V(\tau) : \mu(e) > 0 \text{ for each edge } e \text{ in } \tau\}$.

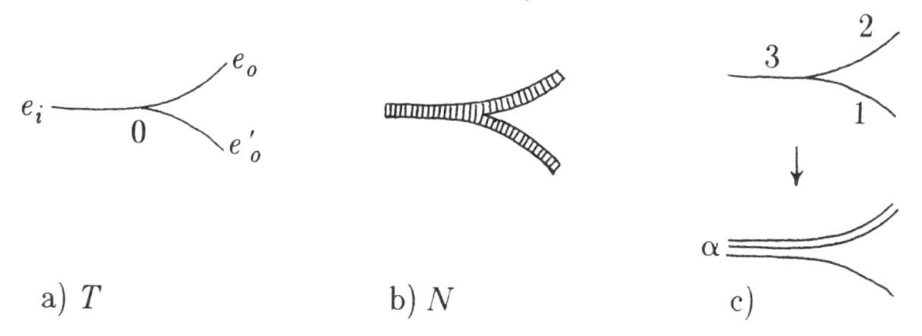

Figure 1.

Train tracks were introduced in [Thurston, 1979] as a convenient tool in studying measured laminations on surfaces. If τ is a digon track on F and (δ,υ) is a measured lamination on F, then (δ,υ) is *carried* by τ if δ can be isotoped into a fibered regular neighbourhood N of τ (the local picture is given in Figure 1b so that the leaves of δ are transverse to the fibers of N. By integrating along fibers, the measure υ determines a transverse measure μ_δ on τ. The track τ is *recurrent* if for each edge $e \in \tau$ there is a simple closed curve α carried by τ such that $\mu_\alpha(e) > 0$.

If α is a union of pairwise disjoint, essential simple closed curves in F which are carried by a train track τ in F, then the weights of μ_α are all integers. (One can recover the isotopy class of α from these weights by gluing parallel copies of the edges together near the vertices (see Figure 1c for the local picture)). If τ is recurrent then this extends to a map $\phi : V(\tau) \to ML(F)$, and one has the following:

Theorem (Thurston). *If τ is a complete, recurrent train track, then this map $\phi : V(\tau) \to ML(F)$ is a homeomorphism onto its image. The restriction of ϕ to $V_+(\tau)$ is a coordinate chart for $ML(F)$.*

Thurston's original proof assumed that the track is also transversely recurrent, but Papadopoulos showed in [Papadopoulos, 1983] how to prove the theorem without that assumption.

We now give Thurston's construction of the local symplectic structure im $ML(F)$. Let τ be a digon track in F. For each vertex $v_j \in \tau$, choose an orientation preserving diffeomorphism of a neighbourhood of v_j in F to the neighbourhood of 0 in Figure 1a. Let $e_{j,o}$ be the edge in τ which corresponds to e_o under the diffeomorphism, and let $e_{j,o}{}'$ be the edge in τ which corresponds to $e_o{}'$. Given transverse measures μ and υ on τ, define the pairing $(\mu,\upsilon)_\tau$ by

$$(\mu,\upsilon)_\tau - \sum_j \mu(e_{j,o})\upsilon(e'_{j,o}) - \mu(e'_{j,o})\upsilon(e_{j,o}).$$

Then $(\ ,\ )_\tau$ is bilinear and skew-symmetric. Furthermore one has the following:

Theorem (Thurston). *If τ is a complete, recurrent train track on F, then $(\ ,\ )_\tau$ is non-degenerate.*

For an exposition of Thurston's proof, see [Harer-Penner, 1986].

We are interested in defining $ML(F)$ and $PL(F)$ when F has components which are tori. Let T be a torus, which we will think of as $\mathbb{R}^2/\mathbb{Z}^2$. Given an essential, simple closed curve α in T, we can isotope α in T so that its lift to $\mathbb{R}^2$ is a line through the origin. Let $(p,q) \in \mathbb{Z}^2$ be a point on the line with p and q relatively prime. We let

$$ML(T) = \mathbb{R}\mathrm{P}^1 \times \mathbb{R}_{\geq 0}/(\mathbb{R}\mathrm{P}^1 \times \{0\}).$$

We will denote the equivalence class of $\mathbb{R}\mathrm{P}^1 \times \{0\}$ by 0. Given an essential, simple closed curve α in T and a weight $r \in \mathbb{R}_+$ on α, let $f(\alpha,r) = ([p:q],\ r(p^2 + q^2)^{-1/2})) \in ML(T)$. Then $ML(T) = \mathrm{cl}\{f(\alpha,r) : \alpha$ is an essential, simple closed curve in T and $r \in \mathbb{R}_+$ }.

Given a closed, orientable surface F such that $F = F_1 \cup \ldots \cup F_k$, where $F_1, \ldots, F_{k-1}$ are distinct components of F which are tori and each component of F_k has genus greater than one, we define $ML(F) = ML(F_1) \times \ldots \times ML(F_k)$ and $PL(F) = (ML(F) - \{0\})/\mathbb{R}_+$. A digon track τ on F_k is a *complete train track* on F if the restriction of τ to F_k is a complete train track on F_k and for each $i \in \{1, \ldots, k-1\}$ the restriction of τ to F_i is homeomorphic to the digon track in Figure 4a. With these definitions, it is easy to check that the above results are still valid when F has components which are tori.

2. Branched surfaces and the symplectic pairing

A *branched surface* B in a manifold M^3 is a closed subset of M which in int(M) is locally modelled on the space U shown in Figure 2a and which near ∂M is locally modelled on the space U^+ in Figure 2b. The *branch locus* of B is $\{ p \in B : B$ is not a manifold in a neighbourhood of $p \}$. The *vertices* of B are the points in the branch locus at which the branch locus is not a manifold. The branch locus is a union of properly immersed curves in M. We define $\partial B = B \cap \partial M$.

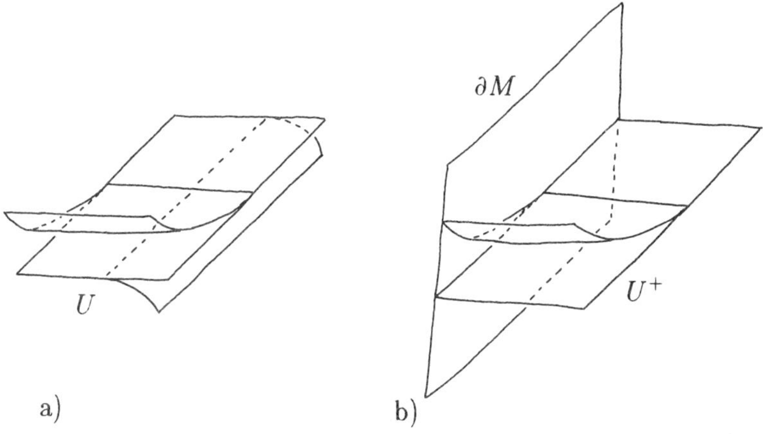

Figure 2.

Analogously to the definition for digon tracks in surfaces, one can define a transverse measure on a branched surface B and the concept of a surface S in M being carried by B (see [Floyd-Oertel, 1984] for more details). If B is a branched surface in M such that ∂B is a digon track in

∂M, then a transverse measure μ on B restricts to a transverse measure $\partial\mu$ on ∂B (the condition that ∂B is a digon track is because we did not define a transverse measure on a general branched 1-manifold).

Proposition. *Let B be a branched surface in a compact, oriented 3-manifold M such that ∂B is a digon track in ∂M. If μ and υ are transverse measures on B, then $(\partial\mu,\partial\upsilon)_{\partial B} = 0$.*

Proof. Choose an orientation for each of the curves in the branch locus of B, and let $v_1, \ldots, v_n$ be the vertices of B. Let v be a vertex in B, and let α and β be the two local curves of the branch locus near v. Let $x_{\alpha,+}$ ($x_{\alpha,-}$) be a point on the positive (negative) side (using the orientation on α) of v on α, and let $x_{\beta,+}$ ($x_{\beta,-}$) be a point on the positive (negative) side of v on β. Let $N_{\alpha,+}$, $N_{\alpha,-}$, $N_{\beta,+}$ and $N_{\beta,-}$ be normal planes to α and β at $x_{\alpha,+}$, $x_{\alpha,-}$, $x_{\beta,+}$, and $x_{\beta,-}$ (respectively), with orientations induced from the orientations on M, α, and β. These normal planes intersect B in train tracks near v, and as in the previous section one can define local intersection pairings $(\mu,\upsilon)_{\alpha,+}$, $(\mu,\upsilon)_{\alpha,-}$, $(\mu,\upsilon)_{\beta,+}$ and $(\mu,\upsilon)_{\beta,-}$. Let $(\mu,\upsilon)_v = (\mu,\upsilon)_{\alpha,+} - (\mu,\upsilon)_{\alpha,-} + (\mu,\upsilon)_{\beta,+} - (\mu,\upsilon)_{\beta,-}$.

An easy calculation shows that $(\mu,\upsilon)_v = 0$.

Thus $(\mu,\upsilon)_B = \sum_{i=1}^{n} (\mu,\upsilon)_{v_i} = 0$.

However, for each immersed curve α in the branch locus, the signed intersection pairings on α coming from the vertices cancel in pairs except for the contribution when the arcs of α go out to ∂M, in which case they give the local intersection pairings for $\partial\mu$ and $\partial\upsilon$ on ∂B. Since the local intersection pairings for $\partial\mu$ and $\partial\upsilon$ cancel for arcs in the branch locus which are disjoint from the vertices, $(\partial\mu,\partial\upsilon)_{\partial B} = (\mu,\upsilon)_B = 0$.

Proposition

3. Getting rid of annuli and digons

Let M be an oriented, Haken 3-manifold. By [Floyd-Oertel, 1984] there are finitely many branched surfaces $B_1, \ldots, B_n$ in M such that

i) the 2-sided surfaces carried with positive weights by the B_i's are exactly the two sided, incompressible, ∂-incompressible surfaces in M, and

ii) the ∂B_i's are digon tracks.

If S_1 and S_2 are surfaces carried by a B_i, then we saw in the previous section that $(\partial S_1, \partial S_2)_{\partial B_i} = 0$. However, the B_i's are digon tracks, and we need train tracks to get coordinate charts for $ML(F)$. In this section we show how to resolve this problem.

We will say that a set C of transverse measures on a digon track τ in F satisfies $(\star)$ if

i) all the weights of the elements of C_i are rational numbers,

ii) C is closed under $\mathbf{Q}_+$ linear combinations, and

iii) if $\mu, \upsilon \in C$ then $(\mu,\upsilon)_\tau = 0$.

Note that if C satisfies $\star$, then the elements of C determine systems of pairwise disjoint, simple closed curves in F with rational weights. Given digon tracks $\tau_1, \ldots, \tau_m$ in F and sets $C_i \subset V(\tau_i)$ each satisfying $\star$, and digon tracks $\eta_1, \ldots, \eta_r$ in F and sets $C_j{}' \subset V(\eta_j)$ each satisfying $\star$, we say $\bigcup_{i=1}^{m} C_i$ and $\bigcup_{j=1}^{r} C_j{}'$ are *compatible* if they determine the same system of pairwise disjoint, simple closed curves in F with rational weights.

The idea is to change τ to get rid of components of $F - \tau$ which are annuli or digons, where a digon is a disk component of $F - \tau$ which has exactly 2 cusps. The operations we will use are shifts, splits, edge collapses, and digon collapses.

A *shift* takes a digon track τ in F to another digon track τ' in F. The track is unchanged outside a neighbourhood of an edge e in τ, and the local picture near e is shown in Figure 3a. A transverse measure μ on τ naturally determines a transverse measure μ' on τ', and if $C \subset V(\tau)$ satisfies $\star$ then $C' \subset V(\tau')$ satisfies $\star$ and C and C' are compatible. (The third condition for $\star$ can be shown by direct calculation or by using Section 2 since there is a branched surface B in $F \times I$ so that $B \cap (F \times \{0\}) = \tau$ and $B \cap (F \times \{1\}) = \tau'$). Note that a shift does not change the numbers or types (topological types plus numbers of cusps) of the complementary components.

In a *split* a digon track τ in F is replaced by two digon tracks τ_1' and τ_2' in F. Each τ_i' is identical to τ outside a neighbourhood of an edge e in τ, and the local pictures near e are given in Figure 3b. A transverse measure μ on τ with $\mu(e_1) \geq \mu(e_2)$ determines a transverse measure μ_1 on τ_1', and a transverse measure μ on τ with $\mu(e_1) \leq \mu(e_2)$ determines a transverse measure μ_2' on τ_2'. If $C \subset V(\tau)$ satisfies $\star$, then the induced sets $C_1' \subset V(\tau_1)$ and $C_2' \subset V(\tau_2')$ both satisfy $\star$ and C and $C_1' \cup C_2'$ are compatible

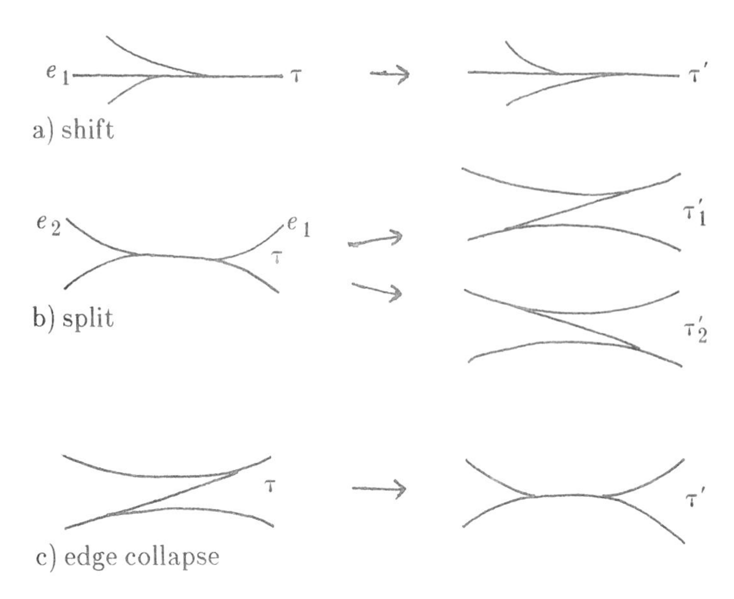

Figure 3.

(once again, the third condition follows since the splits can be extended to branched surfaces in $F \times I$). If $C'_i = \phi$ then we will not make use of the digon track τ'_i. A split also does not change the numbers or types of the complementary components.

An *edge collapse* takes a digon track τ in F to another digon track τ' in F with the same numbers and types of the complementary components. The tracks τ and τ' agree outside a neighbourhood of an edge e in τ, and the local picture near e is given in Figure 3c. An edge collapse is the inverse of one of the two pieces of a split. A transverse measure μ on τ determines a transverse measure μ' on τ', and if $C \subset V(\tau)$ satisfies $\star$ then the induced set $C' \subset V(\tau')$ also satisfies $\star$ and is compatible with C.

A *digon collapse* takes a digon track τ in F to another digon track τ' in F. A digon collapse is obtained as follows: take an embedded digon D in $F - \tau$, fiber $\mathrm{cl}(D) - \mathrm{cusps}(D)$ by intervals such that each interval is transverse to the boundary of D and no interval has vertices of τ on both boundary components, and then collapse each interval to a point. The new track τ' has one less digon, but all the other numbers and types of the complementary regions are the same. The digon collapse is not unique if there are vertices of τ on both components of $\partial D - cusps(D)$. A transverse measure μ on τ determines a transverse measure μ' on τ', and if $C \subset V(\tau)$ satisfies $\star$ then the induced set $C'' \subset V(\tau')$ also satisfies $\star$ and is compatible with C.

Given a digon track τ in F, a *toral digon* is a digon in a toral component F_i of F so that the restriction of τ to F_i is as in Figure 4a. An *annular digon* is a digon in $F-\tau$ as in Figure 4b, and a *Reeb digon* is a digon as in Figure 4c. In the two latter cases we allow vertices of τ on the boundary. If an annular digon or a Reeb digon has further self-intersections on the boundary, then we say it is immersed. given a component D of $F-\tau$ which is a digon, let the complexity $c(D)$ be the number of edges e in τ with $e \subset \mathrm{int}(\mathrm{cl}(D))$. That is, $c(D)$ is the number of edges e which have D on both sides. For example, $c(D) = 1$ if D is an embedded annular digon or Reeb digon.

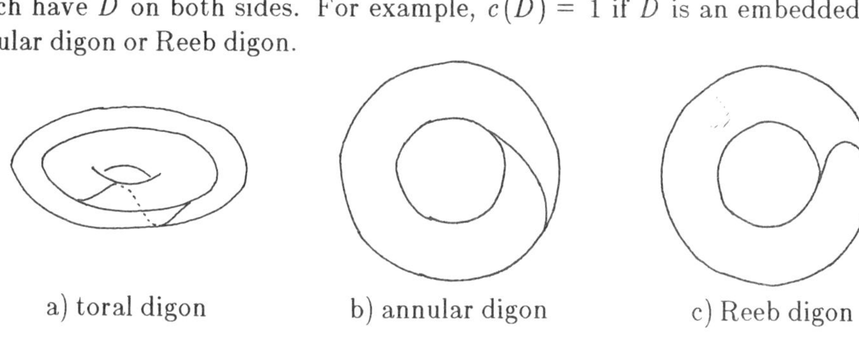

a) toral digon b) annular digon c) Reeb digon

Figure 4.

Proposition. *Let τ be a digon track in F, and C a set of transverse measures on τ which satisfies $\star$. Then there are a finite number of complete, recurrent train tracks $\tau_1, \ldots, \tau_m$ in F and sets $C_i \subset V(\tau_i)$ each satisfying $\star$ so that C and $\bigcup_{i=1}^{m} C_i$ are compatible.*

Proof. Let τ be a digon track in F, and C a set of transverse measures on τ which satisfies $\star$. Suppose τ is not a train track. For convenience, we first get rid of components of $F-\tau$ which are annuli. If A is a component of $F-\tau$ which is an annulus, then we will add an edge e to τ in A to make it into an annular digon, and then do an edge collapse on e to get a digon track. We do this in each complementary component to get a new digon track τ' without complementary components which are annuli and an induced set of transverse measures $C' \subset V(\tau')$ which satisfies $\star$ and is compatible with C.

We proceed by induction on the number, $n(\tau')$, of components of $F-\tau'$ which are digons but are not toral digons. If $n(\tau') = 0$, then τ' is a train track. By passing to the maximal recurrent subtrack of τ' and then adding edges and doing edge collapses, we can get a complete, recurrent train track τ'' in F and induced set $C'' \subset V(\tau'')$ such that C'' satisfies $\star$ and is compatible with C. So we can assume that $n(\tau') = n$ is greater than zero, and that the lemma holds for digon tracks η with $n(\eta) < n$ and no annular components of $F-\tau$.

For this fixed n, we will do an induction on the complexity $c(\tau') = \min\{c(D) : D$ is a component of $F-\tau'$ which is a digon but not a toral digon$\}$. If $c(\tau') = 0$, then there is an embedded digon in $F-\tau'$, and we can do a digon collapse to reduce $n(\tau')$. Thus we can assume that $c(\tau') = r > 0$, and that the lemma holds for digon tracks η in F with no annular components of $F-\eta$, $n(\eta) = n$, and $c(\eta) < r$. Let D be a digon in $F-\tau'$ which is not a toral digon, and let e be an edge in int(cl(D)). Up to symmetry, the possibilities for τ' and D near e are given in Figure 5. If the local picture near e is as in Figures 5a – 5c, then shifting along e produces a digon track η and a set $C'' \subset V(\eta)$ such that C'' satisfies $\star$, C and C'' are compatible, $n(\eta) = n$, and $c(\eta) < (\tau')$. So in this case we are done by induction. If the local picture is as in Figures 5e or 5f, then we can reduce the complexity by splitting along e (though we will create two digon tracks) and then using induction on the two new digon tracks. In cases 5h and 5i we can reduce the complexity (and hence complete the proof by induction) by an edge collapse along e. So we can assume that the local picture is one of cases (d), (g), (j), (k), (l), or (m) in Figure 5. In each of these 6 cases, in our local picture we have already accounted for the two cusps of D.

Suppose the local picture near e is that of Figure 5d. If there is another vertex in the component of τ' containing e, then there is an edge adjacent to e with local picture that of Figure 5c or 5f. In this case we can complete the proof by the above analysis of those cases. Otherwise, e is in a torus component of F. The maximal recurrent sub-track in this component is a circle and we can add an edge and do an edge collapse to get a toral digon (reducing the complexity).

If the local picture near e is that of Figure 5g, then if there are no other vertices of τ' in the component F' of F containing e, the component of τ' in F' is a train track. Since this contradicts our choice of D, there must be another vertex at the other end of one of e_1, e_2, e_3, or e_4. The local picture near this edge is that of Figure 5c, so we can complete the induction step by a shift along that edge.

If the local picture is that of Figure 5j, then after doing an edge collapse along e we are in case 5 (g), which we have just treated.

In cases 5k – 5m, D is a (possibly immersed) Reeb digon. If $c(D) > 1$ let $e' \neq e$ be an edge of τ' with $e' \subset$ int(cl(D)). Then the local picture near e' is one of those in Figure 5a – 5j, so we are done by our previous analysis. So the only remaining case is 5 (k) when D is an embedded Reeb

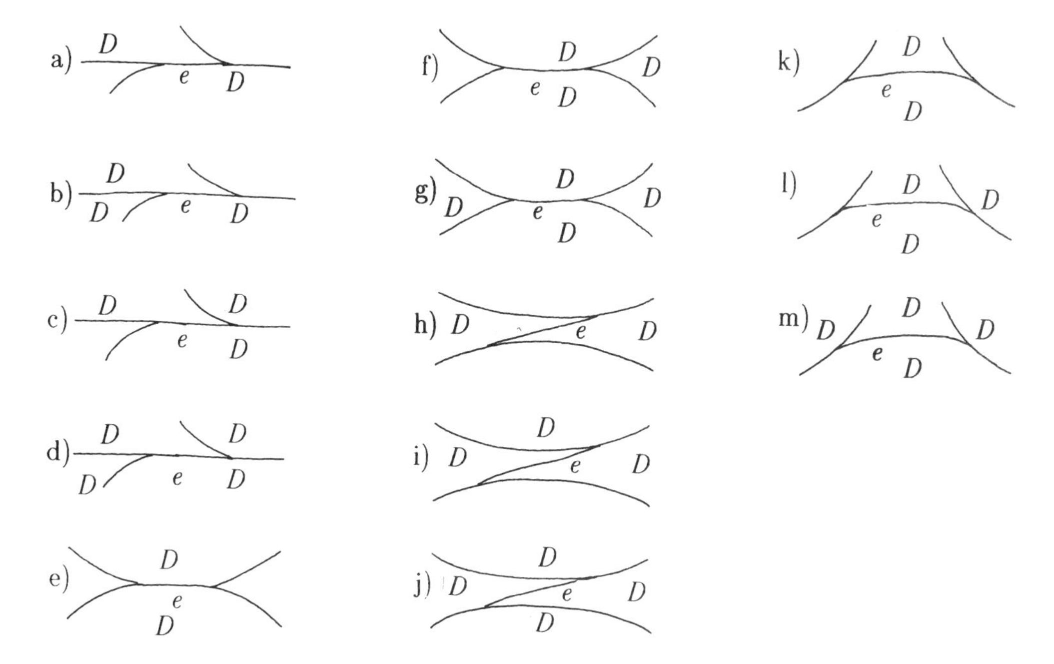

Figure 5.

digon. If D is an embedded Reeb digon, then after possibly doing some shifts and splits along edges in $\mathrm{cl}(D)$, we can assume that we have replaced τ' by a finite number of digon tracks $\eta_1, \ldots, \eta_s$ in F and associated sets $C_i' \subset V(\eta_i)$ each satisfying $\star$ so that C and $\bigcup_{i=1}^{s} C_i'$ are compatible, each η_i has $n(\eta_i) = n$ and no annular components of $F - \eta_i$, and the local picture near D in η_i is that of Figure 6a. Given $i \in \{1, \ldots, s\}$, let η_i^1 and η_i^2 be digon tracks in F such that η_i^1 and η_i^2 are identical with τ' outside a neighbourhood of D, and in a neighbourhood of D the local picture for η_i^1 (η_i^2) is that of Figure 6b (Figure 6c). Given a transverse measure μ on τ', if $\mu(e_1) \leq \mu(e_2)$ then μ determines a transverse measure μ_i^1 on η_i^1 by letting $\mu_i^1 = \mu$ outside our local picture, $\mu_i^1(\alpha) = \mu(e_5)$, $\mu_i^1(\beta) = \mu(e_2) - \mu(e_1)$, $\mu_i^1(\gamma) = \mu(e_4) - \mu(e_3)$, and $\mu_i^1(\delta) = \mu(e_6)$. If $\mu(e_1) \geq \mu(e_2)$ then μ determines a transverse measure μ_i^2 on η_i^2 by letting $\mu_i^2 = \mu$ outside our local picture, $\mu_i^2(\alpha) = \mu(e_5)$, $\mu_i^2(\beta) = \mu(e_1) - \mu(e_2)$, $\mu_i^2(\gamma) = \mu(e_3) - \mu(e_4)$, and $\mu_i^2(\delta) = \mu(e_6)$. The induced sets $C_i^1 \subset V(\eta_i^1)$ and $C_i^2 \subset V(\eta_i^2)$ both satisfy $\star$ and $C_i^1 \cup C_i^2$ is compatible with C'_i. Since $n(\eta_i^1) = n(\eta_i^2) = n-1$, we can proceed by induction to finish the proof.

Proposition

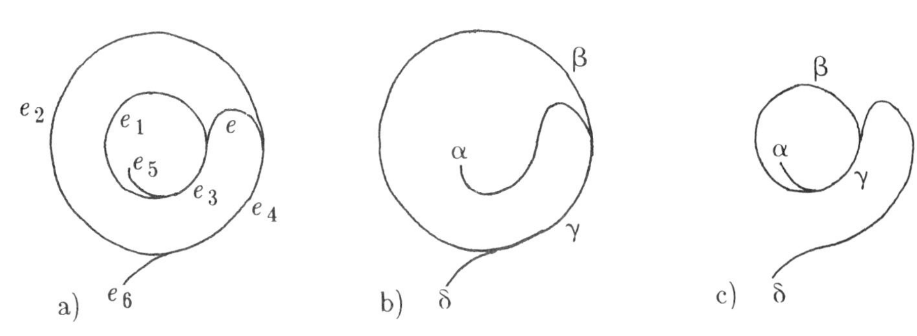

Figure 6.

4. The proof

We are now ready to prove the theorem stated in the introduction.

Proof. Let M be a Haken 3-manifold with incompressible boundary, and choose an orientation for M. By [Floyd-Oertel, 1984] there are finitely many incompressible branched surfaces $B_1, \ldots, B_n$ in M such that the 2-sided surfaces carried by the B_i' s are exactly (up to isotopy) the 2-sided, incompressible, ∂-incompressible surfaces in M. We will choose $B_1, \ldots, B_n$ to be the incompressible branched surfaces constructed in [Floyd-Oertel, 1984] so that each ∂B_i is a digon track in ∂M.

For each i, let $C_i = \{ \partial\mu_i : \mu_i$ is a transverse measure on B_i with all weights positive, rational numbers$\}$. By the proposition in Section 2, each C_i satisfies $\star$. By applying the proposition in Section 3 to each ∂B_i and $C_i \subset V(\partial B_i)$, one sees that there are a finite number $\tau_1, \ldots, \tau_m$ of complete, recurrent train tracks in ∂M and sets $K_j \subset V(\tau_j)$ each satisfying $\star$ so that $\bigcup_{i-1}^{n} C_i$ and $\bigcup_{j=1}^{m} K_j$ are compatible.

Let $\pi : (ML(\partial M) - \{0\}) \to PL(\partial M)$ be the projection, and for each $i \in \{1, \ldots, m\}$ let $\phi_i : V(\tau_i) \to ML(\partial M)$ be the natural map which is a homeomorphism onto its image. Then $\{[\partial S] : S$ is a 2-sided, incompressible, ∂-incompressible surface in $M\} = \pi(\bigcup_{j=1}^{m} \phi_j(K_j))$. Since each K_j satisfies $\star$, for each $j \in \{1, \ldots, m\}$ cl(K_j) is a convex subset of $V(\tau_j)$ which lies in a Lagrangian subspace of $V(\tau_j)$ (technically, for this one should allow transverse measures with negative weights). Hence each cl(K_j) has dimension $\leq \dim(V(\tau_j)/2$. So cl $\{[\partial S] : S$ is 2-sided, incompressible, ∂-incompressible surface in $M\} = \pi(\bigcup_{j=1}^{m} \phi_j(\text{cl}(K_j)))$ is a finite union of closed cells, each of dimension less than one half the dimension of $PL(\partial M)$.

Main Theorem

References

Fathi-Laudenbach-Poenaru, 1979.
A. Fathi, F. Laudenbach, and V. Poenaru, "Travaux de Thurston sur les Surfaces," *Société Mathématique de France*, **66-67**, 1979.

Floyd-Oertel, 1984.
W.J. Floyd and U. Oertel, "Incompressible surfaces via branched surfaces," *Topology*, **23**, pp. 117-125, 1984.

Harer-Penner, 1986.
J. Harer and R. Penner, "Combinatorics of Train Tracks," *Annals of Math. Studies*, Princeton University Press, 1986. (with an Appendix by N. Kuhn) (to appear)

Hatcher, 1982.
A. Hatcher, "On the boundary curves of incompressible surfaces," *Pacific J. Math*, **99**, pp. 373-377, 1982.

Papadopoulos, 1983.
A. Papadopoulos, *Réseaux ferroviares, difféomorphismes pseudo-Anosov et automorphismes symplectiques de l'homologie d'une surface*, 1983. Thèse, Université de Paris-Sud

Thurston, 1977.
W. P. Thurston, *On the geometry and dynamics of diffeomorphisms of surfaces, I,* Princeton University Mathematics Department, 1977. Preprint

Thurston, 1979.
W. P. Thurston, *The Geometry and Topology of 3-Manifolds,* Princeton University Mathematics Department, 1979. Part of this material – plus additional material – will be published in book form by Princeton University Press

W. Floyd,
University of Michigan,
Ann Arbor,
Michigan, 48109,
U.S.A.

On 3-Manifolds Finitely Covered by Surface Bundles

*by David Gabai**

Section 0

Question 0.1. *Given a compact oriented 3-manifold M does it have a finite sheeted covering space $\tilde{M}$ which fibres over S^1.*

This question was asked in 1976 by William Thurston who was interested in the case that M was hyperbolic. In fact his question was upgraded in 1984 to

Conjecture 0.2. *(Thurston) If M is a compact oriented 3-manifold whose interior supports a hyperbolic metric of finite volume, then M has a finite sheeted covering space $\tilde{M}$ which fibres over S^1.*

In this note we collect some necessary conditions on M which have such finite covers. We state the precise answer (first obtained in [Orlik-Raymond, 1969]) for Seifert fibred spaces. We present old and new examples of hyperbolic manifolds which have such finite covers, give (metaphysical) evidence for a positive solution, show that if a cover $\tilde{M}$ fibres and M does not, then

$$\text{rank } H_2(\tilde{M},\partial\tilde{M}) > \text{rank } H_2(M,\partial M)$$

and close with related questions.

* Partially supported by a grant from the National Science Foundation.

Section 1

Proposition 1.1. *Let $p : N \to M$ be a finite sheeted covering map of a compact oriented 3-manifold M such that N fibres over S^1, then*

1) *M is either irreducible or $S^2 \times S^1$ or $RP^3 \# RP^3$*
2) *∂M is a possibly empty union of tori*
3) *$|\pi_1(M)| = \infty$ and if N is an S bundle over S^1 with $\chi(S) < 0$, then $\pi_1(M)$ has exponential growth.*

Proof.

1) N is either $S^2 \times S^1$ or irreducible, hence M is either irreducible or a quotient of $S^2 \times S^1$.
2) Restricting the fibration to ∂N yields a foliation of ∂N, hence each component of ∂N is a torus.
3) $|\pi_1(N)| = \infty$ and $\pi_1(N) \subset \pi_1(M)$. Finally, $\pi_1(S)$ has exponential growth if $\chi(S) < 0$.

$\square$

The following result (using different language) was first proved by P. Orlik and F. Raymond in 1969. See also [Conner-Raymond, 1971], [Neumann-Raymond, 1977].

Theorem 1.2. *Let $\xi : M \to T$ be a Seifert fibration where M is compact oriented and connected. Let $e(\xi)$ denote the rational Euler number of ξ, then M has a finite sheeted covering space which fibres over S^1 if and only if either $e(\xi) = 0$ or $\chi(T) = 0$, where χ denotes the orbifold Euler characteristic* [Thurston, 1979 chap. 13] *of T.*

Proof. First suppose that $\xi : M \to T$ is a circle bundle over an orientable surface T. Let $E(\xi) \in H_2(T, \partial T) = Z$ denote the Euler class of ξ. If M fibres over S^1, then by [Waldhausen, 1967] a fibre S is either transverse to the circle fibration, or is vertical (i.e. a union of fibres).

If S is a vertical torus in M then $M - \mathring{N}(S) = (T - \mathring{N}(C)) \times S^1$ where $\xi(S) = C$. Therefore, M fibres over S^1 with fibre a vertical torus if and only if $T = T^2$. E is the obstruction to finding a section of ξ, thus M fibres over S^1 with fibre transverse to the circles if and only if $E(\xi) = 0$. If $p : M' \to M$ is a finite sheeted cover and $\xi' : M' \to T'$ is the induced circle bundle, then the above discussion and the fact $E(\xi') = 0$ if and only if $E(\xi) = 0$ implies that M fibres if and only if M' fibres.

Now let $\xi : M \to T$ be the Seifert fibred space with T orientable. If $T = S^2$ and ξ has either 1 singular fibre or 2 singular fibres of distinct indices, then $|\pi_1(M)| < \infty$. Also $e(\xi) \neq 0$, else there exists an embedded surface $T \subset M$ such that T is transverse to the fibres so $H_2(M) \neq 0$.

If ξ is not as above, then T viewed as an orbifold (it has cone points of index $\alpha_1, \ldots, \alpha_n$ where α_i is the index of the ith singular fibre of ξ) is good [Thurston, 1979]. Hence, there exists a circle fibration $\xi' : M' \to T'$ such that the diagram commutes,

$$\begin{array}{ccc} M' & \xrightarrow{\xi'} & T' \\ \downarrow p & & \downarrow \pi \\ M & \xrightarrow{\xi} & T \end{array}$$

where p is a covering map and π is an orbifold covering map. ξ' is obtained (see [Bonahon-Siebenmann]) by pulling back ξ where T' is an orbifold without cone points. By definition

$$E(\xi') = \frac{(\text{degree } \pi)^2}{\text{degree } p} \, e(\xi)$$

$$\chi(T') = \text{degree } \pi \; \chi(T)$$

By the first paragraph and covering space theory, M' fibres if and only if some finite sheeted cover of M fibres. The result follows from the circle bundle case and the transformation rules of e and χ.

If T is nonorientable, then the result follows by passing to 2-fold covers and applying the oriented base case.

□

Section 2

Very little is known about Thurston's conjecture beyond the elementary fact that hyperbolic manifolds satisfy the necessary conditions of Proposition 1.1. In fact, until the example of Theorem 2.2 was discovered, the only known example of a nonfibred hyperbolic manifold covered by a bundle was:

Example 2.1. (Thurston 1976). Let N be an orientable [0,1] bundle over a non-orientable surface S. Foliate N so that one leaf (the 0 section of N) is S and the other leaves 2-fold cover S. Let M be the 3-manifold obtained by gluing 2 copies of N along their boundaries. The foliation on N extends to a foliation $\mathcal{F}$ on M. If $p: \tilde{M} \to M$ is the 2-fold cover such that $p^{-1}(T)$ is orientable for every leaf T of $\mathcal{F}$, then $\tilde{M}$ fibres over S^1 with fibres $\{p^{-1}(T) : T \text{ is a leaf of } \mathcal{F}\}$. If $\chi(S) < 0$, then by choosing an appropriate gluing map one can construct a hyperbolic 3-manifold M such that M does not fibre.

Przytycki [Przytycki] has observed that such examples arise as surgeries on certain punctured torus bundles over S^1.

Theorem 2.2. *$\tilde{M} = S^3 - \mathring{N}(\tilde{L})$ 2-fold covers $M = S^3 - \mathring{N}(L)$ where $\tilde{L}$, L are the links of figure 1. $\tilde{M}$ fibres over S^1, M neither fibres, nor has a codimension-1 foliation with only compact leaves. Finally $S^3 - \mathring{N}(L)$ is hyperbolic.*

Proof. By [Oertel, 1984] $S^3 - \mathring{N}(L)$ is either atoroidal and anannular or L is a torus link. Since the components of L have zero linking number the latter cannot hold. By [Thurston, 1983] M is hyperbolic.

The fibre $\tilde{F}$ of $\tilde{M}$, exhibited in figure 2, can be built up by Stallings twisting and Murasugi summing, hence $\tilde{M}$ fibres by. [Stallings, 1975]

To show that M does not fibre, we use the following.

Theorem. ([Thurston, 1986]) *Let M be a compact oriented 3-manifold. The set of elements of $H_2(M,\partial M;\mathbb{R})$ represented by surfaces S such that M fibres over S^1 with fibre S equals the set of lattice points in the cone of unions of open faces in the unit ball of the (Thurston) norm.*

One calculates (as in [Thurston, 1986]) that the unit ball of $H_2(M,\partial M)$ is as in Figure 3, where elements of $H_2(M,\partial M)$ are parametrized by the intersection numbers with meridians of components of L. Therefore, M does not fibre if none of the 4 classes $(\pm 1, \pm 1)$ are represented by fibres. In fact, we need only check the class (1,1) since L is a symmetric link. A Seifert surface for L oriented as in figure 2 represents (1,1). By [Neuwirth, 1965], any minimal genus surface for the oriented link L would be a fibre if (1,1) was representable by a fibre. The surface S of figure 4 is minimal genus, else L would bound a surface of non negative Euler characteristic. Hence, L would either be trivial or non anannular. S is a Murasugi sum of surfaces S_1, S_2 where $S^3 - \mathring{N}(\partial S_i)$ fibres (resp., does not

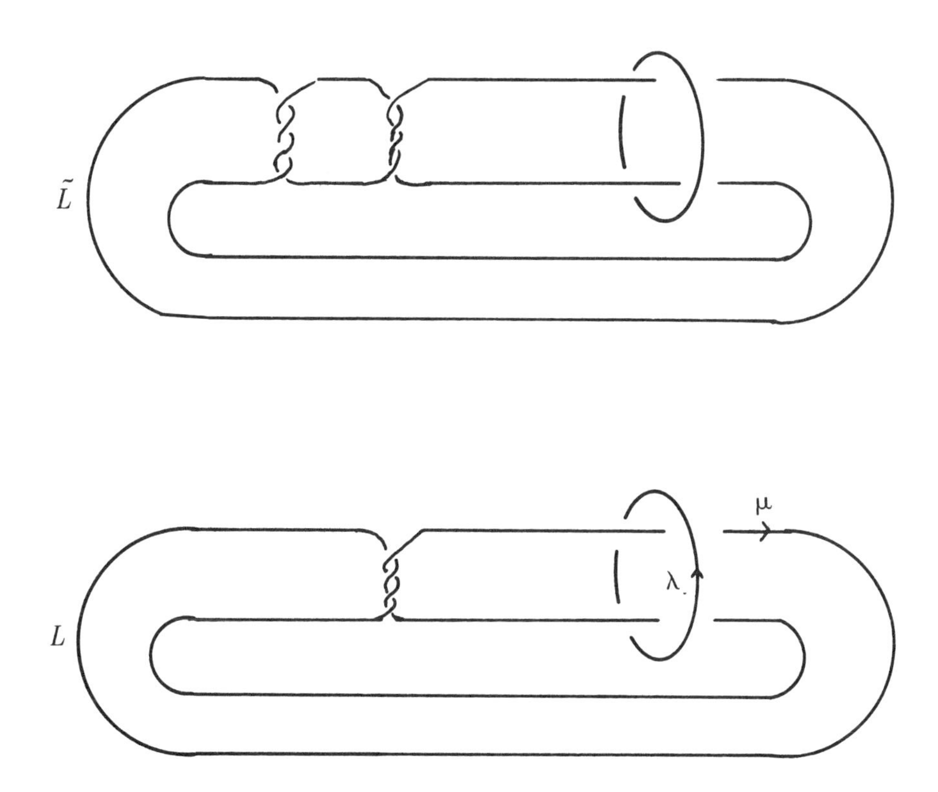

Figure 1.

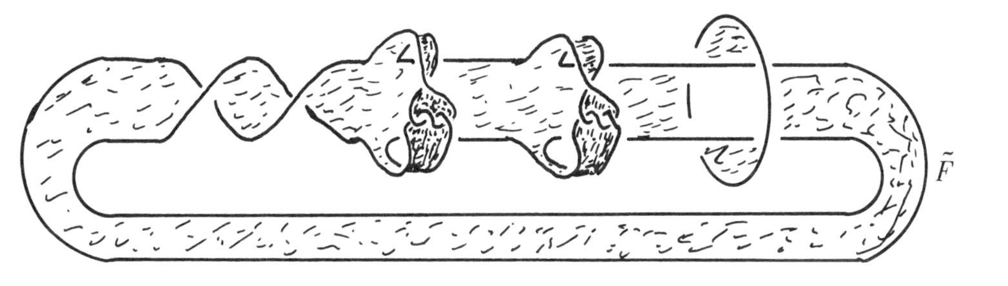

Figure 2.

fibre) with S_1 (resp., S_2). By [Gabai, 1985], L does not fibre with fibre S, hence M does not fibre.

If M had a foliation $\mathcal{F}$ with only compact leaves, then $\mathcal{F}$ would be transverse to ∂M else M would be covered by $T^2 \times I$. Therefore, M either fibres or there exists a 2-fold cover $p : M' \to M$ such that M' fibres where

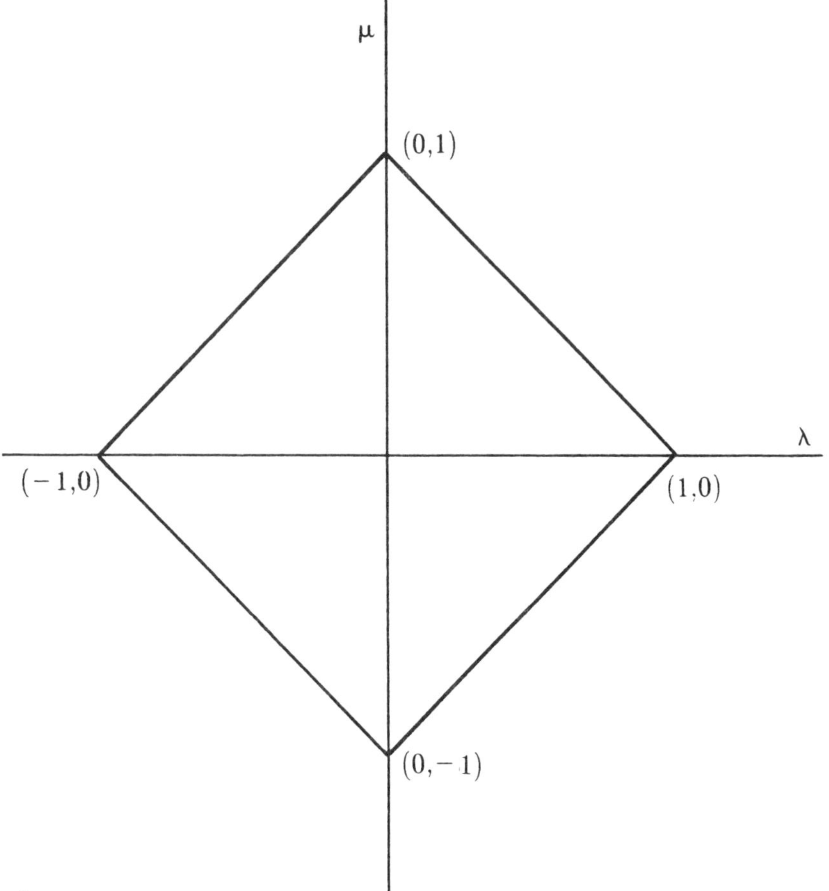

Figure 3.

the fibres are the leaves of $p^*\mathcal{F}$. Since M does not fibre, some fibre F of M' equals $p^{-1}(\partial N(S))$ where S is a non orientable leaf of $\mathcal{F}$. Therefore, $p_*[F] = 0 \in H_2(M,\partial M)$ for every fibre of $p^*(\mathcal{F})$.

Any $\mathbb{Z}_2$ cover $p : M' \to M$ is the 2-fold cyclic cover corresponding to a primitive element (r,s) of $H_2(M,\partial M)$. Say r is odd. If $Q = \partial N(\lambda)$, then $p^{-1}(Q) = \tilde{Q}$ is connected and $p_* : H_1(\tilde{Q}) \to H_1(Q)$ is injective. Any homology class z representing a fibre of M' must satisfy $0 \neq [z \cap \tilde{Q}] \in H_1(\tilde{Q})$. Therefore $p_*(z) \neq 0$.

$\square$

Remarks 2.3.

1) $\tilde{M}$ was a 2-fold cover of M and could be obtained by gluing together 2 copies of M split along a disk D with 2 holes. (D is the disk that λ "bounds".) F was obtained by doing cut and paste with a copy of D (viewed in $\tilde{M}$) and a surface R which had the property that $0 \neq [R] \in H_2(\tilde{M},\partial\tilde{M})$, $0 = p_\star[R] \in H_2(M,\partial M)$. In a similar way one can find examples of closed fibred 3-manifolds which 2-fold cover non fibred ones and do not arise as in Example 2.1.

2) If L is a non split non alternating link of ≤ 9 crossings, then by [Gabai] with 4 exceptions $M = S^3 - \mathring{N}(L)$ fibres over S^1. By taking 2 or 4-fold covers as in Theorem 2.2 (which analyzes an exceptional case) one finds manifolds which fibre over S^1.

3) Theorem 5.5 of [Gabai, 1983] asserts that if M satisfies the necessary conditions of Theorem 1.1 and $H_2(M,\partial M) \neq 0$, then M almost fibres over S^1 in the sense that M has a foliation without Reeb components, and after removing a finite number of leaves the remaining leaves are the fibres of a fibration over S^1.

Lemma 2.4. *Let M be a compact oriented 3-manifold. If $p : \tilde{M} \to M$ is a degree $p < \infty$ cover and $0 \neq z \in H_2(M,\partial M;\mathbb{Z}) = H^1(M;\mathbb{Z})$ is not represented by a surface S such that M fibres over S^1 with fibre S, then $p^\star(z) \in H^1(\tilde{M};\mathbb{Z})$ is not represented by a surface T such that $\tilde{M}$ fibres over S^1 with fibre T.*

Proof. We prove the contrapositive. We can assume that $M \neq S^2 \times S^1$. Let S be a Thurston norm minimizing surface such that $[S] = z$. If $\tilde{S} = p^{-1}(S)$, then $[\tilde{S}] = p^\star(z)$. By Theorem 6.13 of [Gabai, 1983] $\tilde{S}$ is a norm minimizing surface for $p^\star(z)$. By [Neuwirth, 1965], M (resp. $\tilde{M}$) fibres over S^1 with fibre S (resp. $\tilde{S}$) if and only if z (resp. $p^\star(z)$) is represented by a fibre. If $\tilde{M}$ fibres with fibre $\tilde{S}$, then the product $\tilde{M} - \mathring{N}(\tilde{S})$ is a finite sheeted cover of $M - \mathring{N}(S)$. Therefore each component N of $M - \mathring{N}(S)$ is irreducible and for each component J of $\partial N - \partial M, \pi_1(J) \to \pi_1(N)$ is an isomorphism. By [Stallings, 1962], N is a product, hence M fibres over S^1. $\square$

Corollary 2.5. *Let M be a compact oriented 3-manifold and $p : \tilde{M} \to M$ a finite sheeted cover. If $\tilde{M}$ fibres and M does not fibre over S^1, then*

$$\text{rank } H_2(\tilde{M},\partial\tilde{M};\mathbb{Z}) > \text{rank } H_2(M,\partial M;\mathbb{Z})$$

Proof. If rank $H_2(\tilde{M},\partial\tilde{M};\mathbb{Z})$ = rank $H_2(M,\partial M;\mathbb{Z})$ then $p^\star : H^1(M;\mathbb{R}) \to H^1(\tilde{M};\mathbb{R})$ is an isomorphism. Therefore the cone of each open face of the unit ball of the Thurston norm of $H_2(\tilde{M},\partial\tilde{M};\mathbb{R}) = H^1(\tilde{M};\mathbb{R})$ contains an element $p^\star(z)$ for $z \in H_2(M,\partial M;\mathbb{Z})$. By Lemma 2.4 $p^\star(z)$ is not represented by a fibre, hence by Thurston's theorem quoted earlier $\tilde{M}$ does not fibre.

□

Section 3

In the light of Corollary 2.5 we ask

Question 3.1 Let M be a hyperbolic 3-manifold. Does there exist a finite cover $\tilde{M} \to M$ such that rank $H_2(\tilde{M},\partial\tilde{M}) >$ rank $H_2(M,\partial M)$? Do such covers exist even in the special case when M fibres over S^1?

It would be interesting to identify a fibre of $\tilde{M}$ as an immersed surface F in M.

Question 3.2 How can one decide if an immersed surface S in M lifts to a fibre in some finite cover of M? How does one decide if it lifts to an embedded surface?

To help answer question 3.2 we recall the old

Question 3.3 To what extent are 3-manifold groups LERF, i.e. subgroup separable? In particular if $G = \pi_1$ (surface group) $\subset \pi_1(M)$ and $C \in \pi_1(M) - G$, when does there exist a subgroup H of finite index in $\pi_1(M)$ such that $G \subset H$, $C \notin H$?

Question 3.3 has been studied by [Scott, 1978] who in particular showed that compact Seifert fibred spaces are LERF.

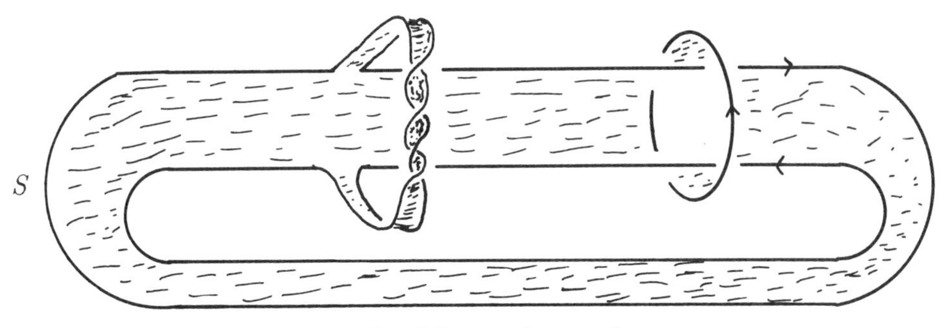

is a Murasugi sum of

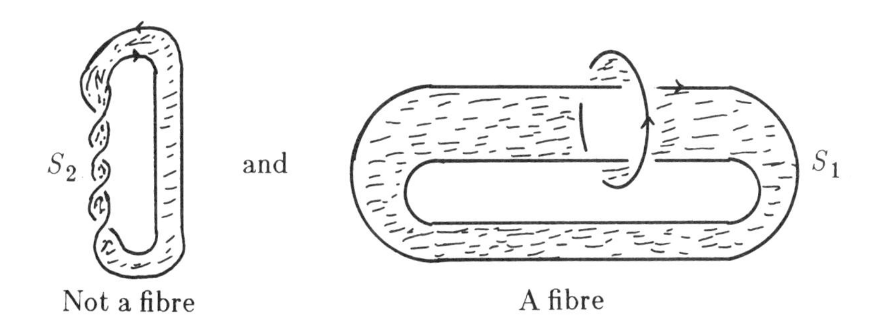

Figure 4.

References

Bonahon-Siebenmann.
F. Bonahon and L. Siebenmann, *Seifert 3-orbifolds and their role as natural crystalline parts of arbitrary compact irreducible 3-orbifolds,* London Mathematical Society Lecture Notes, C.U.P..

Conner-Raymond, 1971.
P. Conner and F. Raymond, "Injective operations of the toral groups," *Topology*, **10**, pp. 283-96, 1971.

Gabai.
D. Gabai, "Detecting Fibred Links in S^3," *Comm. Math. Helv.*.

Gabai, 1983.
D. Gabai, "Foliations and the topology of 3-manifolds," *Journal of Differential Geometry*, pp. 445-503, 1983.

Gabai, 1985.
D. Gabai, "The Murasugi sum is a natural geometric operation II," *Contemporary Mathematics*, **44**, pp. 93-100, AMS, 1985.

Neumann-Raymond, 1977.
W. Neumann and F. Raymond, "Seifert manifolds, plumbing, μ invariant, and orientation-reversing maps," *Lecture notes in Math*, **664**, pp. 162-195, Springer, 1977.

Neuwirth, 1965.
L. Neuwirth, "Knot Groups," *Annals of Math. Studies*, **56**, 1965.

Oertel, 1984.
U. Oertel, "Closed incompressible surfaces in complements of star links," *Pac.J.Math.*, **111**, pp. 209-230, 1984.

Orlik-Raymond, 1969.
P. Orlik and F. Raymond, "On 3-manifolds with local SO(2) actions," *Quart. J. of Math.*, **20**, pp. 143-60, 1969.

Przytycki.
J. Przytycki, *Non Orientable Incompressible Surfaces in Punctured Torus Bundles over S^1*. preprint

Scott, 1978.
P. Scott, "Subgroups of Surface groups are almost Geometric," *J. London Math. Soc. (2)*, **17**, pp. 555-565, 1978.

Stallings, 1962.
J.R. Stallings, "On fibering certain 3-manifolds," in *Topology of 3-manifolds*, pp. 95-100, Prentice Hall, 1962.

Stallings, 1975.
J.R. Stallings, "Constructions of Fibred Knots and Links," *Proc. Symp. Pure Math. AMS*, **315-319**, 1975.

Thurston, 1979.
W. P. Thurston, *The Geometry and Topology of 3-Manifolds,* Princeton University Mathematics Department, 1979. Part of this material – plus additional material – will be published in book form by Princeton University Press

Thurston, 1983.
W. P. Thurston, "Three Dimensional Manifolds, Kleinian Groups, and Hyperbolic Geometry," *Proc. Symp. Pure Math.*, **39**, pp. 87-114, 1983.

Thurston, 1986.
W. P. Thurston, "A Norm on the Homology of 3-Manifolds," *Memoirs of the AMS*, **339**, 1986.

Waldhausen, 1967.
Waldhausen, F., "Eine Klasse von 3-dimensionalen Mannigfaltigkelten. I & II," *Inv. math*, **3 & 4**, 1967.

D. Gabai,
Caltech,
Pasadena,
California 91125,
U.S.A.

On Surfaces in One-Relator 3-Manifolds

Klaus Johannson

1. Introduction

Attaching one 2-handle along some simple closed curve in the boundary of a handlebody M gives a 3-manifold whose fundamental group is a one-relator group. It is the aim of this paper to study incompressible surfaces in such 3-manifolds. It turns out that the method used also applies when M is supposed to be boundary-reducible, and so we will include this more general case in our considerations. This situation suggests the following definition:

Definition. A *one-relator 3-manifold* is defined to be an (orientable) 3-manifold obtained from some (compact) boundary-reducible 3-manifold, M, by attaching a 2-handle E, along some simple closed curve, k, in ∂M. We denote such a one-relator 3-manifold by $M^+(k)$, or simply by M^+. Then M^+ contains a proper arc, t, such that $(M^+ - E)^-$ is the boundary-reducible 3-manifold M, where E is a regular neighbourhood of t in M^+. Note that the pair (M, k) as well as the pair (M^+, t) can be considered as a *presentation* of M^+ as one-relator 3-manifold. The presentation is called *non-trivial* if the arc t is essential in M^+ (i.e. does not bound, together with an arc in ∂M^+, a non singular disc in M^+). Almost all Seifert fibre spaces with non-empty boundary are examples of 3-manifolds with non-trivial presentation as one-relator 3-manifold (see Section 4).

This paper is devoted to the study of incompressible surfaces in $M^+(k)$. In Section 2 we formulate and discuss our Main Theorem whose proof takes up Section 3. In Section 4 we begin a study of homeomorphisms of M^+. In Section 5 we give an application of our Main Theorem to Dehn-surgeries, and in Section 6 we finally discuss a more group-theoretical application, namely the question of decomposability of groups into free products.

2. Formulation of the Main Theorem

Throughout this paper we are working in the PL-category.

Given an incompressible surface, S^+, in the one-relator 3-manifold M^+, one would in general not expect that S^+ can be isotoped out of the 2-handle E and so into M (although this is sometimes the case; see Proposition 4.2). Nevertheless, there are more drastic procedures (than just isotopy) which also do not enlarge the complexity of S^+, but which eventually force S^+ into M. This is the content of our main result. To give a more precise statement, define the complexity, $c(S)$, of a surface S simply to be the "enlarged" first Betti number, i.e. $c(S) := \beta_1(S) + \text{genus}(S)$. Furthermore, it is to be understood that surfaces in 3-manifolds are orientable. Then we have the following:

Theorem A. *Let $M^+(k)$ be a one-relator 3-manifold, and suppose that M is irreducible and that each component of $\partial M - k$ is incompressible in M. If there is an incompressible, non-separating surface, S^+, in $M^+(k)$, then there is an incompressible, non-separating surface, $S^\star$, in $M^+(k)$, with $c(S^\star) \leq c(S^+)$ and which is entirely contained in the submanifold M.*

In addition.

1) Suppose that k separates a once-punctured torus T_0 in ∂M. If $\partial S^+ \subset \partial M - T_0$, then $S^\star$ may also be chosen so that, in addition, $\partial S^\star \subset \partial M - T_0$.

2) If S^+ is an annulus, then $S^\star$ can be chosen to be an annulus so that at least one component of ∂S^+ can be isotoped in ∂M^+ into $\partial S^\star$. This stronger assertion also holds for the annuli in the situation of addition 1.

Remark. It will be apparent from the proof that, in the case when S^+ is a disc or a 2-sphere, we do not need to suppose that S^+ is non-separating. The case when S^+ is a disc or a 2-sphere was first dealt with in [Przytycki] for handlebodies M, and later extended by [Jaco, 1984] to the case when M is supposed to be boundary-reducible. Our method gives an independent proof. Furthermore, [Floyd] solves the case when S^+ is an annulus or torus.

To illustrate Theorem A, let us first point to some of its immediate consequences. One of them is the following

Corollary 1. *Let M and k be given as in Theorem A. Let S be a non-separating surface in M with $\partial S \subset \partial M - k$. Suppose S has minimal complexity (as a surface in M). Then S is an essential surface in $M^+(k)$.*
Remark. See [Johannson, 1979] for the term "essential".

Proof. If S is inessential in M^+, then by cutting and pasting along compression discs, we obtain from S a non-separating, essential surface in M^+. Then, by Theorem A, there is a similar surface in M^+ which is entirely contained in M and whose complexity is not larger than that of the previous one. But compressing surfaces along discs reduces their complexities and so we get a contradiction to our minimality condition on S.

□

As a result of this application of Theorem A, we may note that the study of non-separating, incompressible surfaces in M^+ is essentially the same as that in M.

Our next application is to knots in handlebodies. Recall that one may distinguish between knotted and unknotted curves in 3-manifolds. As usual, a proper curve in 3-manifold M is called unknotted (or inessential) if it bounds, possibly together with an arc in ∂M, a non-singular disc in M. A well-known application of the Alexander Theorem says that a classical knot, k, in S^3 is knotted if and only if the knot space $S^3 - k$ contains a closed, incompressible surface. A similar result for arcs in handlebodies is obtained from Theorem A:

Corollary 2. *Let t be an arc in the handlebody M with $t \cap \partial M = \partial t$. Then t is knotted if and only if $\partial M - t$ contains a closed, incompressible surface.*

Proof. One direction is obvious since handlebodies have no closed, incompressible surfaces at all. For the other direction, let $U(t)$ denote a regular neighbourhood in M. Now, as a handlebody, M contains non-separating discs. It follows from Theorem A that either a component of $\partial M - U(t)$ is compressible in $M - U(t)$, or $\partial(M - U(t))^-$ is incompressible in $M - t$. In the latter case, we are done. In the first case, we split M along compression discs (contained in $M - t$). After finitely many of such steps, we finally obtain a 3-ball, $\tilde{M}$, from M containing the arc t. Then, by Alexander's Theorem, $\partial(\tilde{M} - U(t))^-$ is incompressible in $\tilde{M} - t$ if and only if t is knotted in $\tilde{M}$. Corollary 2 then follows since $\tilde{M}$ is obtained from M by splitting along discs, only.

□

We close this section with an application to the placement-problem for 3-manifolds. This problem asks for a method of resolving intersections of pairs of submanifolds whose dimensions add up to the dimension of the underlying manifold. A specific case of this problem is the case of a proper arc, t, (knotted or not), together with an essential disc, in an irreducible 3-manifold, N, with non-empty boundary. If there is an essential disc in N, as previously given, one might also ask whether there is such a disc in N which avoids the arc t, entirely. With the help of Theorem A, this problem can now be translated into algebra as follows:

Corollary 3. *Let N be an irreducible 3-manifold, and let t be an arc in N with $t \cap \partial N = \partial t$. If there is any essential disc in N, then there is such a disc disjoint to t, if and only if $\pi_1(N - t)$ is a non-trivial free product.*

Proof. If there is any essential disc in N disjoint to t, then, by a result of [Stallings, 1962], $\pi_1(N - t)$ actually is a non-trivial free product. For the other direction, let $U(t)$ again be a regular neighbourhood of t in N, and denote $M := (N - U(t))^-$. Now, $\pi_1(N - t)$ is supposed to be a non-trivial free product, and so, by another well known result of Stallings [Stallings, 1971], M is boundary-reducible. Furthermore, M is irreducible since N is. But, by hypothesis, N is boundary-reducible, and so it follows from Theorem A that $(\partial M - U(t))^-$ is compressible in M. This gives the required disc in M.

□

3. Proof of Theorem A

Let $k^\star$ be any closed curve in ∂M and suppose that each component of $\partial M - k^\star$ is incompressible in M (and that $k^\star$ separates a punctured torus from ∂M, provided we are in the proof of the addition of Theorem A). We associate to each such curve a complexity, $c(k^\star)$, defined as follows: First consider the one-relator 3-manifold $M^+(k^\star)$, uniquely associated to

$k^\star$. Given a surface, $S^\star$, in $M^+(k^\star)$, we denote by $d(S^\star)$ the number of all components of $E^\star \cap S^\star$ (here $E^\star$ denotes the 2-handle attached to M along $k^\star$), and we define

$$c(k^\star) := \min\left(c(S^\star), d(S^\star) \right),$$

where the minimum is taken (with respect to the lexicographical order) over all surfaces, $S^\star$, in $M^+(k^\star)$ satisfying the hypothesis of Theorem A. In particular, $S^\star$ is orientable, incompressible and non-separating. Moreover, $\partial S^\star$ is contained in the complement of the punctured torus in ∂M separated by $k^\star$, provided we are in the proof of addition 1 of Theorem A. One component of ∂S^+ can be isotoped in ∂M^+ into $\partial S^\star$, if we are in the proof of addition 2 of Theorem A. Note that these extra conditions have to be taken into account whenever we are estimating complexities, $c(k^\star)$.

The proof of Theorem A will be by an induction on the complexity $c(k^\star)$ – the first step of the induction being trivial.

For the induction step fix a simple closed curve, k, in ∂M as above and suppose that Theorem A holds for all those curves which have strictly smaller complexities than k. Observe that the complexity $c(k)$ can be realized by a surface in $M^+ = M^+(k)$. In other words, there is a surface, S^+, in M^+, satisfying the hypothesis of the Theorem, with

$$c(k) = \left(c(S^+), d(S^+) \right).$$

If $d(S^+) = 0$, we are done. Thus, in order to fix ideas, we assume the converse. We are going to show that the assumption $d(S^+) \neq 0$ leads to contradictions.

First observe that S^+ may be isotoped in M^+ so that the above holds and that, in addition, $E \cap S^+$ consists only of discs (here and in the following E denotes the 2-handle attached to M along k). To achieve this, simply choose an unknotted arc, t, in the 2-handle E, joining the two components of $(\partial E - \partial M)^-$, being transverse to S^+ and intersecting each component of $E \cap S^+$ in at most one point. The claim then follows since a regular neighbourhood of t can be expanded to the entire 2-handle E, using an ambient isotopy of M^+.

Let A denote the regular neighbourhood

$$A := E \cap \partial M$$

of k in ∂M, and let us introduce the following subsurfaces of S^+:

$$B := E \cap S^+, \text{ and}$$

$$S := (S^+ - B)^-.$$

Then $B \neq \emptyset$ and, by our previous adjustment of S^+, B is a non-empty system of discs. Furthermore, $\partial B = A \cap \partial S$ consists of simple closed curves in the annulus A, all parallel to the components of ∂A, whose number equals the number of the components of $E \cap S^+$. In particular, S is connected. Moreover, we observe

Lemma 3.1. *The surface S is incompressible in M.*

Proof. Let D be a disc in M with $D \cap S = \partial D$. Then ∂D bounds a disc, D', in S^+ since S^+ is incompressible in M^+. Replacing D' by D does not change the complexity $c(S^+)$, but it diminishes $d(S^+)$ if $D' \cap B \neq \emptyset$. Thus, by induction hypothesis, it follows that $D' \cap B = \emptyset$, and this in turn proves the claim.

□

After these remarks we organize the material of the actual proof of Theorem A in four sections under the following headlines:

A.1 The intersection patterns

A.2 Good compression discs

A.3 Constructing a new one-relator 3-manifold

A.4 The conclusion of the proof

A.1 The intersection patterns

Recall from our hypothesis on M, that M is boundary-reducible. In other words, there is at least one essential (= non boundary-parallel) disc in M. Let us choose such a disc, D, so that, in addition,

$$\mathcal{B} := D \cap S$$

is a system of curves (general position) whose number is minimal. This disc has to intersect the attaching curve k since no component of $\partial M - k$ is compressible in M. Hence $\mathcal{B} \neq \emptyset$ since, by induction hypothesis, the intersection $E \cap S^+$ is non-empty. Observe that the system $\mathcal{B}$ can be considered as an intersection pattern in D as well as in S, and we shall have to consider both points of view. But to analyze the intersection patterns more carefully, we first have to introduce a few notations.

Definition. Let D be a disc. A *system, $(I, \mathcal{R})$, of collections* (of type n) in ∂D is a set, $\mathcal{R}$, of points in ∂D, together with a system, I, of pairwise disjoint intervals in ∂D, such that each component of I contains precisely n points from $\mathcal{R}$.

Furthermore, the set, $\mathcal{R}_i$, of points in the i-th interval, I_i, of I is called a "collection" (from $\mathcal{R}$). A point, z, in $\mathcal{R}_i$ is a *boundary-point* of $\mathcal{R}_i$ (or $\mathcal{R}$), if z separates an interval of I_i which meets $\mathcal{R}_i$ in z, only. Two collections, $\mathcal{R}_i$ and $\mathcal{R}_j$, from $\mathcal{R}$ are *neighbouring collections*, if there is an arc, b, in ∂D such that $b \cap \mathcal{R}$ consists of boundary-points of $\mathcal{R}_i$ or $\mathcal{R}_j$. Now, let Θ be a system of arcs, t, in D with $t \cap \partial D = \partial t$. Then Θ is called a *good arc-system* in D (with respect to $\mathcal{R}$), if the following holds:

(1) $\mathcal{R} \subset \partial\Theta$ and $\partial\Theta \subset \mathcal{R} \cap (\partial D - I)$, where $\partial\Theta$ denotes the set of end-points of Θ.

(2) $\partial t \not\subset \mathcal{R}_i$, for every t from Θ and every i.

(3) If b is an arc in ∂D with $b \cap \Theta = \partial b$ contained in one component of Θ, then $b \cap \mathcal{R} = \partial b$.

If Θ is a good arc-system in D, then a collection, $\mathcal{R}_0$, from $\mathcal{R}$ is called a *special collection* (with respect to Θ), provided there are neighbouring collections $\mathcal{R}_1$, $\mathcal{R}_2$ from $\mathcal{R}$ (possibly $\mathcal{R}_1 = \mathcal{R}_2$) such that any point (resp. boundary-point) of $\mathcal{R}_0$ will be joined by an arc from Θ with a point (resp. boundary-point) of $\mathcal{R}_1$ or $\mathcal{R}_2$.

Lemma 3.2. *Let $\mathcal{R}$ be a system of collections in ∂D, and suppose Θ is a good arc-system in D. Then $\mathcal{R}$ has at least one special collection, $\mathcal{R}_0$, with respect to Θ.*

Proof. Either Lemma 3.2 is true, or there is at least one arc, t, from Θ which does not join two neighbouring collections from $\mathcal{R}$. Let t be chosen so that, in addition, t is minimal. More precisely, let t be chosen so that it separates a disc, D_1, from D with the property that any arc from $D_1 \cap \Theta$, different from t, joins neighbouring collections. Of course, this choice is always possible since D is a disc.

By our choice of t, either $D_1 \cap \Theta = t$, or the interior of the arc $D_1 \cap \partial D$ contains at least one collection, $\mathcal{R}_0$, entirely. In the first case, considering the possible positions of the arc $D_1 \cap \partial D$, we get a contradiction to one of the properties of good arc-systems, and in the second case, $\mathcal{R}_0$ has to be a special collection.

$\square$

It will be shown in this section that $\mathcal{B} = D \cap S$ can be considered as a good arc-system in D. For this purpose, we first need the following two elementary consequences of our choice of D.

Lemma 3.3. *The system $\mathcal{B} = D \cap S$ consists of arcs and each arc from $\mathcal{B}$ is essential in S.*

Proof. Recall from Lemma 3.1 that the surface S is incompressible in M. Furthermore, recall that D is chosen so that $D \cap S$ is minimal. Hence the claim follows by applying the innermost-disc-argument twice.

□

Lemma 3.4. *Let A be the annulus $A = E \cap \partial M$. If b is a component of $D \cap A$, then b is an arc joining both of the components of ∂A in such a way that the number of points of $b \cap \partial S$ equals the number of components of $A \cap \partial S$.*

Proof. Suppose the converse. Then we find a subarc, b', in b, whose end-points are contained in one component of $A \cap \partial S$, say $\partial_1 S$. An appropriate isotopy of D in M, which pushes b' across $\partial_1 S$, either creates a closed curve in $D \cap S$, or decreases the number of arcs from $D \cap S$. By Lemma 3.3, this contradicts our minimal choice of D.

□

A *special compression disc* for S is a disc, D_0, in M whose boundary is the union of two arcs, $b_1, b_2 \subset \partial D_0$, with $b_1 \cap b_2 = \partial b_1 = \partial b_2$, such that

(1) $b_1 = D_0 \cap \partial M$ and $b_2 = D_0 \cap S$,

(2) $b_1 \cap A$ is empty or connected, and

(3) b_2 is an essential arc in S.

Lemma 3.5. *The surface S admits no special compression discs.*

Proof. Suppose there is a special compression disc, D_0, for S and let $\partial D_0 = b_1 \cup b_2$ be the decomposition of ∂D_0 as given by its definition.

Case 1 $b_1 \cap A = \emptyset$.

In this case the end-points of b_1 are contained in ∂S^+. Thus pushing b_2 across D_0 into ∂M and splitting S^+ along b_2, we obtain from S^+ a 2-manifold $S^\star$. At least one of the components of $S^\star$ has to have the same topological properties as S^+ and, in particular, it cannot have strictly smaller complexity than S^+ (see our choice of S^+) and, if we are in the proof of addition 1, observe that $b_1 \cap T_0 = \emptyset$ since $\partial S^+ \subset \partial M - T_0$ and $b_1 \cap A = \emptyset$. This implies that b_2 has to be inessential in S^+ and so separates a disc, D_1, from S^+. Since, by definition, b_2 is essential in S, each such disc has to contain at least one component of B. Now, $D_0 \cup D_1$ is a disc in M which is boundary-parallel or not. But the first case is impossible for the intersection $E \cap S^+$ cannot be diminished, using an isotopy of S^+ in M^+, and also the second case is impossible because of our minimal choice of S^+. This contradiction proves Lemma 3.5 in Case 1.

Case 2 $b_1 \cap A \neq \emptyset$.

By the very definition of special compression discs, the intersection $t := b_1 \cap A$ is connected. Furthermore, the arc b_2 is isotopic to t, using an isotopy in D_0 which fixes one end-point of b_2 and which leaves the other one in b_1. It follows that there is an ambient isotopy in M^+ which pushes b_2 into E and which is constant outside of a regular neighbourhood of D_0 in M^+. Choosing such an isotopy carefully, it pushes S^+ into a surface, $S^\star$, with

$$S^\star \cap A = ((S \cap A) - U(t)) \cup t_1 \cup t_2,$$

where t_1, t_2 denote two copies of t in the regular neighbourhood $U(t)$ of t in A. Hence, there is an arc, b, in A, joining the two components of ∂A, and which intersects $S^\star$ in a number of points strictly smaller than the number of components of $A \cap \partial S$ and so of $E \cap S^+$. A regular neighbourhood of b in E can be expanded to E, using an ambient isotopy of M^+. By our choice of b the product of all these isotopies applied to S^+ reduces the number of components of $E \cap S^+$. But this contradicts our minimal choice of S^+. This contradiction proves Lemma 3.5 in Case 2.

Now, finally, consider $I := A \cap \partial D$ as a system of intervals in ∂D, and let $\mathcal{R} := I \cap \partial S$. Then, by Lemma 3.4, it follows that $(I, \mathcal{R})$ is a system of collections.

$\square$

The result of the previous discussion, i.e. of Lemmas 3.2–3.5 is the following:

Lemma 3.6. *The system $\mathcal{B} = D \cap S$ is a good arc-system with respect to $\mathcal{R}$. In particular, $\mathcal{R}$ contains at least one special collection.*

A.2. Good compression discs

After having established the non-existence of special compression discs for S, we now have to face another sort of compression disc which we call *good compression discs*.

To define the latter notion, let us call an arc in S *recurrent* (in S) if both its end-points are contained in one component of B $(= E \cap S^+)$. Then a *good compression disc* for S is a disc, D_0, in M whose boundary is the union of two arcs, $b_1, b_2 \subset \partial D_0$, with $b_1 \cap b_2 = \partial b_1 = \partial b_2$ and $b_2 \subset D_0 \cap S$, such that

(1) $b_1 = D_0 \cap \partial M$, and $b_1 \cap A$ is a regular neighbourhood of ∂b_1 in b_1,

(2) $D_0 \cap S$ consists of arcs, joining the two components of $b_1 \cap A$ and none of which is recurrent in S, and

(3) the number of components of B is less than or equal to twice the number of arcs in $D_0 \cap S$.

A candidate for a good compression disc is obtained as follows. Let $\mathcal{R}_0$ be a special collection of $\mathcal{R}$. The existence of such a collection is given by Lemma 3.6. Denote by $\mathcal{R}_1$ and $\mathcal{R}_2$ (possibly $\mathcal{R}_1 = \mathcal{R}_2$) the two neighbouring collections of $\mathcal{R}_0$. Let $\mathcal{B}_1$, resp. $\mathcal{B}_2$, be the set of all arcs from $\mathcal{B}$, joining $\mathcal{R}_0$ with $\mathcal{R}_1$, resp. $\mathcal{R}_2$. Recall from the definition of special collections that there are arcs from $\mathcal{B}_1$, resp. $\mathcal{B}_2$, which separate discs, D_1, resp. D_2, from D with

$$D_i \cap \mathcal{B} = \mathcal{B}_i \, , \, i = 1,2.$$

Let us suppose that the indices are chosen so that card $\mathcal{B}_1 \geq$ card $\mathcal{B}_2$. Then the disc D_1 is at least a candidate for a good compression disc. To show that, under the previous hypothesis, the disc is indeed a good compression disc (for S), we first have to deal with the recurrent arcs in $\mathcal{B}_1 \cap \mathcal{B}_2$. The following result is crucial for our purpose.

Lemma 3.7. *Let $\mathcal{B}_1$ and $\mathcal{B}_2$ be given as above. Suppose $\mathcal{B}_1$ (or $\mathcal{B}_2$) contains some recurrent arc, say b_0, and let B_0 denote that component of B containing ∂b_0. Then the following holds :*

(1) *Each arc from $\mathcal{B}_1$ (resp. $\mathcal{B}_2$) is recurrent.*

(2) *The arc b_0 is inessential in $(S^+ - B_0)^-$.*

Proof. (1): Since b_0 is recurrent, the end-points of b_0 are contained in one component, say $\partial_0 S$, of $A \cap \partial S$. Recall that $\mathcal{B}_1$, (resp. $\mathcal{B}_2$) consists of arcs all parallel to b_0 (see definitions of special collections and good arc-systems). Let b_1 be the arc from $\mathcal{B}_1$ (resp. $\mathcal{B}_2$) which lies next to b_0. If b_1 is not recurrent, it follows from Lemma 3.4 that the two end-points of b_1 have to lie in different components of $A - \partial_0 S$. Since b_0 and b_1 are parallel in D, this implies that S has to be one-sided. This, however, contradicts the fact that S^+, and so S, is an orientable surface in an orientable 3-manifold. So, b_1 is recurrent and our claim follows successively.

Proof. (2): We first check that we can restrict ourselves to the case when b_0 is innermost in D. To see this, recall that b_0 is a recurrent arc from $\mathcal{B}_1$, say, and so, by (1) above, all arcs from $\mathcal{B}_1$ are recurrent. At least one arc, say b_m, from $\mathcal{B}_1$ is innermost in D, in the sense that b_m separates a disc from D which meets $\mathcal{B}$ in b_m alone. Let B_0 and B_m be the two components of B containing ∂b_0 and ∂b_m, respectively. Then ∂B_0 and ∂B_m are disjoint curves in the annulus A which are parallel in A. Since also b_0 and b_m are parallel (in D), we find appropriate arcs, β_0 and β_m, in ∂B_0 and ∂B_m, respectively, with $\partial\beta_i = \partial b_i$, $i = 0, m$, such that the closed curves $b_0 \cup \beta_0$ and $b_m \cup \beta_m$ are freely homotopic in M^+. Hence $b_0 \cup \beta_0$ is contractible in M^+ if and only if $b_m \cup \beta_m$ is. Since S^+ is incompressible in M^+ it follows that $b_i \cup \beta_i$, $i = 0$ and m, is contractible (in M^+) if and only if b_i is inessential in $(S^+ - B_i)^-$. So, b_0 is inessential in $(S^+ - B_0)^-$ if and only if b_m is inessential in $(S^+ - B_m)^-$, proving our claim.

By what we have verified so far, we may suppose that b_0 separates a disc, D_0, from D with $D_0 \cap \mathcal{B} = b_0$. Now recall that B_0 is that component of $E \cap S^+$ which contains ∂b_0 and that B_0 is a disc which splits the 2-handle E into two 3-balls. One of them is containing $E \cap D_0$ (since S^+ is 2-sided). Then a solid torus, N, is obtained by taking the union of this 3-ball with a regular neighbourhood of D_0 in M^+ (recall that $A \cap D_0$, and so $E \cap D_0$, has precisely two components). The intersection $N \cap \partial M^+$ as well as $N \cap S^+$ is an annulus and, more precisely, $N \cap S^+$ is the union $B_0 \cup U(b_0)$, where $U(b_0)$ is the regular neighbourhood of b_0 in S^+.

Now, let $U(N \cup S^+)$ denote the regular neighbourhood of $N \cup S^+$ in M^+. Then, by the previous properties of N, one component of

$$(\partial U(N \cup S^+) - \partial M^+)^-$$

is a copy of S^+ and the other two, denoted by S_1^+ and S_2^+ (possibly equal), are copies of the components of $S^+ - (B_0 \cup U(b_0))^-$.

If we are *not* in the proof of addition 1 of Theorem A, the proof of Lemma 3.7 is now easily completed as follows: Since S^+ is supposed to be non-separating, also one of S_1^+ or S_2^+, say S_1^+, is non-separating in M^+. Moreover, if we are in the proof of addition 2 of Theorem A, observe that S_1^+ has to contain at least one component of ∂S^+ since S^+ is then supposed to have minimal complexity and disconnected boundary. But S_1^+ meets E in strictly fewer components than S^+ does. Thus, by our minimal choice of S^+, it follows that the complexity (= enlarged Betti number) of S_1^+ has to be strictly larger that that of S^+. Now, the union $B_0 \cup U(b_0)$ can be considered as the regular neighbourhood of some simple closed curve in S^+. However, splitting a surface along a simple closed curve does not in general increase the complexity, except in the case when this curve is contractible. It follows that the core of $B_0 \cup U(b_0)$ is contractible in S^+, and that therefore b_0 has to be inessential in $(S^+ - B_0)^-$.

If we *are* in the proof of addition 1 of Theorem A, we have to argue a little more carefully. For in this case the curve k (the attaching curve of the 2-handle) separates a punctured torus, T_0, from ∂M, and we suppose ∂S^+ to be contained in $\partial M - T_0$, while on the other hand the boundary of the surface S_1^+ constructed above need not be contained in $\partial M - T_0$. But if the latter is the case, it follows from our choice of N that then the annulus $N \cap \partial M^+$ is contained in some torus-component, say $\partial_0 M^+$, of ∂M^+. Moreover, $E \cap \partial_0 M^+ \subset N \cap \partial_0 M^+$, and so $E \cap (\partial_0 M^+ - N)^- = \emptyset$.
Thus

$$S' = S_1^+ \cup (\partial_0 M^+ - N)^- \cup S_2^+$$

is a surface in M^+ which is non-separating since S^+ is and whose boundary lies in $\partial M - T_0$ since $\partial S' = \partial S^+ \subset \partial M - T_0$.

Now, assume that b_0 is essential in $(S^+ - B_0)^-$. Then it follows that $c(S') = c(S^+)$. But, by our choice of S', the surface S' meets E in strictly fewer components than S^+ does. By our minimal choice of S^+ and by the previously mentioned properties of S', the surface S' has to be compressible. Compressing S' along discs, we eventually end up with an incompressible, non-separating surface, $S^\star$, with $\partial S^\star = \partial S' \subset \partial M - T_0$, which might meet E in more components than S^+ does. However, $c(S^\star) < c(S')$ $= c(S^+)$ and recall that we chose S^+ minimal with respect to the pair $(c(S^+), d(S^+))$ (and to the lexicographical order). Thus the existence of the surface S' gives the required contradiction again, proving that b_0 is inessential in $(S^+ - B_0)^-$.

To formulate our next result, we first introduce the notion of a "good subset". For this purpose let $\mathcal{B}_1$ and $\mathcal{B}_2$ again be given as in Lemma 3.7. Then a subset $\mathcal{B}^\star \subset \mathcal{B}_1 \cup \mathcal{B}_2$ is called a *good subset* (in $\mathcal{B}_1 \cup \mathcal{B}_2$), if the following holds :

(1) the set $\mathcal{B}^\star$ contains all recurrent arcs from $\mathcal{B}_1 \cup \mathcal{B}_2$,

(2) each component of B must contain at least one end-point of $\mathcal{B}^\star$, and

(3) there is a disc in S^+ which contains the union $B \cup \mathcal{B}^\star$.

Lemma 3.8. *There is a good subset in* $\mathcal{B}_1 \cup \mathcal{B}_2$.

Proof. To $\mathcal{B}_i$, $i = 1, 2$, we associate a set $\mathcal{B}_i^\star$ as follows:

If $\mathcal{B}_i$ contains some recurrent arc, then define $\mathcal{B}_i^\star := \mathcal{B}_i$. In the other case, let $B_i^\star$ be the union of all those components of B containing at least one end-point of $\mathcal{B}_i$. Then let $\mathcal{B}_i^\star$ be a minimal set satisfying the property that each component of $B_i^\star$ contains at least one end-point of $\mathcal{B}_i^\star$.

Define $\mathcal{B}^\star := \mathcal{B}_1^\star \cup \mathcal{B}_2^\star$. It is easily checked that $\mathcal{B}^\star$ satisfies (1) and (2) of a good subset in $\mathcal{B}_1 \cup \mathcal{B}_2$ (see Lemmas 3.4 and 3.7 and recall that $\mathcal{R}_0 \subset \mathcal{B}_1 \cup \mathcal{B}_2$). Furthermore, each component of $\partial U(B \cup \mathcal{B}^\star)$ in contractible in S^+, where $U(B \cup \mathcal{B}^\star)$ is a regular neighbourhood in S^+ (see Lemma 3.7). It follows that $B \cup \mathcal{B}^\star$ is contained in a system of discs in S^+, and so in one large disc since S^+ is connected.

□

As a consequence of Lemma 3.8 we obtain

Lemma 3.9. *At least one arc in* $\mathcal{B}_1 \cup \mathcal{B}_2$ *is not recurrent. In particular, the attaching curve* k *has to be non-separating in* ∂M.

Proof. Suppose the converse. Then, by Lemma 3.8, $\mathcal{B}_1 \cup \mathcal{B}_2 \cup B$ is contained in a disc. Since each arc from $\mathcal{B}_1 \cup \mathcal{B}_2$ is recurrent and since each component of B contains at least one end-point of $\mathcal{B}_1 \cup \mathcal{B}_2$, an inductive argument shows that at least one arc in $\mathcal{B}_1 \cup \mathcal{B}_2$ has to be inessential in S. But this contradicts Lemma 3.3, i.e. our choice of the disc D.

To prove the addition, let b be an arc in $\mathcal{B}_1 \cup \mathcal{B}_2$ which is not recurrent. b splits D into two discs. Let D_1 be that one of those discs which does not contain any collection of $\mathcal{R}$ entirely, and denote $b_1 := D_1 \cap \partial D$. Then b_1 is an arc in ∂M and since b is not recurrent it follows that b_1

intersects ∂A in precisely two points and each component of ∂A in precisely one point. Thus A, and so k, is non-separating in ∂M.

□

Lemma 3.10. *If* card $\mathcal{B}_1 \geq$ card $\mathcal{B}_2$, *then no arc from* $\mathcal{B}_1$ *is recurrent. In particular, the surface* S *admits a good compression disc.*

Proof. Suppose the contrary. Then, by Lemma 3.7, $\mathcal{B}_1$ consists of recurrent arcs. Then, by Lemmas 3.9 and 3.7, no arc from $\mathcal{B}_2$ is recurrent. Denote by B' the union of all components of B containing some end-point of $\mathcal{B}_1$. It follows that card $\mathcal{B}_1$ equals the number of components of B' (see Lemma 3.4). Moreover, it follows that $\mathcal{B}_2$ has a very special pattern in S. Indeed, recalling Lemma 3.4, our special choice of $\mathcal{B}_1$ and $\mathcal{B}_2$, and the fact that card $\mathcal{B}_1 \geq$ card $\mathcal{B}_2$, a moment's reflection shows that each component of B' (resp. $B - B'$) contains at most (resp. precisely) one end-point of $\mathcal{B}_2$ and that no two components of B' and no two components of $B - B'$ are joined by an arc from $\mathcal{B}_2$. In particular, any good subset in $\mathcal{B}_1 \cup \mathcal{B}_2$ has to contain $\mathcal{B}_1 \cup \mathcal{B}_2$ itself. Hence, by Lemma 3.8 and the very definition of good subsets, $\mathcal{B}_1 \cup \mathcal{B}_2 \cup B$ is contained in a disc, D^+, in S^+. Denote by B^+ a regular neighbourhood of $B \cup \mathcal{B}_2$ in S^+. Then each component of B^+ contains the end-points of at least one arc from $\mathcal{B}_1$, and $\mathcal{B}_1 \subset (D - B^+)^-$ and $B^+ \cup \mathcal{B}_1 \subset D^+$. Hence, by an inductive argument (as in Lemma 3.9), this implies that at least one arc in $\mathcal{B}_1$ is inessential in $(D^+ - B^+)^-$, and so in $(S^+ - B^+)^-$. This arc is either inessential in $(S^+ - B)^- = S$, too, or separates an annulus from S containing an arc from $\mathcal{B}$ which is inessential in S. To see the latter recall the properties of $\mathcal{B}_2$ described above and observe that, by Lemma 3.4, every component of ∂B contains the same number of end-points of $\mathcal{B}$. However, the existence of an arc from $\mathcal{B}$ which is inessential in S contradicts Lemma 3.3 (i.e. our choice of the disc D).

□

Lemma 3.11. *The number of components of* $A \cap \partial S = \partial B$ *is even.*

Proof. Assume the converse. Then the special collection $\mathcal{R}_0$ has an odd number of points (see Lemma 3.4). In particular, $\mathcal{R}_0$ has a midpoint. It follows that the arc from $\mathcal{B}_1 \cup \mathcal{B}_2$, containing this midpoint, is recurrent. Let $\mathcal{B}_1$ be that one of the two sets $\mathcal{B}_1$, $\mathcal{B}_2$ containing this particular arc. Then card $\mathcal{B}_1 \geq$ card $\mathcal{B}_2$, and we either get a contradiction to Lemma 3.7 or to Lemma 3.10.

A.3. Constructing a new one-relator 3-manifold

Recall from Lemma 3.10. that the surface S admits a good compression disc, D_0. Let $b_1 := D_0 \cap \partial M$. Then, by property (1) and (2) of good compression discs, $b_1' := (b_1 - A)^-$ is an arc in $(\partial M - A)^-$ which joins the two boundary curves of the annulus A $(= E \cap \partial M)$. Define

$$k^\star := \partial U(A \cup b_1'), \quad \text{and}$$

$$A^\star := U(k^\star),$$

where the regular neighbourhoods are taken in ∂M. Then $k^\star$ is connected. More precisely, $k^\star$ is a simple closed curve in ∂M which separates the once punctured torus $T_0 = U(A \cap b_1')$ from ∂M with $k \subset T_0$ and $\partial S^+ \subset \partial M - T_0$. The existence of such a simple closed curve in ∂M gives rise to the definition of the new one-relator 3-manifold,

$$M^\star := M^+(k^\star),$$

which is the 3-manifold obtained from M by attaching a 2-handle, $E^\star$, along $k^\star$. But before verifying that $M^\star$ satisfies the hypothesis of Theorem A, we first modify S^+ as to obtain an appropriate surface $S^\star$ in $M^\star$.

Consider the arcs from $D_0 \cap S$. By property (2) of the good compression disc D_0, they all join the two components of $b_1 \cap A$, i.e. they are all parallel in D_0. Let $c_1, c_2, \ldots, c_m$ be an enumeration of these arcs beginning with the innermost arc, c_1, in D_0. Let n denote the number of components of $A \cap \partial S = \partial B$. Then, by Lemma 3.11, n is even and, by property (3) of good compression discs (see page 166), $m \geq n/2$. Consider the first $n/2$ arcs $c_1, c_2, \ldots, c_{n/2}$ only. Considering b_1 in ∂M, it is easily checked that the components of B fall into two disjoint subsets with the property that each component of one subset will be joined by the arcs from $c_1, c_2, \ldots, c_{n/2}$ with precisely one component of the other subset. In other words, the components of B will be paired by $c_1, c_2, \ldots, c_{n/2}$. In particular,

$$B^\star := B \cup U\Big(\bigcup_{1 \leq i \leq n/2} c_i\Big)$$

has precisely $n/2$ components, where the regular neighbourhood is taken in

S^+. Pushing the arcs $c_1, c_2, \ldots, c_{n/2}$, one after the other, across D_0 into b_1, we find that $S' := (S - B^\star)^-$ can be considered as a surface in M with $\partial B^\star \subset A^\star$. Moreover, $S^\star$ is incompressible in M since S is, and each component of $\partial B^\star$ is parallel to the components of $\partial A^\star$. Now, $\partial B^\star$ is also the boundary of a system, B', of discs in the 2-handle $E^\star$, and we denote by $S^\star$ the union

$$S^\star := S' \cup B'.$$

Then $S^\star$ is a surface in $M^\star$ homeomorphic to S^+ with $d(S^\star) < d(S^+)$.

The further relevant properties of $S^\star$ will be considered in the next section. Here we use $S^\star$, or rather S', to finally show that $M^\star$ indeed satisfies the hypothesis of Theorem A. For this purpose we have to show that each component of $\partial M - k^\star$ is incompressible in M. Let $\partial_0 M$ denote the component of ∂M containing T_0. We claim that $(\partial_0 M - T_0)^-$ cannot be a disc. For a moment suppose the converse. Then it follows that S' has to be a disc since S' is incompressible in M and since the components of $A^\star \cap \partial S'$ are parallel to $k^\star$. Since $\partial S'$ also bounds a disc in ∂M and since M is irreducible, S' and so S has to be boundary-parallel. Thus alo S^+ is boundary-parallel; but this contradicts our hypothesis on S^+. Hence $(\partial_0 M - T_0)^-$ has to be different from the disc. Now, $(\partial M - T_0)^- \subset \partial M - k$ and so each component of $(\partial M - T_0)^-$ is incompressible in M since each component of $\partial M - k$ is. Thus, in order to show that each component of $\partial M - k^\star$ is indeed incompressible in M, it only remains to show the following Lemma.

Lemma 3.12. *Let M be any (orientable) 3-manifold with non-empty boundary. Let T_0 be a torus with hole in the component $\partial_0 M$ of ∂M. Then T_0 is incompressible in M, if $(\partial_0 M - T_0)^-$ is incompressible in M and not a disc.*

Proof. Assume the converse. Then there is a disc, D, in M, with $D \cap \partial M = \partial D$, such that ∂D is contained in T_0 but not contractible in T_0.

Since $(\partial_0 M - T_0)^-$ is neither a disc nor compressible in M, ∂D cannot be boundary-parallel in T_0. Since T_0 is a punctured torus, it follows that ∂D has to be non-separating in T_0. In this case, however, there is some simple closed curve, t, in T_0 which intersects ∂D in precisely one point. Using this curve t, we define

$$D^\star := (\partial U(t \cup D) - \partial M)^-,$$

where the regular neighbourhood is taken in M. Then $D^\star$ is a disc in M

with $D^\star \cap \partial M = \partial D^\star \subset T_0$. Furthermore, $\partial D^\star$ is separating in T_0 since it bounds the regular neighbourhood of $t \cup \partial D$ in T_0. Hence, replacing D by $D^\star$, we obtain a contradiction as above.

$\square$

We now have all the tools together to conclude the proof of Theorem A. This will be carried out in the next section.

A.4. The conclusion of the proof

In order to conclude the proof of Theorem A, it still remains to carry out the inductive step, i.e. we have to show Theorem A for the specified curve k (under the assumption that Theorem A holds for all curves with smaller complexity). For this consider the one-relator 3-manifold $M^+(k^\star)$ as constructed in the previous section, and estimate the complexity $c(k^\star)$ of the curve $k^\star$ (see the beginning of the proof for the definition of $c(k^\star)$). We claim that $c(k^\star) < c(k)$. To see this consider the surface $S^\star$ in $M^+(k^\star)$ as also given in section 3. Recall from the construction of $S^\star$ that both $S^\star$ and S^+ are homeomorphic and that, moreover, $d(S^\star) < d(S^+)$, i.e. $(c(S^\star), d(S^\star)) < (c(S^+), d(S^+))$. Now, $S^\star$ satisfies the hypothesis of Theorem A, except possibly that it might be compressible. But $S^\star$ is non-separating since S^+ is, and so, compressing $S^\star$ along discs, if necessary, we eventually end up with some surface in $M^+(k^\star)$ satisfying the hypothesis of Theorem A completely. Since the process of compressing along discs reduces the (enlarged) first Betti number, our claim that $c(k^\star) < c(k)$ finally follows. Hence, by the induction hypothesis there is an incompressible, non-separating surface, S', in $M^+(k^\star)$ satisfying the conclusion of Theorem A with respect to the curve $k^\star$. In particular, $S' \subset M$. Furthermore, recall from Section 3.3 that $k^\star$ separates a punctured torus, T_0, from ∂M and that $\partial S^\star \subset \partial M - T_0$. Thus without loss of generality also $\partial S' \subset \partial M - T_0$ (addition of Theorem A). Since the original curve k is contained in T_0 (see section 3.3), this implies that S' is also contained in $M^+(k)$. Furthermore, by our minimal choice of $c(S^+)$, the surface S' is not only incompressible in $M^+(k^\star)$, but also in $M^+(k)$ since compressing along discs reduces complexity and since $c(S') \leq c(S^\star) = c(S^+)$. Now, by Lemma 3.9, k has to be non-separating and so we cannot be in the proof of addition 1 (or, more precisely, the assumption that k is separating is already inconsistent with the assumption $d(S^+) \neq 0$). Thus, the existence of the surface S' shows that Theorem A with addition 1 holds for the curve k, and so, inductively, for all curves satisfying the hypothesis of Theorem A. □

In order to check that addition 2 of Theorem A is also true, start reading the proof again, this time under the additional hypothesis that Theorem A and addition 1 are already proven. Then, for every curve k in ∂M satisfying the hypothesis of Theorem A, the one-relator 3-manifold $M^+(k)$ is irreducible and boundary-irreducible (see remark). Thus, every non-separating annulus in M^+ is already incompressible. For example, the surface $S^\star$ obtained in the very last step above has to be an annulus which is incompressible in $M^+(k^\star)$. Furthermore, recall that $\partial S^\star = \partial S^+$ and check that only in the proof of Lemma 3.5 an isotopic deformation of S^+ is applied and only to show that certain special compression discs give rise to such isotopies. However, $S^\star$ admits no special compression discs at all as S^+ (after deformation in M^+) does not. Thus the existence of the required annulus follows by an application of a similar induction as used in the course of the proof of Theorem A, but we omit the details. □

On Homeomorphic Images of Surfaces in One-Relator 3-Manifolds

The main result of this section is Theorem B.

As already mentioned in this introduction, a presentation of a 3-manifold, N, as one-relator 3-manifold is given by some specified arc, t, in N such that $(N - U(t))^-$ is boundary-reducible. It follows that every 3-manifold, N, which admits an essential annulus, A, $A \cap \partial N = \partial A$, is a one-relator 3-manifold in a non-trivial way. To see this simply consider an essential arc, t, in the annulus, A. Then t is essential in N and $(N - U(t))^-$ is obviously boundary-reducible since $(A - U(t))^-$ is a disc. Thus almost all I-bundles over surfaces, almost all Seifert fibre spaces with non-empty boundary, all knot spaces of product-knots and knots with companions etc. are examples of 3-manifolds which are one-relator 3-manifolds in a non-trivial way.

Considering Seifert fibre spaces, for example, it is easy to construct examples for which the presentation as one-relator 3-manifold is not unique. To give a concrete example, let X be the S^1-bundle over the 2-sphere with three holes. Let A_1, resp. A_2, be vertical, essential annuli in X

which have their boundary curves in the same, resp. different, components of ∂X. Then the essential arcs in A_1, resp. A_2 give rise to two non-equivalent presentations.

However, on the other hand, there are other 3-manifolds which definitely have, even in the strong sense, a unique presentation as one-relator 3-manifold. Examples are given by the following result.

Proposition 4.1. *Let X be a product I-bundle over some closed, orientable surface of genus ≥ 1. Let $h : X \to X$ be any homeomorphism. Then for any two non-trivial one-relator presentations $(X,\ t_1)$ and $(X,\ t_2)$ of X, we have that $h(t_1)$ is isotopic to (t_2).*

Proof. Let $(X,\ t)$ be any non-trivial one-relator presentation of X. Then, by the very definition, t is an essential arc in X such that $(X - U(t))^-$ is boundary-reducible. Since h can be isotoped into a fibre preserving homeomorphism (see [Waldhausen, 1968]), it suffices to show that t is isotopic to some fibre of X.

Case 1. Suppose t is contained in a 3-ball, E, in X such that $(\partial E - \partial X)^-$ is a disc.

We are going to show that this case cannot occur. Denote by $\tilde{X}$ the manifold obtained from $(X - U(t))^-$ by splitting along the disc $(\partial E - \partial X)^-$. Since X is not a 3-ball, it follows that no component of $\tilde{X}$ is a 3-ball. But, by our choice of t, $(X - U(t))^-$ is boundary-reducible, and so also at least one component of $\tilde{X}$ (innermost-disc-argument). Now, $(X - E)^-$ is homeomorphic to X, and I-bundles over closed surfaces are known to be boundary-irreducible. Hence $(E - U(t))^-$ has to be boundary-reducible. Now, the boundary of $(E - U(t))^-$ is a torus and $(E - U(t))^-$ is irreducible since X is irreducible as an I-bundle over a closed orientable surface different from S^2. Thus $(E - U(t))^-$ is boundary-reducible if and only if it is a solid torus. It follows that t is unknotted, i.e. inessential, in the 3-ball E. Since $E \cap \partial X$ is a disc, it follows that t is inessential in X. This is a contradiction to our hypothesis on t.

Case 2 . Suppose t is contained in a 3-ball, E, in X such that $(\partial E - \partial X)^-$ is a vertical annulus in X.

The annulus $(\partial E - \partial X)^-$ is incompressible in $X - t$, for otherwise we are in the situation of Case 1 and get a contradiction as described there. In particular, the arc t in the 3-ball E has to join the two components of $E \cap \partial X$. A moment's reflection shows that then t is indeed isotopic to a fibre of X, provided t is inessential in E (recall that the annulus $(\partial E - \partial X)^-$ is supposed to be vertical). Now, $(X - U(t))^-$ is boundary-reducible and so there is an essential disc, D, in $(X - U(t))^-$. But this disc cannot be isotoped in $(X - U(t))^-$ out of E since X is boundary-irreducible. Hence, by the innermost-disc-argument, it follows that $(E - U(t))^-$ is boundary-reducible and so, as in Case 1, t is inessential in E. This proves proposition 4.1 in Case 2.

Case 3. Suppose t is contained in a solid torus, W, in X such that $(\partial W - \partial X)^-$ is an incompressible annulus.

Since $(\partial W - \partial X)^-$ is incompressible, observe that there is at least one non-separating annulus, A, in X which cannot be isotoped out of W. But A can be isotoped into an annulus, A', with $t \cap A' = \emptyset$ (this follows from addition 2 of Theorem A). Now, $(\partial W - \partial X)^-$ is boundary-parallel (X is a product I-bundle) and we are not in Case 1. Hence, all closed curves can be removed from $A' \cap (\partial W - \partial X)^-$, by an isotopy of A' which does not meet t (innermost-disc-argument). This implies the existence of a meridian disc in W which is disjoint to t. Splitting W along this disc, we see that we are in the situation of Case 1.

Case 4. Suppose neither one of Case 1, 2, or 3 holds.

By our hypothesis on X, there is a set of vertical, non-separating annuli, $A_1, A_2, \ldots, A_n$, in X such that

(1) $(X - \cup\, U(A_i))^-$ consists of 3-balls, and that

(2) $A_i \cap A_{i+1}$ consists of precisely one arc, for all i such that $1 \leq i \leq n-1$ (take the pre-images of some appropriate set of curves in the base of X).

By property (1), the arc t cannot be contained in $(X - \cup\, U(A_i))^-$ since we are not in Case 2. Let m, $m \geq 1$, be the largest integer with the property that $A_1, \ldots, A_m$ are pairwise isotopic to vertical annuli $A'_1, \ldots, A'_m$ with $t \cap (\cup_{1 \leq i \leq m} A'_i) = \emptyset$. Now, observe that under the hypothesis of property (2), property (1) is preserved under fibre preserving isotopies in X. It follows that m has to be strictly smaller than n.

Define Z_m to be the union of $\bigcup_{1 \leq i \leq m} U(A'_i)$ with all ball-components of its complement in X. Then the arc t is contained in $(X - Z_m)^-$ since we are not in Case 2. Furthermore, Z_m is a vertical I-bundle in X and $(\partial Z_m - \partial X)^-$ consists of essential annuli.

Now, consider the annulus A_{m+1}. It follows from addition 2 of Theorem A that A_{m+1} can be isotoped into an annulus, A'_{m+1}, with $t \cap A'_{m+1} = \emptyset$. By general position, we may suppose that $Aprimesup\, m+1 \cap (\partial Z_m - \partial X)^-$ consists of curves only. Since X is irreducible and boundary-irreducible and since X is a product I-bundle, it follows that all closed curves and all essential arcs from $A'_{m+1} \cap (\partial Z_m - \partial X)^-$ can be removed by an isotopy of A'_{m+1}. Since we are neither in Case 1 nor in Case 3, this isotopy may be chosen so that it does not meet t (innermost-disc-argument). Thus we may suppose that $A'_{m+1} \cap (\partial Z_m - \partial X)^-$ consists of essential arcs in A'_{m+1}. Then an ambient isotopy of X is easily constructed so that afterwards $A'_1, \ldots, A'_m, A'_{m+1}$ are vertical and that, in addition, $t \cap \cup_{1 \leq i \leq m+1} A'_i = \emptyset$. This, however is a contradiction to our maximal choice of the integer m (recall that in X two vertical annuli are isotopic iff they are fibre-preserving isotopic).

This proves that Case 4 cannot occur.

□

By what we have seen so far, there are classes of 3-manifolds with, and others without, unique one-relator presentation. This problem arises to characterize these classes internally. However, nothing is known in this direction. Even the answer to the following stronger form of the question remains open: when is a one-relator presentation (M^+, t) of M^+ determined by the complement $(M^+ - U(t))^-$ alone? We can deduce from Theorem A the necessary existence of some hierarchy of M^+ missing t, such that its image under each homeomorphism of M^+ also misses t. This property suggests the study of homeomorphic images of incompressible surfaces in M^+ in its own right. In the remainder of this section we study the case when the attaching curve k of the 2-handle E is separating, or dually, when the arc t joins different components of ∂M^+. Our result is the following.

Theorem B. *Let M, k and $M^+(k)$ be given as in Theorem A. Let $h^+ : M^+ \to M^+$ be a homeomorphism. Furthermore, suppose that k is separating in ∂M, and that M^+ is atoroidal.*

Then there is an incompressible, non-boundary-parallel surface, S, in M^+ such that S as well as $(h^+)^2(S)$ is contained in M, up to isotopy. If, in addiction, every Seifert fibre space in M^+ is the product S^1-bundle over the disc, then more precisely also $h^+(S)$ is contained in M, up to isotopy. **Remarks**

1) Recall that M^+ is a Haken 3-manifold (by Theorem A) and that a Haken 3-manifold, M^+, is called *atoroidal* if every incompressible torus in M^+ is boundary-parallel. A Seifert fibre space, X, in M^+ is called *essential* in M^+ if $(\partial X - \partial M^+)^-$ consists of annuli or tori which are incompressible and not boundary-parallel in M^+. It is

further to be understood that $\partial X \cap \partial M^+$ is saturated in the Seifert fibration of X.

2) Observe that the atoroidal-property survives under splitting along incompressible surfaces.

Before proving Theorem B, we state and prove the following result which has a stronger hypothesis, but which also has a stronger conclusion. This result is crucial for the proof of Theorem B. In order to formulate it, let M, k and E again be given as in Theorem A. Denote by $\mathbf{m}$ the completed boundary-pattern of M induced by the regular neighbourhood of k in ∂M (see [Johannson, 1979] for the relevant terminology).

Proposition 4.2. *Suppose that k is separating in ∂M, and that every essential annulus in $(M, \mathbf{m})$ is boundary-parallel in M. Then any non-boundary-parallel, incompressible surface in $M^+(k)$ whose complexity is minimal, can be isotoped out of the 2-handle E (and so into M).*

Remark. See [Johannson, 1979] for the notion "essential", and recall that the term "complexity" refers to the enlarged first Betti number.

Proof. The proof is similar to the proof of Theorem A. But its argument can be simplified considerably whenever (as in our case) the attaching curve is separating. It will turn out that in this particular case essentially only those constructions remain in the proof of Theorem A which can be carried out by isotopies. In order to check this more carefully, let us follow Theorem A again (and let us use the notations introduced there).

Let S^+ be any non-boundary-parallel, incompressible surface in M^+ whose complexity is minimal. We have to show that S^+ can be isotoped out of E. Suppose that S^+ is *isotoped* in M^+ so that $E \cap S^+$ is a system of discs, parallel to the components of $(\partial E - \partial M)^-$, whose number is minimal (see the beginning of the proof of Theorem A). Assume $E \cap S^+ \neq \emptyset$ (for otherwise we are done).

By an argument of Lemma 3.5, the surface

$$S := M \cap S^+$$

admits no special compression disc (for the definition see Section 3.1). To see this simply check that the reductions used in Lemma 3.5 can in fact be carried out by isotopies of S^+, provided that M^+ is boundary-irreducible. That M^+ is indeed boundary-irreducible follows from Theorem A, Remark. Again, the non-existence of a special compression disc implies the existence of at least one essential disc, D, in M such that

$$\mathcal{B} := D \cap S$$

is a good arc-system in D and essential in S. In particular, the system $\mathcal{R} := \partial D \cap \partial S$ in ∂D admits a special collection, say $\mathcal{R}_0$, with respect to $A \cap \partial D$, where $A = E \cap \partial M$ (see Section 3.1).

Let $\mathcal{B}_0$ denote the sub-system of all those arcs from $\mathcal{B}$ which have one end-point in $\mathcal{R}_0$. Since $\mathcal{R}_0$ is a special collection, we find two arcs, b_1, b_2 (possibly equal) in $\mathcal{B}_0$ which separate discs, D_1, D_2, respectively, from D such that $\mathcal{B}_0 \cap (D_1 \cup D_2) = \mathcal{B}_0$. Since k is separating in ∂M, it is easily checked that $\mathcal{B}_0$ consists only of recurrent arcs.

The latter implies that the regular neighbourhood, U, of $B \cup \mathcal{B}_0$ in S^+ is a system of annuli in S^+ (here B denotes the system of discs $E \cap S^+$). Furthermore, no component of $(S^+ - U)^-$ contains a disc since $\mathcal{B}_0$ consists of arcs which are essential in S (see above). It follows that no component of $(S^+ - U)^-$ has larger complexity S^+.

To continue the proof observe that $(S^+ - U)^-$ corresponds, in a canonical way, to a homeomorphic 2-manifold, S', in M^+ with $S' \cap \partial M^+ = \partial S'$ and also contained in M. To obtain S' recall that $\mathcal{B}_0 \subset D_1 \cup D_2$. Then split S^+ along $\mathcal{B}_0$ and push the arcs from $\mathcal{B}_0$ in the appropriate order along D_1, resp. D_2, first into ∂M and then out of $E \cap \partial M$. By the properties of S^+, it follows that each component of this 2-manifold, S', is incompressible in M^+, and so in M, and that at least one component, say S'_1, of S' is not boundary-parallel in M^+. Denote by S_1^+ the component of $(S^+ - U)^-$ corresponding to S'_1. Then $c(S_1^+) = c(S'_1) \geq c(S^+)$, and $c(S_1^+) \leq c(S^+)$ (see above).

By what we have seen so far, $(S^+ - U)^-$ contains a component, S_1^+, which is homeomorphic to S^+. Hence all components of $(S^+ - U)^-$, different from S_1^+, are annuli and at least one such annulus, say S_2^+, contains a component of ∂S^+.

Let S'_2 be that component of S' corresponding to S_2^+. Then S'_2 is an admissible annulus in $(M, \mathbf{m})$ which is incompressible in M. We claim that S'_2 is boundary-parallel in M. If S'_2 is essential in $(M, \mathbf{m})$, this is our hypothesis. If not, this claim follows since $\partial M - k$ is incompressible in M and so $\mathbf{m}$ is a useful boundary-pattern (see [Johannson, 1979]). Since S'_2 is boundary-parallel in M, we easily find, in particular, a compression disc, D_0, for S'_2 which satisfies all properties of a special compression disc, except that it might meet S in more than one component. Then the existence of an honest special compression disc follows from the innermost-disc-argument. But there are no special compression discs for S (see above), i.e. our assumption $E \cap S^+ \neq \emptyset$ leads to contradictions.

□

Proof of Theorem B. We first prove Theorem B in special cases:

Lemma 4.3. *Theorem B holds, if M^+ is an I-bundle or Seifert fibre space.*

Proof. If M^+ is an I-bundle, observe that ∂M^+ has to have two components since the attaching curve k is supposed to be separating in ∂M. Thus M^+ is a product I-bundle and Lemma 4.3 follows from 4.1.

Suppose that M^+ is a Seifert fibre space. Let $h^+ : M^+ \to M^+$ be any homeomorphism and let t be the core of the 2-handle E. Then, by [Waldhausen, 1967] we may suppose that h^+ is fibre preserving, and, by our hypothesis on k, the arc t joins two different boundary-components of M^+. Since M^+ is atoroidal, the sum of boundary-components and exceptional fibres of M^+ cannot be larger than 3. Since ∂M^+ has at least two components, M^+ has at most one exceptional fibre. There is at least one non-separating, incompressible annulus, A, in M^+. By Theorem A, the annulus A can be chosen so that, in addition, $A \cap t = \emptyset$. Furthermore, by a result of Waldhausen [Waldhausen, 1967] we may suppose that A is vertical in M^+ (for it is isotopic to a vertical one).

If the orbit surface of M^+ is the annulus, then $h^+(A)$ is isotopic to A and we are done (recall that h^+ is fibre-preserving).

If the orbit surface of M^+ is not the annulus, then M^+ has to be the product S^1-bundle over the 2-sphere with three holes (see our discussion above). Splitting M^+ along A we obtain the S^1-bundle, $\tilde{M}$, over the annulus. Now, it follows that $(\tilde{M} - U(t))^-$ is boundary-reducible since $(\tilde{M} - U(t))^-$ is. Hence, again by Theorem A, there must exist some non-separating, vertical annulus, B, in $\tilde{M}$ with $B \cap t = \emptyset$. Without loss of generality, $\partial B \subset \partial M^+$ and so A and B are non-separating, vertical annuli in M^+ which are not isotopic in M^+. Since h^+ is fibre-preserving, it is easily checked that $h^+(A)$ or $h^+(B)$ is isotopic in M^+ to A or B, and we are done.

□

We are now in the position to prove Theorem B in its full generality. For this observe that $M^+(k)$ is a Haken 3-manifold. This is an immediate consequence of Theorem A, Remark. Now, recall that in any Haken 3-manifold the characteristic submanifold exists and is unique up to isotopy (see [Johannson, 1979]). The idea is now to apply the theory of characteristic submanifolds to our problem.

By Proposition 4.2, we may suppose that the manifold $(M, \mathbf{m})$ admits an essential annulus which is *not* boundary-parallel in M, for otherwise nothing is to be shown any more. This property implies that the characteristic submanifold of M^+ is non-trivial, in the sense that it is non-empty *and* not the regular neighbourhood of components of ∂M^+:

Lemma 4.4. *The characteristic submanifold, V^+, of M^+ is non-trivial, and either M^+ is an I-bundle or Seifert fibre space, or at least one component of $(\partial V^+ - \partial M^+)^-$ is an annulus.*

Proof. We are first going to show that V^+ is non-trivial. Recall that characteristic submanifolds contain all possible essential annuli and tori, up to isotopy (see [Johannson, 1979]). Furthermore, $(M, \mathbf{m})$ is supposed to contain at least one essential annulus, A, which is not boundary-parallel in M. Thus in order to prove that $V^+ \neq \emptyset$, it suffices to show that the annulus A is essential in M^+, too.

Assume the converse. Since M^+ is a Haken 3-manifold, it follows that either A is parallel in M^+ to an annulus, A', in ∂M^+, or that the two boundary curves of A bound discs, say A'_1, A'_2, in ∂M^+. For convenience consider the 2-handle E as a small neighbourhood of some appropriate arc, t, in M^+. If A is parallel to A', then note that, by hypothesis, the annulus A' cannot lie in ∂M. Hence A' has to contain at least one disc from $E \cap \partial M^+$. But $E \cap A = \emptyset$ and so it follows that A' has to contain the other disc from $E \cap \partial M^+$, too. This, however, is impossible since the attaching curve k separates ∂M. Therefore the components of ∂A bound discs in ∂M^+. These discs cannot be contained in ∂M since A is incompressible in M and so they each contain a disc from $E \cap \partial M^+$. Since k is separating, the disc A'_1 cannot lie in A'_2 and vice versa, and so $A'_1 \cap A'_2 = \emptyset$. Since M^+ is irreducible, A separates a 3-ball, N, from M^+ and this 3-ball contains E. Now, $\partial(N - E)^-$ is a torus. Furthermore, M^+ is boundary-irreducible, and $M$is irreducible and boundary-reducible. It follows that $(N - E)^-$ has to be a solid torus. Since we obtain N from $(N - E)^-$ by attaching the 2-handle E and since N is a 3-ball, the circulation number of the annulus $E \cap \partial M$ with respect to the solid torus $(N - E)^-$ is ± 1. This implies that A is boundary-parallel in M, contradicting our choice of the annulus A.

By what we have proved so far, the characteristic submanifold V^+ of M^+ is non-trivial. Now, V^+ consists of I-bundles and Seifert fibre spaces which are essential in M^+. In particular, $(\partial V^+ - \partial M^+)^-$ consists of annuli and tori which are essential in M^+. Furthermore, by the completeness of characteristic submanifolds [Johannson, 1979], $(\partial V^+ - \partial M^+)^-$ cannot consist of boundary-parallel tori. Thus the second half of Lemma 4.4 follows from the hypothesis that M^+ is atoroidal.

□

Lemma 4.4 will now be our starting point.

Let V^+ be the characteristic submanifold of M^+. Let V_0^+ be the union of all those components of V^+ which are not regular neighbourhoods of components of ∂M^+. Denote by A^+ the frontier

$$A^+ := (\partial V_0^+ - \partial M^+)^-.$$

By Lemmas 4.3 and 4.4, we may suppose that A^+ is non-empty. In general, A^+ is a system of essential annuli *and* tori in M^+. But in our situation, A^+ consists of annuli alone since M^+ is atoroidal and V^+ is complete. Since M^+ is a one-relator 3-manifold, recall the existence of a proper arc, t, in M^+ with $M = (M^+ - U(t))^-$. More precisely, t is the core of the 2-handle E. Given this arc t, we are looking for some essential annulus, A, in M^+ with $A \cap t = \emptyset = (h^+)^2(A) \cap t$, respectively $h^+(A) \cap t = \emptyset$, up to isotopy. Here, $h^+\colon M^+ \to M^+$ denotes the homeomorphism given in Theorem B.

We may suppose that t is (properly) isotoped in M^+ so that t is in general position with respect to A^+ and that, in addition, t intersects A^+ in a minimal number of points. Now, recall that h^+ can be isotoped so that afterwards $h^+(V^+) = V^+$ (uniqueness of the characteristic submanifold). In particular, $h^+(A^+) = A^+$. Thus Theorem B follows immediately, provided $t \cap A^+ = \emptyset$.

Thus we may suppose the converse. Then the set of all components of A^+ naturally falls into two disjoint subsets, say αA^+ and βA^+, consisting of all those annuli from A^+ whose intersection with the arc t is empty or non-empty, respectively. The aim is now to show that the set αA^+ is strictly larger than βA^+, for then there has to be at least one annulus $A \in \alpha A^+$ with $h^+(A) \in \alpha A^+$ and Theorem B follows immediately. However, instead of proving this (which might not be true in general), we will show a slightly weaker result which still suffices for our purposes.

We first study the arc system $t \cap V_0^+$, and it will turn out (see Lemma 4.5) that the embedding type of this system in V_0^+ cannot be very complicated.

To begin with we choose an essential disc, D, in M. This choice is possible since M is boundary-reducible. Moreover, as in the proof of Lemma 4.2 (and Theorem A as well) this disc may be chosen in such a way that

$$\mathcal{B} := D \cap A^+$$

is a (non-empty) good arc-system in D and essential in $(A^+ - E)^-$ (for notations see again the proof of Theorem A). In particular, the system $\mathcal{R} := \partial D \cap E \cap A^+$ has a special collection, $\mathcal{R}_0$. the subsystem, $\mathcal{B}_0$, of all those arcs from $\mathcal{B}$ which have one end-point in $\mathcal{R}_0$ consists of recurrent arcs, only (recall that k is separating).

We will utilize the fact that to any arc from $\mathcal{B}_0$ there is associated a pair of annuli constructed as follows: Given any arc, b_i, from $\mathcal{B}_0$, let D_i denote the disc separated from D by b_i with $D_i \cap \mathcal{B} \subset \mathcal{B}_0$, and let $b'_i := D_i \cap \partial D$. Now, let B_i be the component of $E \cap A^+$ containing the two end-points of b_i (recall that b_i is recurrent and note that $E = U(t)$). Then ∂B_i is a simple closed curve in the annulus $E \cap \partial M$ which separates this annulus into two annuli. Since A^+ is 2-sided in M^+, the arc b'_i can intersect only one of the latter two annuli. The intersection of b'_i with this particular annulus consists of two essential arcs whose union splits the annulus into two squares. Denote by A_i the union of D_i with one of these squares, and by A'_i the union with the other. We call A_i, A'_i the *annuli associated to the arc* b_i. Note that A_i and A'_i only differ by one of the components, say Q, separated by B_i from E, i.e. A_i and A'_i are "parallel", and we call Q the *parallelity region of* A_i and A'_i.

The existence of the annuli A_i, A'_i associated to the arcs from $\mathcal{B}_0$ is crucial in the proof of our next result.

Lemma 4.5.

(1) *If* $t \cap A^+ \neq \emptyset$, *then at least one component of* V_0^+ *is a Seifert fibre space.*

(2) *If* X *is a Seifert fibre space from* V_0^+, *then the following holds :*

  (a) $t \cap X$ *consists of at most two arcs, and*

  (b) *there are disjoint, essential annuli,* B_1, B_2 *(possibly* $B_1 = B_2$*) in* X *joining* $(\partial X - \partial M^+)^-$ *with* $X \cap \partial M^+$ *such that* $U(B_i)$, $i = 1, 2$, *contains precisely one component of* $t \cap X$.

Remark. $U(B_i)$ denotes the regular neighbourhood of B_i in X.

Proof. To prove 4.5, we are going to utilize the completeness of characteristic submanifolds [Johannson, 1979]. It is known that the completeness gives certain restrictions to the "length" and possible positions of annuli in M^+ and these have relevance for our purpose. To describe the relevant restrictions, suppose we are given a nested sequence of annuli

$$A_1 \subset A_2 \subset \ldots A_n \,, n \geq 1 \,,$$

such that

(1) $((A_i - A_{i-1})^-, i = 1, \ldots, n)$, is an essential annulus either in V^+ or in $(M^+ - V_0^+)^-$, and that

(2) one component of ∂A_1 is contained in ∂M^+ and the other one in $(\partial V_0^+ - \partial M^+)^-$.

We claim that A_1 is contained in a Seifert fibre space from V_0^+, that $A_n = A_1$ or $A_n = A_2$, and that $A_n = A_2$ is possible only if A_2 meets an I-bundle from V_0^+. To see this recall that, as one of the properties of characteristic submanifolds, any component of $(M^+ - V_0^+)^-$, containing an essential annulus not parallel to $(\partial V_0^+ - \partial M^+)^-$, is a product I-bundle over the annulus (it cannot be the I-bundle over the torus since $(\partial V_0^+ - \partial M^+)^-$ consists of annuli alone). It follows from this property that A_1 has to be contained in some component, X_1, of V^+ (and not in $(M^+ - V^+)^-$). Any component of V_0^+ is either an I-bundle or a Seifert fibre space and we may suppose that A_1 is either vertical or horizontal in X_1 (see e.g. [Johannson, 1979 5.6]). If X_1 is an I-bundle, A_1 cannot be vertical since one boundary curve of A_1 is contained in $(\partial X_1 - \partial M^+)^-$ and not in a lid of X_1. In this case, X_1 is the I-bundle over the annulus or Möbius band and hence it also admits a fibration as a Seifert fibre space. This proves the first part of our claim. If $A_n \neq A_1$, then $(A_2 - A_1)^-$ is contained in a component of $(M^+ - V_0^+)^-$ and so, by (1), this component has to be annulus $\times$ I. Thus, by completeness of V^+, at least one component of V_0^+ meeting this particular component of $(M^+ - V_0^+)^-$ has to be an I-bundle, and cannot be a Seifert fibre space. Hence, by the previous argument, $A_n = A_2$ and this proves the second part of our claim.

Now, consider the subsystem $\mathcal{B}_0$ of $\mathcal{B} = D \cap A^+$ (see above). The arc-system $\mathcal{B}_0$ is, more precisely, the union $\mathcal{B}_0 = \mathcal{B}_1 \cup \mathcal{B}_2$ (possibly $\mathcal{B}_1 = \mathcal{B}_2$), where $\mathcal{B}_i$, $i = 1, 2$, consists of all those arcs from $\mathcal{B}_0$ which join $\mathcal{R}_0$ with a given collection neighbouring $\mathcal{R}_0$. It is easily checked that the annuli associated to the arcs from $\mathcal{B}_i$, $i = 1$ and 2, define a nested sequence of annuli, $A_1, \ldots, A_n$, as considered previously. To show that all $(A_i -$

$A_{i-1})^-$ are indeed essential, recall our minimality condition on $t \cap A^+$ (look for special compression discs, for these would give rise to isotopies as described in Lemma 3.5). It follows from our claim above that $\mathcal{B}_1$ as well as $\mathcal{B}_2$ has at most two arcs and so $\mathcal{R}_0$ consists of at least one, two or four points (it cannot consist of three points since $\mathcal{B}_0 \subset (\partial V^+ - \partial M^+)^-$). Hence $E \cap V^+$ consists of at most three components and, again by our claim above, precisely those components of $E \cap V^+$ which meet ∂M^+ are contained in Seifert fibre spaces from V^+. These components also give "parallelity regions" for pairs of annuli associated to arcs from $\mathcal{B}_0$. Hence and since E is a regular neighbourhood of t in M^+, Lemma 4.5 follows.

□

Let W^+ be the union of all Seifert fibre spaces from V_0^+.
If $V_0^+ \cap \partial t = \emptyset$, then Theorem B follows from Lemma 4.5 since $A^+ \neq \emptyset$. We suppose the converse. Then, by Lemma 4.5, W^+ is non-empty. Every homeomorphism of M^+ which preserves V^+ also preserves W^+. Therefore we may suppose that the homeomorphism $h^+ : M^+ \to M^+$ is isotoped so that $h^+(W^+) = W^+$ (uniqueness of the characteristic submanifold).

We now distinguish between two cases.

Case 1. No component of W^+ is the S^1-bundle over the annulus.

For every component, X of W^+, let $n_\alpha(X)$, resp. $n_\beta(X)$, be the number of all annuli from αA^+, resp. βA^+, contained in X, and denote

$$n_\alpha := \sum n_\alpha(X), \text{ resp.} n_\beta := \sum n_\beta(X),$$

where the sums are taken over all components of W^+. There is an annulus from αA^+ which is mapped under h^+, resp. $(h^+)^2$, to an annulus from αA^+ again, provided $n_\alpha > n_\beta$, resp. $n_\alpha = n_\beta$. Thus it suffices to show that, for every component, X, from W^+,

$$n_\alpha(X) \geq n_\beta(X), \text{ or } n_\alpha(X) > n_\beta(X),$$

according to whether we are in the proof of Theorem B or its addition.

Let X be any component of W^+. First we claim that X is a solid torus. To see this recall that X is a Seifert fibre space. Take a boundary component of X, or rather some torus in the interior of X and parallel to this boundary component. Observe that the latter torus cannot be boundary-parallel in M^+ since V^+ is complete and since we are in Case 1. It follows that the torus is compressible (M^+ is atoroidal). From the irreducibility of M^+ we then conclude that X is a solid torus proving our claim.

Now, $(\partial X - \partial M^+)^-$ has to be a non-empty system of annuli, i.e. $X \cap A^+ \neq \emptyset$. Thus $n_\alpha(X) > n_\beta(X) = 0$, if $t \cap X = \emptyset$. Therefore we may suppose that $t \cap X \neq \emptyset$. Let B_1, B_2 (possibly $B_1 = B_2$) be the two disjoint, essential annuli in X as given in Lemma 4.5. Without loss of generality, the annuli B_1, B_2 are vertical in X (see [Johannson, 1979, 5.6]) and therefore they project down to essential arcs, b_1, b_2, in the orbit-disc, F, of X (which is a disc with exceptional point, if X has an exceptional fibre). Furthermore, observe that each component of $X \cap \partial M^+$ contains at most one component of $\partial B_1 \cup B_2$ since the attaching curve k is supposed to be separating in ∂M (i.e. t joins different components of ∂M^+). Checking the possible configurations of $b_1 \cup b_2$ in the (possibly punctured) disc F (with respect to its boundary-pattern induced by $X \cap \partial M^+$; see [Johannson, 1979]), it is now easy to see that $n_\alpha(X)$ and $n_\beta(X)$ satisfy the required inequalities.

Case 2. At least one component of W^+ is the S^1-bundle over the annulus.

If ∂t is contained in $V^+ - V_0^+$, the Theorem B follows since then, by Lemma 4.5, $t \cap W^+ = \emptyset$, but $A^+ \cap W^+ \neq \emptyset$. Thus we may suppose the converse. We may then follow the argument of Case 1 again. Indeed, a moment's reflection shows we find the required surface either as a component of $(\partial W^+ - \partial M^+)^-$, applying a similar counting argument as used in Case 1, or, by a direct (and simple) search of appropriate, essential annuli in those components of W^+ which are S^1-bundles over the annulus. □

5. An Application to Dehn-Surgeries

In this chapter we are going to apply Theorem A to the question under which conditions Dehn-surgeries lead to Haken 3-manifolds.

Let us recall the definition of Dehn-surgeries. For this let us be given a closed 3-manifold, M, together with a simple closed curve, k, in M and some specified non-trivial, simple closed curve, t, in $\partial U(k)$. Here $U(k)$ denotes the regular neighbourhood of k in M. Furthermore, let V be the solid torus, and let s be some specified meridian curve in ∂V. We say that a 3-manifold, N, is obtained from M by a Dehn-surgery along k, or, more precisely, by the $(k,\ t)$-*Dehn-surgery*, if N is homeomorphic to a 3-manifold obtained from $(M - U(k))^-$ by attaching V to it, via a homeomorphism $h: \partial V \to \partial (M - U(k))^-$ with $h(s) = t$.

In recent time, it has been shown by various authors that Dehn-surgeries along knots in S^3 and certain Stallings fibrations do not necessarily lead to Haken 3-manifolds. Theorem C below can be regarded as some result in the other direction.

To formulate Theorem C suppose that the curve $k \subset M$ is chosen so that $\tilde{M} := (M - U(k))^-$ is an irreducible and boundary-irreducible 3-manifold (e.g. k a non-trivial knot in S^3). Furthermore, let t_0 be that non-trivial curve in $\partial\tilde{M}$ which is zero-homologous in $\tilde{M}$. It is well-known that such a curve t_0 always exists and that it is unique, up to isotopy in $\partial\tilde{M}$. More precisely, t is always the boundary of some orientable and incompressible surface in $\tilde{M}$.

Theorem C. *Under the previous conventions, either the (k, t_0)-Dehn-surgery, or all $(k,\ t')$-Dehn-surgeries lead to Haken 3-manifolds, where t' denotes the simple closed curve in $\partial\tilde{M}$ whose intersection number with t_0 is neither* 0 *nor* ± 1.

Proof. By our choice of t_0, the curve t_0 is boundary of some (orientable) incompressible surface, S, in $\tilde{M}$. The regular neighbourhood, $U(S)$, of S in $\tilde{M}$ carries a product structure, $S \times I$, with $(S \times I) \cap \partial\tilde{M} = \partial S \times I$. Consider the manifold

$$N := (\tilde{M} - U(S))^-.$$

Case 1. N is boundary-irreducible.

Let t' be any simple closed curve in $\partial\tilde{M}$ whose intersection number with t_0 is different from 0 and ± 1. Denote by $M^\star$ the manifold obtained from M by $(k,\ t')$-Dehn-surgery. We claim $M^\star$ is a Haken 3-manifold, i.e. we have to show that $M^\star$ is irreducible and contains at least one incompressible surface.

Consider the complement, N_0, of N in $M^\star$. Then N_0 is the union of the product I-bundle $U(S) = S \times I$ and the solid torus $V = (M^\star - \tilde{M})^-$, which intersect themselves in an annulus $A := U(S) \cap V = \partial S \times I$. Finally, note that t_0 is the core of the annulus A since it is the boundary of the surface S.

Lemma 5.1. *The annulus A is incompressible and boundary-incompressible in N_0.*

Proof. The curve t_0 does not bound a disc in $U(S)$, for $\tilde{M}$ is boundary irreducible and so S not a disc. It also does not bound a disc in V, for otherwise both t_0 and t' would be zero-homologous in V, and so their intersection number would be zero. This proves that A is incompressible in N_0. Now, assume A is boundary compressible in N_0. Then there is a disc, D, either in $U(S)$ or in V such that $D \cap A$ is an arc joining the two boundary curves of A. In particular, the intersection number of t_0 and ∂D is ± 1. Hence D has to be contained in V, for t_0 is zero-homologous in $U(S)$. Since ∂D intersects t_0 in precisely one point, it cannot bound a disc in ∂V, and so it is a meridian disc in V. But any two meridian discs in a solid torus are isotopic [Waldhausen, 1967] so t' and ∂D are homologous, and so the intersection number of t_0 and t' would be ± 1. This again contradicts our choice of t', proving Lemma 5.1.

□

Now, by [Alexander, 1924] and [Waldhausen, 1967], $U(S)$ and V are irreducible, and so is N_0. Hence, again by [Waldhausen, 1967], also $M^\star$ is irreducible, provided ∂N is incompressible in $M^\star$. Moreover, in the latter case, $M^\star$ is also sufficiently large and hence a Haken 3-manifold. Since we are in Case 1, ∂N is incompressible in $M^\star$, if N_0 is boundary-irreducible.
To see that N_0 is indeed boundary-irreducible, let D be any disc in N_0 with $D \cap \partial N_0 = \partial D$. It remains to show that D is boundary-parallel in N_0. For this purpose, let D be isotoped so that $D \cap A$ consists of curves whose number is as small as possible. Since A is incompressible and since $U(S)$ as well as V is irreducible, $D \cap A$ consists of arcs (innermost-disc-argument).

Suppose $D \cap A = \emptyset$. In this case, D is contained either in $U(S)$ or in $V - A$. But S is incompressible in $U(S) = S \times I$ and so D is either boundary-parallel, or it has to lie in V. Now, the disc D cannot be a meridian disc in V, for otherwise the complement of D in V would be a 3-ball containing A which contradicts the fact that A is incompressible. Hence, in any case, D is boundary-parallel, if $D \cap A = \emptyset$.

Suppose $D \cap \neq \emptyset$. In this case there is at least one arc of $D \cap A$ which separates a disc, D_1, from D not containing any other curve from $D \cap A$. In particular, ∂D_1 meets only one boundary curve of A, for A is boundary-incompressible. Furthermore, D_1 is contained either in V or in $U(S)$ and, by what we have just seen, D_1 can be isotoped in one of these manifolds (modulo boundary) so that afterwards it does not meet A any more. Hence, by the argument of the previous paragraph, D_1 is boundary-parallel, and we get a contradiction to our minimality condition on $D \cap A$.

This proves Theorem C in Case 1.

Case 2. N is boundary-reducible.

Denote by $M^\star$ the manifold obtained from M by (k,t_0)-Dehn-surgery. Then the solid torus $V = (M^\star - \tilde{M})^-$ contains a disc, D, with $D \cap \partial V = \partial D$ and $\partial D = \partial S$. Define $S^+ := S \cup D$, and denote by $U(S^+)$ the regular neighbourhood of S^+ in $M^\star$. It suffices to show that $(M^\star - U(S^+))^-$ is irreducible and boundary-irreducible, for in this case S^+ is incompressible in $M^\star$, i.e. $M^\star$ is sufficiently large, and, by [Waldhausen, 1967] $M^\star$ is irreducible since $U(S^+)$ and $N^\star := (M^\star - U(S^+))^-$ are.

Now, $N^\star = (M^\star - U(S^+))^-$ is the union of the irreducible but boundary-reducible (we are in Case 2) manifold N and a 3-ball attached to N along the annulus $A = (\partial\tilde{M} - U(S^+))^- = N \cap \partial\tilde{M}$. This means that we are in the situation of Theorem A. It remains to show that $(\partial N^\star - A)^-$ is incompressible in N. But $(\partial N - A)^-$ is equal to $U(S) \cap \partial N$ and therefore it consists of two copies of S. Now, S is supposed to be incompressible in $\tilde{M}$, and so our claim follows.

This proves Theorem C in Case 2.

□

Appendix: On Decomposition of Groups

In certain situations (see e.g. corollary 3 in Section 2 of this paper) the question arises whether a given group G can be decomposed into a (non-trivial) free product $A \star B$ of groups. However, as for so many problems in combinatorial group theory, also this question is known to be recursively undecidable in general [Rabin, 1958]. Therefore, for special classes of groups only, one can hope to answer the question of their indecomposability. For example, it follows from geometric methods that surface groups, knot groups etc. are indecomposable. For one-relator groups this depends on the choice of the relator ([Shenitzer, 1955] has given some concrete sufficient conditions, by purely algebraic reasoning). In light of Theorem A, Remark, still other examples can be produced.

Here is a general procedure (which easily can be generalised to some other cases for the relators).

Proposition 6.1. *Let G be a group which is a free product $G = G_1 \star G_2 \star \ldots \star G_n$, where each G_i is the fundamental group of a Haken 3-manifold, M_i. Let w_i be a word in the presentation of G_i whose conjugacy class is represented by a simple closed curve in ∂M_i. Then adding $w_1 w_2 \ldots w_n = 1$ to the relators of G gives an indecomposable group G'.*

Proof. Simply check that G' is the fundamental group of a 3-manifold obtained from the M_i's by boundary-connected-sum construction and an additional attaching of a 2-handle. Show that the attaching curve represents (the conjugacy class of) $w_1 w_2 \ldots w_n$ and can be chosen so that it satisfies the hypothesis of Theorem A. Then the proposition follows from Theorem A, remark.

□

Since knot spaces are special examples of Haken 3-manifolds, we obtain the following corollary as an immediate consequence of our previous proposition.

Corollary 6.2. *Let G_1, G_2 be two knot groups with Wirtinger presentations $\langle A_i \mid R_i \rangle$, $i = 1, 2$, then*

$$\langle A_1 \cup A_2 \mid R_1 \cup R_2 \cup a_1^{\pm 1} a_2^{\mp 1} \rangle,$$

for all $a_1 \in A_1$ and $a_2 \in A_2$, is not a (non-trivial) free product $A \star B$ of groups.

In other words, by simply adding e.g. $a_1 = a_2$, for all $a_1 \in A_1$ and $a_2 \in A_2$, to the relations of $G_1 \star G_2$ we always can destroy the decomposition.

Finally note that a one-relator group is the fundamental group of a handlebody, M, with a 2-handle attached along a loop, k, in the boundary, ∂M. Here k represents the conjugacy class of the relator, and we may distinguish between simple and non-simple relators according as they can be represented by simple closed curves $k \subset \partial M$ or not. For simple relators we obtain from Theorem A (remark) Shenitzer's criterion for one-relator groups. This reads for simple relators, represented by a simple closed curve, $k \subset \partial M$, as follows:

A one-relator group with simple relator is indecomposable, if $\partial M - k$ is incompressible in the handlebody, M.

In this context observe that, for handlebodies, M, and simple closed curves, $k \subset \partial M$, there is a useful criterion for the question, whether $\partial M - k$ is incompressible in M. Simply cut ∂M along some fixed system of meridian curves into a 2-sphere with holes S, and consider the trace $k' := k \cap S$

of k in S. Then $\partial M - k$ is incompressible in M, if (S, k') is irreducible, in the sense that every curve, t, in S with $t \cap \partial_i S = \partial t$, for some boundary curve, $\partial_i S$, of S, and $t \cap k' = \emptyset$, separates a disc, D, from S with $D \cap k' = \emptyset$. This last condition can always be checked in a finite number of steps.

References

Alexander, 1924.
Alexander, J.W., "On the subdivision of 3-space by a polyhedron," *Proc. Nat. Acad. Sci.*, **10**, pp. 6-8, 1924.

Floyd.
W.J. Floyd, *Hyperbolic manifolds obtained by adding a 2-handle to a 3-manifold with compressible boundary.* preprint

Jaco, 1984.
W. Jaco, "Adding a 2-handle to a 3-manifold: an application of Property R," *Proc. A.M.S.*, **92**, pp. 288-292, 1984.

Johannson, 1979.
K. Johannson, "Homotopy equivalences of 3-manifolds with boundary," *Springer-Verlag Lecture notes in Math*, **761**, 1979.

Przytycki.
Przytycki, J., *Incompressible surfaces in 3-manifolds.* thesis

Rabin, 1958.
Rabin, M.D., "Recursive unsolvability of group theoretic problems," *Ann. of Math.*, **67**, pp. 172-194, 1958.

Shenitzer, 1955.
Shenitzer, A., "Decomposition of a group with a single defining relation into a free product," *Proc. Amer. Math. Soc.*, **6**, pp. 273-279, 1955.

Stallings, 1962.
J.R. Stallings, "On fibering certain 3-manifolds," in *Topology of 3-manifolds*, pp. 95-100, Prentice Hall, 1962.

Stallings, 1971.
Stallings, J.R., *Group theory and three dimensional manifolds,* Yale Univ. Press, 1971.

Waldhausen, 1967.
Waldhausen, F., "Eine Klasse von 3-dimensionalen Mannigfaltigkelten. I & II," *Inv. math*, **3 & 4**, 1967.

Waldhausen, 1968.
Waldhausen, F., "On irreducible 3-manifolds which are sufficiently large," *Annals of Math.*, **87**, pp. 56-88, 1968.

K. Johannson,
Rice University,
Texas 77001,
USA.

Determining Knots by Branched Covers

*Sadayoshi Kojima**

1. Introduction

A branched cover of a knot in the 3-sphere has been one of the fundamental objects in knot theory. There are several distinct knots whose n fold cyclic branched covers are homeomorphic. See for example [Birman-González-Acuña-Montesinos, 1976], [Takahashi, 1977], for $n = 2$ and [Viro, 1979], [Nakanishi, 1981], [Sakuma, 1981] for another n. On the contrary, our purpose is to prove

Theorem. *For each prime knot K in S^3, there exists $N_K > 0$ such that two prime knots K, K' are equivalent (in the weakest sense) if their n fold cyclic branched covers are homeomorphic for some $n >$ $\max(N_K, N_{K'})$.*

Corollary. *Two prime knots which have infinitely many homeomorphic cyclic branched covers are equivalent (in the weakest sense).*

This answers Problem 1.27 in [Kirby, 1978] by Goldsmith. We remark that there is no universal bound of N_K, which can be known from the examples by Nakanishi [Nakanishi, 1981] and Sakuma [Sakuma, 1981]. However, it is still unknown whether infinitely many branched covers are needed to determine the knot.

The theorem is not valid for non prime knots, but we can claim the same assertion for each prime factor of a knot as its corollary. The precise statement and examples are in the last section.

The proof is based on the Jaco-Shalen Johannson decomposition of branched covers and characteristicity of the shortest geodesic in hyperbolic

* Partially supported by the Sakkokai Foundation

Dehn surgery theory. The next section is for preliminaries. We study a similar property of unbranched covers in Section 3 and then prove the theorem in Section 4.

2. Preliminaries

We will use the torus decomposition in the sense of Jaco-Shalen [Jaco-Shalen, 1979] and Johannson [Johannson, 1979], and Thurston's uniformization theorem [Morgan-Bass, 1984]. They assert for the manifolds which show up in this paper that every one is uniquely decomposed by tori into geometric pieces each of which either is a Seifert fibred space or admits a complete hyperbolic structure of finite volume in its interior. This is the form of the uniformization we use hereafter.

Let us recall some terminologies in knot theory. Let K be an unoriented knot in oriented S^3. Two knots are equivalent (in the weakest sense) if there is a homeomorphism of S^3 which takes one knot to the other. By $E(K)$, we mean the exterior $S^3 - \text{int } N(K)$ of K. Note that $\partial E(K)$ is incompressible unless K is trivial. Let $E_n(K)$ denote the n fold cyclic cover of $E(K)$ and let $B_n(K)$ denote the n fold cyclic branched cover of S^3 branched along K. $B_n(K)$ is topologically the union of $E_n(K)$ and the solid torus.

Let us think of some properties of the torus decomposition of $E(K)$ in the sense of Jaco-Shalen Johannson. First assume that K is non trivial. Then a decomposed piece of $E(K)$ either is a Seifert fibred space over a hyperbolic orbifold or admits a complete hyperbolic structure of finite volume in its interior. Since a splitting torus in the decomposition is in S^3, it is separating. Hence the graph associated to the decomposition is a tree. Furthermore, since any embedded torus in S^3 bounds a solid torus, the torus in the decomposition splits $E(K)$ into some knot exterior and the exterior of a knot in the solid torus.

By the length of the decomposition of $E(K)$, we mean the maximal length of a sequence of proper exclusions: $E(K) = E_1 \supset E_2 \supset \cdots \supset E_S \supset \emptyset$, where each E_j is a nontrivial knot exterior split by some torus of the decomposition. The decomposition of $E_n(K)$ in the sense of Jaco-Shalen Johannson is clearly obtained by lifting tori of the decomposition of $E(K)$. We define the length of the decomposition of $E_n(K)$ to be the length of the decomposition of $E(K)$. It stands for the minimal number of intersections between the tori of the decomposition and a path from the boundary to a component having a connected boundary.

There is a usual definition of a system of meridian and longitude on $\partial E(K)$. We will define such a canonical system of simple loops on an incompressible torus in $E_n(K)$ which covers an incompressible torus in $E(K)$. We define them to be the usual system of the knot defined by the core of the solid torus in S^3 bounded by T. For an incompressible torus in $E_n(K)$ which covers T, just take the simple loops which cover the loops defined in $E(K)$. We will call them also a system of meridian and longitude. Since S^3 is oriented, it induces the orientation on each incompressible torus in $E(K)$ as a boundary of a knot complement and hence on each lift in $E_n(K)$ and $B_n(K)$. We adopt the convention that the ordered pair (meridian, longitude) is compatible with this orientation on each torus. Note that we did not specify the direction of loops and we did specify the order. Hence a direction of one of them defines the other's. We say a homeomorphism preserves meridian-longitude systems if it maps a meridian to a meridian and a longitude to a longitude.

Now let J be a family of incompressible tori in $E(K)$ to decompose it in the sense of Jaco-Shalen Johannson and let M be the component of $E(K) - \text{int } N(J)$ which contains $\partial E(K)$. This section is in fact for studying several geometric properties of M.

We first think of the case when M is a Seifert fibred space. The Seifert fibred spaces which show up later have a really good uniqueness property with only a few exceptions. Namely, a fibration is topologically unique when it is defined over a hyperbolic orbifold. Moreover, if $\partial \neq \emptyset$, a fibration itself is unique up to isotopy. Hence the direction of a regular fibre can be specified. A proof of these facts can be found for instance in [Jaco, 1977].

Lemma 1. *If M is a Seifert fibred space, then K is either*

(1) *a (p, q)-torus knot: p and q are positive coprime integers with $p > q > 1$,*

(2) *a (p, q)-cable of a nontrivial knot: p and q are coprime integers with $q \geq 2$ (p may be negative) or*

(3) *a composite knot.*

Furthermore if K is prime, a Seifert fibration of M extends to a tubular neighbourhood of K.

Proof. The first half is almost the same as Lemma 4.1 of [Gordon, 1983]. The latter half is then obvious since a Seifert fibration of M for this case is given by a saturated restriction of some Seifert fibration of S^3 or $S^1 \times D^2$. □

Let M_n be the pre-image of M in $E_n(K)$ and let $\overline{M}_n$ be the n fold cyclic branched cover of $M_n \cup N(K)$ branched along K, which is embeded-ded in $B_n(K)$. This is sometimes called the completion of M_n.

Lemma 2.

(1) *If K is the (p,q)-torus knot, then $E_n(K)$ has a Seifert fibration with (q,n) singular fibres of index $p/(p,n)$ and (p,n) singular fibres of index $q/(q,n)$. Its orbit orbifold has Euler characteristic $=$ $(pq,n)(1/p+1/q-1)$. A regular fibre represents $pq/(pq,n) \cdot$ meridian $+\ n/(pq,n) \cdot$ longitude. This extends to a fibration on $B_n(K)$ by adding a neighbourhood of a singular fibre of index $n/(pq,n)$. The Euler characteristic of the orbit orbifold is $(pq,n)(1/p+1/q+1/n-1)$.*

(2) *If K is the (p,q)-cable of the nontrivial knot, then M_n has a Seifert fibration with orbit Euler characteristic $= (pq,n)(1/q-1)$. A regular fibre on each boundary of M_n except for $\partial E_n(K)$ represents $p/(p,n) \cdot$ meridian $+\ qn/(pq,n) \cdot$ longitude. This also extends to a fibration on $\overline{M}_n$ with orbit Euler characteristic $= (pq,n)(1/q+1/n-1)$.*

(3) *If K is composite, then M_n is a Seifert fibred space homeomorphic to M.*

Proof. Fairly easy combinatorics! □

The next lemma is to find a constant of the theorem when M is Seifert. The estimate is not best possible and can certainly be sharpened, however we adopt it because of a simple argument.

Lemma 3. *Let K be a (p,q)-cable of a (possibly trivial) knot and suppose $n > |pq|$. Then*

(1) *the quotient space of the semi-free $\mathbb{Z}/n\mathbb{Z}$ action on $\overline{M}_n$ admits a Seifert fibration so that the image of the fixed point set consists of fibres.*

(2) *$\overline{M}_n$ cannot be homeomorphic to any one of M_n's of cable knots.*

Proof. Let us prove (1) first. Since $n > |pq| \geq 6$, $\overline{M}_n$ is a Seifert fibred space over a hyperbolic orbifold and hence any finite group action on $\overline{M}_n$ is supposed to be fibre-preserving by Meeks-Scott [Meeks-Scott, 1984]. In particular, since a semi-free $\mathbb{Z}/n\mathbb{Z}$ action with $n > 6$ does not reflect any fibre, the Seifert fibration on $\overline{M}_n$ descends to one on the quotient space so that the image of the fixed point set consists of fibres.

To see (2), suppose that $\overline{M}_n$ is homeomorphic to M_n of a (p',q')-cable of a (possibly trivial) knot. Then since the index of the fibre in $\overline{M}_n$ turned out by the completion is $n/(pq,n)$, it is equal to either $p'/(q',n)$, $q'/(q',n)$ or 1. However any one of them is coprime to n, and therefore $n/(pq,n)$ must be equal to 1. This contradicts the assumption, $n > |pq|$.

□

We are now concerned with the case when int M admits a complete hyperbolic structure. By uniformity of the thin part of complete hyperbolic manifolds observed by Kazhdan-Margulis, we can find the "shortest" closed geodesic in a complete hyperbolic manifold with short geodesics. Recall that a closed geodesic is short if its length is less than the Margulis constant. The shortest one is actually the shortest among all the closed short geodesics and corresponds to the axis of some loxodromic transformation.

The following lemma is an implicit corollary of Thurston's hyperbolic Dehn surgery

Lemma 4. *If* int M *admits a complete hyperbolic structure of finite volume, then there exists* $N_1 > 0$ *such that for* $n > N_1$, int $\overline{M}_n$ *admits a complete hyperbolic structure with respect to which the branched covering translations act isometrically, and in which the branched circle* $\tilde{K}$ *is the unique shortest closed geodesic. Furthermore, the length of this geodesic approaches zero if* n *goes to infinity.*

Proof. Think first of incomplete hyperbolic structures on int M defined by Thurston's hyperbolic Dehn surgery deformation which deforms to be incomplete only on a neighbourhood of the cusp corresponding to K. Then the structure on the x-axis of the deformation space yields topologically the original manifold $\overline{M}$ by completion. For the integral point $(n,0)$ in the deformable range, the completed structure has singularities at K with cone angles of $2\pi/n$ and hence it induces a non singular hyperbolic structure on the n fold branched cover $\overline{M}_n$. Then Thurston's theorem shows that there is $N_0 > 0$ such that the deformation range contains integral points $(n,0)$

for all $n > N_0$. Since the structures on integral points $(n,0)$ converges algebraically to the complete structure, the length of the geodesic attached by the completion (note that this is well defined!) approaches zero when n goes to infinity.

Let us find N_1. Let ϵ be the minimum of the Margulis constant and the infimum of the length of closed geodesics in int M. Notice that ϵ is a positive number and is attained by the shortest geodesic if it is strictly less than the Margulis constant. Let N be a small horoball neighbourhood of the cusp corresponding to K. Then the infimum of the length of closed geodesics in int $M - N$ turns out a continuous function parametrised by hyperbolic Dehn surgery deformation around K. Thus there is a constant $N_1 > N_0 > 0$ such that for any $n > N_1$, the infimum of the length of closed geodesics in int $M - N$ with respect to a hyperbolic structure at $(n,0)$ is greater than $\epsilon/2$, and simultaneously the length of the geodesic attached by the completion is less than $\epsilon/2$. These are the properties of N_1.

Let us then show that N_1 defined above does have the property in the statement. The n fold branched cover int $\overline{M}_n$ is the union of the n fold unbranched cover of int $M - N$ and the n fold branched cover $\overline{N}_n$ of N branched along K. For $n > N_1$, the latter part forms a neighbourhood of a short geodesic $\tilde{K}$ attached by the completion. Think of the thin part of int $\overline{M}_n$ in the sense of Margulis, which is being a disjoint union of neighbourhoods of cusps and short geodesics. The short geodesics of int $\overline{M}_n$ which is supposed to be in this part do not intersect $\partial\overline{N}_n$ since $\overline{N}_n$ forms a small neighbourhood of the short geodesic $\tilde{K}$. Hence they are in int $\overline{M}_n - \overline{N}_n$ except for $\tilde{K}$. Now since int $\overline{M}_n - \overline{N}_n$ is an unbranched cover of int $M - N$, the infimum of the length of closed geodesics there is greater than $\epsilon/2$. Hence $\tilde{K}$ turns out to be the unique shortest geodesic in int $\overline{M}_n$. □

Corollary 5. *Suppose that K is prime. If either M is Seifert and $n > 2$ or M is hyperbolic and $n > N_0$, then a family of tori which decomposes $B_n(K)$ in the sense of Jaco-Shalen Johannson consists of all preimages of tori which decompose $E(K)$. Here N_0 is a constant given in the proof of Lemma 4.*

Proof. The torus decomposition of $E_n(K)$ is obtained by lifting the decomposition of $E(K)$ because geometric structures are lifted to covering spaces. Hence we only need to study geometric effect of gluing a solid torus to M_n. However if K is prime, then this device extends a geometric structure on M_n to $\overline{M}_n$ by lemma 2 and 4.

∎

In view of this lemma, it may be reasonable to define the length of the torus decomposition of $B_n(K)$ by the length of $E_n(K)$ if n is as in the statement.

Lemma 6. *If* int M *admits a complete hyperbolic structure of finite volume, then there exists* $N_2 > N_1 > 0$ *such that for* $n > N_2$*, a cyclic group* $\mathbb{Z}/n\mathbb{Z}$ *acts isometrically and semi-freely on* $\overline{M}_n$ *only by covering translations of the branched cover.*

Proof. Since int M admits a complete hyperbolic structure of finite volume, its isometry transformation group is a finite group and is isomorphic to $\operatorname{out}\pi_1(M)$. Now let N_2 be the maximum of N_1 and $|\operatorname{out}\pi_1(M)|$ and suppose that for some $n > N_2$, a cyclic group $\mathbb{Z}/n\mathbb{Z}$ acts isometrically and semi-freely on $\overline{M}_n$. Then since n is at least N_1, the branched circle $\tilde{K}$ is the unique shortest closed geodesic and hence the action must leave it invariant. Suppose that the action in question fixes it. Then it must be the action of covering translations since it acts isometrically on $\overline{M}_n$ and hence it coincides with the action of covering translations at least on a neighbourhood of $\tilde{K}$ and on everywhere by analyticity. Thus we are done.

So let us assume that it does not fix $\tilde{K}$. Since the action is semi-free, it effectively rotates $\tilde{K}$. Now our claim is that this action induces an effective $\mathbb{Z}/n\mathbb{Z}$ action on M. To see this it suffices to show that every element of our $\mathbb{Z}/n\mathbb{Z}$ action normalizes the group of covering translations. But they actually commute because they can be lifted to a loxodromic transformation and an elliptic transformation of the universal cover with a common axis. Thus we get an effective $\mathbb{Z}/n\mathbb{Z}$ action on $M \cup N(K)$ leaving K invariant and hence on M. However this is a contradiction since any finite group of order greater than $|\operatorname{out}\pi_1(M)|$ cannot act effectively on M.

∎

3. Unbranched Covers

This section is to prove the following proposition.

Proposition 7. *Let K and K' be knots in S^3. If $E_n(K)$ is homeomorphic to $E_n(K')$ for some $n \geq 1$ preserving meridian-longitude systems on the boundaries, then K is equivalent to K' (in the weakest sense).*

Proof. When $n = 1$, the statement is obvious. For $n > 1$, it suffices to show that there is an equivariant homeomorphism of $E_n(K)$ to $E_n(K')$ with respect to the action of covering translations. We will see this by induction on the length of the decomposition of $E_n(K)$ in the sense of Jaco-Shalen Johannson.

Suppose that the length of the decomposition of $E_n(K)$ is one. Then $E_n(K)$ either is a Seifert fibred space or admits a complete hyperbolic structure of finite volume in its interior. When it is Seifert, K and K' must be torus knots of type (p, q) and (p', q') respectively since otherwise we cannot have $E_n(K)$ admitting a Seifert fibration. Since a Seifert fibration on $E_n(K)$ is topologically unique, the Euler characteristic of the orbit space as an orbifold is a topological invariant. Thus because $E_n(K)$ is homeomorphic to $E_n(K')$, we have the identity,

$$(pq,n)(1/p + 1/q - 1) = (p'q', n)(1/p' + 1/q' - 1).$$

Also a given homeomorphism must preserve the classes of regular fibres by the uniqueness of fibrations. In particular, the number of intersections of a fibre and a meridian on each boundary is identical, and we have

$$n/(pq,n) = n/(p'q',n).$$

Then it is not hard to show that $(p, q) = (p', q')$ and we are done.

Now consider the case when int $E_n(K)$ is hyperbolic. Then a given homeomorphism of $E_n(K)$ to $E_n(K')$ is homotopic to a map which is isometric on their interior by Mostow's rigidity theorem. It turns out to be an equivariant homeomorphism, because it preserves meridian-longitude systems and it sends a covering translation of int $E_n(K)$ to a covering translation of int $E_n(K')$ on neighbourhoods of cusps and hence on everywhere by analyticity.

Now suppose that the length of the decomposition of $E_n(K)$ is more than 1 and that there is a homeomorphism of $E_n(K)$ to $E_n(K')$ which preserves meridian-longitude systems on the boundaries. Then by the uniqueness of the decomposition, M_n must be mapped to M'_n, the geometric piece of $E_n(K')$ containing $\partial E_n(K')$, because they are only components facing the boundaries. Now M_n either is a Seifert fibred space or admits a complete hyperbolic structure of finite volume in its interior.

Let us then think of the case when M_n is Seifert. There are two alternative possibilities, (2) and (3) of Lemma 1. We may assume that a given homeomorphism is a fibre preserving map with respect to Seifert fibrations on M_n and M'_n. For the case (3), our map restricted to M_n is isotopic to an equivariant map since the covering translations fix each fibre. Furthermore since it preserves meridian-longitude systems on each boundary of M_n, we are done by induction. Thus assume that K is a (p, q)-cable of a nontrivial knot. Then K' must be a (p', q')-cable of some nontrivial knot. Again since a Seifert fibration is topologically invariant, we have the identity,

$$(pq,n)(1/q - 1) = (p'q',n)(1/q' - 1).$$

Since our homeomorphism preserves the fibres by the uniqueness of fibrations and also preserves meridian-longitude systems on $\partial E_n(K)$ and $\partial E_n(K')$ by assumption, we have two identities,

$$pq/(pq,n) = p'q'/(p'q',n) \quad \text{and}$$

$$n/(pq,n) = n/(p'q',n).$$

Then it is not to hard to check that $(p, q) = (p', q')$. Since the quotients are the same, there is an equivariant homeomorphism of M_n to M'_n. It in turn preserves meridian-longitude systems on each boundary of M_n and M'_n. It preserves meridians on each boundary, since it is equivariant, and it preserves fibres by the uniqueness of fibrations, and hence longitudes.

So if this map extends to a homeomorphism of $E_n(K)$ to $E_n(K')$, we are done by induction. To see this, notice that the given homeomorphism of $E_n(K)$ to $E_n(K')$ preserves meridian-longitude systems on each boundary of M_n and M'_n, because on each boundary in question, it preserves longitudes by homological reason and fibres for the uniqueness of fibrations, and therefore meridians. Thus we had a homeomorphism of $E_n(K) - M_n$ to $E_n(K') - M'_n$ which preserves meridian-longitude systems on each boundary. This is a required map to extend an equivariant map of M_n to M'_n.

Let us now think of the case when int M_n admits a complete hyperbolic structure of finite volume. The argument is almost the same as before. A given homeomorphism restricted to M_n is homotopic to a map which is isometric on their interiors by Mostow's rigidity theorem, and it turns out to be an equivariant map of M_n to M'_n. This is because it sends a covering translation of int M_n to that of int M'_n on some neighbourhoods of cusps corresponding to the knots and hence on everywhere by analyticity. On each boundary of M_n and M'_n, it preserves meridians since it is equivariant and longitudes by homological reason. Since it was homotopic to a given map in M_n, a given map of $E_n(K)$ to $E_n(K')$ is homotopic to a homeomorphism which preserves meridian-longitude systems on each boundary of M_n and M'_n, and which is equivariant on M_n. Thus the induction does work out.

$\square$

4. Proof of Theorem

Let us state the main theorem again.

Theorem. *For each prime knot K, there exists $N_K > 0$ such that K is equivalent to a knot K' (in the weakest sense) if $B_n(K)$ is homeomorphic to $B_n(K')$ for some $n > max(N_K, N_{K'})$.*

Proof. If K is trivial, let N_K be 1 and we are done by the solution of the Smith conjecture [Morgan-Bass, 1984], though we have not defined N_K in general yet. Thus assume that K is non-trivial. It suffices to show that there is an equivariant homeomorphism of $B_n(K)$ to $B_n(K')$ with respect to the actions of branched covering translations.

The first case we consider is when K is a (p, q)-cable of some (possibly trivial) knot. Then for any $n > 2$, the length of $B_n(K)$ does make sense. Define N_K to be the product of pq and the maximum among indices of singular fibers of Seifert pieces in $E(K) - M$, and let K' be another knot such that $B_n(K')$ is homeomorphic to $B_n(K)$ for some $n > N_K$. It is worth noting that we do not need N_K' in this case.

Suppose that the length of the decomposition is 1. Namely, K is the (p, q)-torus knot and $B_n(K)$ is a Seifert fibred space. Since (S^3, K') is the quotient orbifold of a semi-free $\mathbb{Z}/n\mathbb{Z}$ action on $B_n(K') \approx B_n(K)$, $S^3 - K'$ admits a Seifert fibration by Lemma 3, (1). Thus we may assume that K' is a torus knot of type, say, (p', q'). Since a Seifert fibration in this case is topologically unique, its orbit Euler characteristic is a topological invariant. Hence we have

$$(pq,n)(1/p+1/q+1/n-1) = (p'q',n)(1/p'+1/q'+1/n-1).$$

The indices of singular fibres are $p/(p,n)$, $q/(q,n)$ and $n/(pq,n)$. Notice that these are pairwise coprime. Hence we have the second identity by multiplying these three integers,

$$pqn/(pq,n)^2 = p'q'n/(p'q',n)^2.$$

Now cancelling n on both sides and replacing each term by $(pq/(pq,n))/(pq,n)$ and $(p'q'/(p'q',n))/(p'q',n)$, we get the identity of irreducible ratio and hence $(pq,n) = (p'q',n)$. Then it is not hard to see that $(p,q) = (p',q')$.

Assume next that the length of the decomposition is more than 1. That is to say, K is a (p,q)-cable of some nontrivial knot. Let us first see that any homeomorphism of $B_n(K)$ to $B_n(K')$ maps $\overline{M}_n$ to $\overline{M}'_n$. By the definition of N, the singular fibre in $\overline{M}_n$ turned out by the completion attains the unique maximal index in the Seifert part of $B_n(K)$. Hence the image of $\overline{M}_n$ by any homeomorphism of $B_n(K)$ to $B_n(K')$ is the unique geometric piece in $B_n(K')$ which contains such a characteristic fibre. In particular, it is invariant by the branched covering translations of $B_n(K')$. Now, suppose it is not $\overline{M}'_n$. Then it does not contain the branched circle of $B_n(K')$, and hence it must be homeomorphic to one of M_n's of cable knots by Lemma 3, (1). But this is impossible by Lemma 3, (2).

We now have a given homeomorphism of $B_n(K)$ to $B_n(K')$ which maps $\overline{M}_n$ to $\overline{M}'_n$. Since K is prime, $\overline{M}_n$ and hence $\overline{M}'_n$ have a Seifert structure described in (2) of Lemma 2. Then by Lemma 3, (1), M' admits a Seifert fibration and we may assume that K' is a cable knot of the type, say, (p',q'). Again since a Seifert fibration is topologically invariant, we have the identity,

$$(pq,n)(1/q+1/n-1) = (p'q',n)(1/q'+1/n-1).$$

Also since our homeomorphism preserves fibres by the uniqueness of fibrations and longitudes by homological reason, it preserves meridian-longitude systems on each boundary of M_n and M'_n. Hence we have two identities,

$$p/(n,p) = p'/(n,p') \quad \text{and}$$

$$qn/(pq,n) = q'n/(p'q',n).$$

Then it is not hard to see $(p,q) = (p',q')$. Thus since the quotients are the same, there is an equivariant homeomorphism of $\overline{M}_n$ to $\overline{M}'_n$. It in turn preserves meridian-longitude systems on each boundary of M_n and M'_n by the same reason as in the proof of Proposition 7.

So if this map extends to a homeomorphism of $B_n(K)$ to $B_n(K')$, then we are done by Proposition 7. But this is now obvious since a given homeomorphism of $B_n(K) - \overline{M}_n$ to $B_n(K') - \overline{M}'_n$ preserves meridian-longitude systems on each boundary by the same reason as in the proof of Proposition 7.

We are now concerned with the proof when int M admits a complete hyperbolic structure of finite volume. We first need to define N_K. Let ϵ_n be the infimum of the length of closed geodesics in hyperbolic components of $B_n(K)$. Then by Lemma 4, there is $N_3 > N_1 > 0$ such that for $n > N_3$, this infimum is uniquely attained by the branched set, where N_1 is a constant of Lemma 4 for M, To see this, note that the infimum restricted to hyperbolic components other than M_n is not less than the infimum in hyperbolic components of $E(K)$. Now let N_K be the maximum of N_2 and N_3, where N_2 is a constant of Lemma 6 for M. Notice that for any $n > N_k$, the length of the decomposition of $B_n(K)$ does make sense. We have then defined N_K for all prime knots.

Let K' be another knot such that $B_n(K')$ is homeomorphic to $B_n(K)$ for some $n > \max(N_K, N_{K'})$. When the length of the decomposition is 1. That is to say, $B_n(K)$ is hyperbolic. Then a given homeomorphism of $B_n(K)$ to $B_n(K')$ is homotopic to an isometry by Mostow's rigidity theorem. This turns out to be an equivariant homeomorphism by Lemma 6.

Thus assume that the length of the decomposition of $B_n(K)$ is more than 1. First we need to see that a given homeomorphism maps $\overline{M}_n$ to $\overline{M}'_n$ of $B_n(K')$. Since a given homeomorphism is homotopic to an isometry on the hyperbolic part by Mostow's rigidity theorem, the image of $\overline{M}_n$ must contain the unique shortest closed geodesic in the hyperbolic part of $B_n(K')$. Since the covering translations on $B_n(K')$ may be assumed to act isometrically on the hyperbolic part, it must leave the unique shortest closed geodesic invariant and hence the component containing it. However by Lemma 6, such a semi-free action there must fix the shortest geodesic and hence this component is nothing but $\overline{M}'_n$.

Moreover we have shown in Lemma 6 that the isometry of $\overline{M}_n$ to $\overline{M}'_n$ is actually equivariant. Hence it preserves meridians on each boundary. Of course it preserves longitudes on each by homological reason. Since it was homotopic to a given map, we get a homeomorphism of $B_n(K)$ to $B_n(K')$

which preserves meridian-longitude systems on each boundary of $\overline{M}_n$ and $\overline{M}'_n$, and which is equivariant on $\overline{M}_n$. Hence the proof is completed by Proposition 7.

□

Corollary. *Let K and K' be knots in S^3. If $B_n(K)$ is homeomorphic to $B_n(K')$ for infinitely many $n > 1$, then the set of prime factors of K is equal to that of K'. In particular, if K is prime, then K is equivalent to K' (in the weakest sense).*

Proof. First of all, K and K' are uniquely decomposed as $K_1 \# \cdots \# K_s$ and $K'_1 \# \cdots \# K'_t$ into their prime factors. Then for any $n > 1$, $B_n(K)$ and $B_n(K')$ are homeomorphic to $B_n(K_1) \# \cdots \# B_n(K_s)$ and $B_n(K'_1) \# \cdots \# B_n(K'_t)$ respectively, each factor of which is prime as a 3-manifold. Since $B_n(K)$ is homeomorphic to $B_n(K')$ for some n, we have $s = t$ by the uniqueness of prime decomposition. Furthermore since they are homeomorphic for infinitely many n, we can conclude by rearranging the order and by taking a subsequence that for each $j = 1, \ldots, s$, $B_n(K_j)$ is homeomorphic to $B_n(K'_j)$ for infinitely many n. Thus the question comes down to the case of prime knots. However this is a direct corollary to the main theorem.

□

We conclude this paper by remarking the following.

Remark 1. N_K in Theorem is not universal. That is to say, given $n > 0$, there is a prime knot K such that N_K is greater than n. Such an example can be found in the papers by Nakanishi [Nakanishi, 1981] and Sakuma [Sakuma, 1981] as mentioned in the introduction. Sakuma informed me that their examples all are hyperbolic except one case. Thus their example shows that N_1 of Lemma 4 is also not universal. Geometrically this means that the branched set in $B_n(K)$ for their knot is indeed a closed geodesic but not the shortest one. However Thurston announced that there is a universal N_0 for knot complements and it is 3 by Dunbar's list of geometric orbifolds other than hyperbolic case.

Lemma 4 holds not only for int M which appears in our argument but also quite in general. For example, if we take the n fold cover of M_n for $n > N_1$, then the shortest closed geodesic in any branched cover of $\overline{M}_n$ branched along $\tilde{K}$ turns out the branched set itself. This shows that N_1 for M_n is less than 1.

Remark 2. There are several ways to construct inequivalent non prime knots all of whose cyclic branched covers are homeomorphic. For instance, if we take non-invertible knots K_1 and K_2 and think of the knots $K_1 \# K_2$ and $K_1 \# -K_2$. Then clearly they have homeomorphic branched covers for any n, however they are not equivalent in any sense. It is due to the phenomenon which arises by changing the (ambient and intrinsic) orientations of the prime factors of a composite knot.

References

Birman-González-Acuña-Montesinos, 1976.
Birman, J., González-Acuña, F., and Montesinos, J., "Heegaard splitting of prime 3-manifolds are not unique," *Michigan Math. J.*, **23**, pp. 97-103, 1976.

Gordon, 1983.
C. McA. Gordon, "Dehn surgery and satellite knots," *T.A.M.S.*, **275**, pp. 687-708, 1983.

Jaco, 1977.
W. Jaco, "Lectures on 3-manifold topology," *CBMS regional conference series in mathematics*, **43**, 1977.

Jaco-Shalen, 1979.
Jaco, W. and Shalen, P., "Seifert fibered spaces in 3-manifolds," *Mem. Amer. Math. Soc.*, **220**, 1979.

Johannson, 1979.
K. Johannson, "Homotopy equivalences of 3-manifolds with boundary," *Springer-Verlag Lecture notes in Math*, **761**, 1979.

Kirby, 1978.
Kirby, R., "Problems in low-dimensional manifold theory," *Proc. Symp. Pure Math.*, **32**, pp. 273-312, 1978.

Meeks-Scott, 1984.
Meeks, W. and Scott, P., *Finite group actions on 3-manifolds*, 1984. preprint

Morgan-Bass, 1984.
J. Morgan and H. Bass, *Proceedings of the Smith Conjecture Symposium: Columbia University 1979.*, Academic Press, 1984.

Nakanishi, 1981.
Nakanishi, T., "Primeness of links," *Math. Semi. Notes*, **9**, pp. 415-440, 1981.

Sakuma, 1981.
Sakuma, M., "Periods of Composite links," *Math. Semi. Notes*, **9**, pp. 445-452, 1981.

Takahashi, 1977.
Takahashi, M., "Two knots with the same 2-fold branched covering space," *Yokohama Math. J.*, **25**, pp. 91-99, 1977.

Viro, 1979.
Viro, O., "Nonprojecting isotopies and knots with homeomorphic coverings," *J. Soviet Math.*, **12**, pp. 86-89, 1979. translated from Zapiski Nauchnykh Seminarov Leningradskogo Otdeleniya Mathematicheskogo Instituta im. V. A. Steklova AN SSSR 66 (1976) pp.133-147

S. Kojima,
Department of Mathematics,
Tokyo Metropolitan University,
Fukasawa,
Setagaya,
Tokyo,
158 JAPAN.

Hyperbolic 3-Manifolds with Equal Volumes but Different Chern-Simons Invariants

Robert Meyerhoff

1. Theorem and proof

Theorem. *Given any rational number in the circle* $\mathbb{R}/(\frac{1}{2}\mathbb{Z})$ *there exist cusped hyperbolic 3-manifolds* M *and* N *with equal volumes whose Chern-Simons invariants differ by that rational number.*

Proof. We will construct cusped hyperbolic 3-manifolds with high-order symmetry and then use this symmetry to prove the theorem. All hyperbolic 3-orbifolds and 3-manifolds will be assumed complete and orientable.

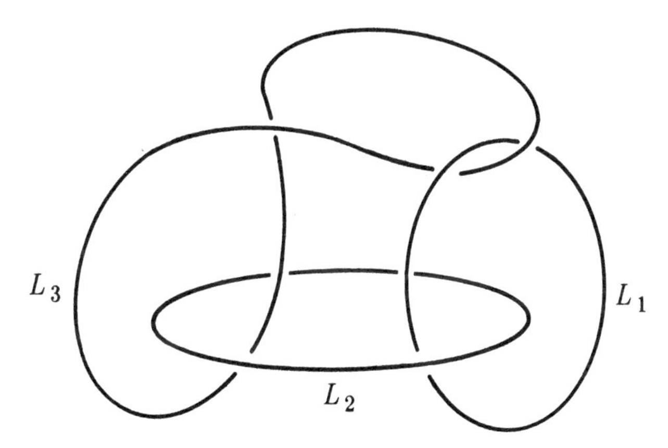

Figure 1.

Consider the 3-component link in the 3-sphere described in Figure 1. This link complement has a hyperbolic structure (see [Thurston, 1979 section 6.8]).

Further, there exists a positive integer A such that hyperbolic Dehn surgery of type $(p,0)$ along L_1 yields a hyperbolic 3-orbifold Q_p for all integers p greater than A (see [Thurston, 1979 Theorem 5.8.2]). Consider the obvious embedded disks D_1 and D_2 spanning L_1 and L_2 respectively in Figure 1. Using D_1 we see that the p-fold cyclic cover of Q_p branched over L_1 is a 3-manifold M_p (see [Rolfsen, 1976 section 5c]). M_p inherits a hyperbolic structure from Q_p, and it has p-fold symmetry – in particular, D_2 has lifted to a disk D'_2 with p-fold rotational symmetry about the point x where D'_2 intersects L'_1. See Figure 2 for the picture in the case $p = 3$. (Lifted objects will generally be denoted by the use of primes.)

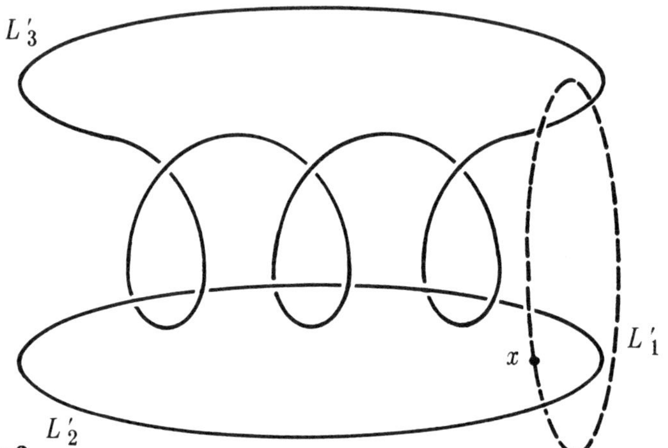

Figure 2.

We now use D'_2 to construct the desired manifolds M and N in the statement of the theorem (assume the rational number is of the form q/p). Cut the manifold $M_p = M$ along D'_2. Since D'_2 has p-fold symmetry about x, we now twist D'_2 by $4\pi q/p$ and reglue, obtaining a new 3-manifold N. N has a hyperbolic structure: we will show that all points n in N have hyperbolic ball neighbourhoods. If n is not on D'_2 then there is an associated point m in M which has a hyperbolic ball neighbourhood B_m not intersecting D'_2. B_m carries over to a hyperbolic ball in N by the twisting and regluing procedure. If n is a point in D'_2 then there is an associated point m in M which has a hyperbolic ball neighbourhood B_m which is cut by D'_2 into two pieces. Project B_m to Q_p and lift back to $M = M_p$. This yields p copies of B_m each intersected by D'_2 in precisely the same way (assuming $m \neq x$, the symmetry point). Cutting, twisting and regluing yields a hyperbolic ball neighbourhood of n in N. Thus N has a hyperbolic structure – the preceding type of argument can be used to show that the structure is complete (see proposition 3.7 of [Thurston, 1979] for a convenient notion of completeness) – and M and N have equal volumes.

What are the Chern-Simons invariants of M and N? It is difficult to compute Chern-Simons invariants; however, by using the "torsion formula" for the Chern-Simons invariant we can compute the difference $CS(M) - CS(N)$. (See sections 2 and 3 for information about the Chern-Simons invariant, special singularities, and the torsion formula; or see [Meyerhoff, 1986].)

We need a particular type of (highly symmetric) frame field on $M = M_p$. By obstruction theory, there exists an orthonormal frame field (see [Meyerhoff, 1986] Theorem 4.3 and Corollary) on Q_p with "special singularities" at $L_1, L_2, L_3, \ldots, L_6$ (see Figure 3); geometrically, at L_1 the angle is $2\pi/p$. This frame field lifts to a special singular frame field on M, where the special singularities are at the lifts of L_1, L_2, L_3 and L_4, and at the p lifts of L_5 and L_6. This frame field carries over to a special singular frame field on N. It is often useful to think of these frame fields as sections s of the orthonormal frame bundle $F(M)$ for the Riemannian manifold M.

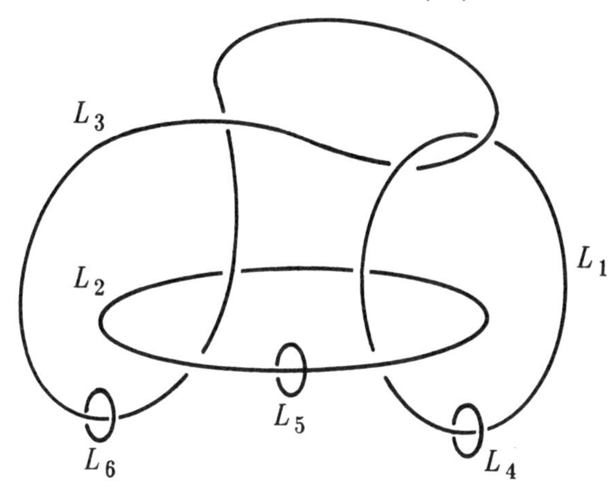

Figure 3.

According to the torsion formula, the Chern-Simons invariant of M is computed by integrating a certain 3-form in $F(M)$ over the image of the section s, and then adding to this $\pi/4$ times the sums of the torsions of the singular curves L'_1, L'_4, L'_5 and L'_6. The cusps L'_2 and L'_3 contribute nothing. The integration term is unchanged in passing from M to N; however, the torsion contributions may be different.

The torsion of a curve K in M is computed by choosing an orthonormal frame field on K with e_1 tangent to K and then integrating the connection form ω_{23} over the associated section of $F(M)$ and reducing this mod 2π. If we choose such a frame field for L_5 while in Q_p and then lift to

$M = M_p$, we see that the frame fields along L'_5 are highly symmetric. In particular, they induce frame fields along the image curves in N after cutting, twisting, and regluing. This is not the case for a frame field along L'_1, because cutting-twisting-regluing produces a broken frame field with a $4\pi q/p$ gap. The torsions of L'_4 and L'_6 are unaffected by cutting-twisting-regluing, because they do not intersect the cutting disc. Thus, the only change in torsion between M and N is contributed by L'_1, and this contribution is $4\pi q/p$. That is,

$$CS(M) - CS(N) = \frac{1}{4\pi} 4\pi q/p = q/p.$$

□

Corollary. *The Chern-Simons invariant for complete hyperbolic 3-manifolds takes on a dense set of values in the circle* $\mathbb{R}/\frac{1}{2}\mathbb{Z}$. This corollary was first proved in. [Meyerhoff, 1981]

Remarks:

1) The volume of a complete hyperbolic 3-manifold can be computed by decomposing the manifold into tetrahedra and computing the volume of each of these, and then summing; the above theorem implies that there is no such "tetrahedral formula" for the Chern-Simons invariant for cusped hyperbolic manifolds. This follows by triangulating (with a vertex at x) the disk D_2 in Q_p, then extending this triangulation over Q_p and lifting to $M = M_p$.

2) The fact that M and N in the above theorem are non-homeomorphic is shown by their differing Chern-Simons invariants. This fact can also be determined by studying the structures of their cusp tori (see [Thurston, 1979 section 6.7] or [Riley, 1970 section 4]), which differ by a shear.

2. The Chern-Simons Invariant: Definitions and Main properties

Given a section, s, of the positively oriented orthonormal frame bundle, $F(M)$, of a closed, orientable Riemannian 3-manifold M, the *Chern-Simons integral* is defined as $(1/8\pi^2)\int Q \pmod 1$ where the integration is performed over $s(M)$ and Q, the *Chern-Simons form*, is the 3-form

$$\omega_{12} \wedge \omega_{13} \wedge \omega_{23} + \omega_{12} \wedge \Omega_{12} + \omega_{13} \wedge \Omega_{13} + \omega_{23} \wedge \Omega_{23}.$$

Here, the ω_{ij}'s are the connection forms and the Ω_{ij}'s are the curvature forms. Such a section always exists for a closed 3-manifold (see, [Steenrod, 1951 section 39.9]).

It can be shown (see [Chern-Simons, 1974]) that the Chern-Simons integral is independent of the section, s. Thus, the Chern-Simons integral is actually an invariant of M, called the *Chern-Simons invariant*, $CS(M)$. Chern and Simons also show that $CS(M)$ is a conformal invariant of the Riemannian 3-manifold M. The 3-manifolds we will be concerned with will have constant sectional curvature -1, i.e., they will be hyperbolic 3-manifolds. By Mostow's theorem complete hyperbolic 3-manifolds of finite volume have unique hyperbolic structures and therefore the Chern-Simons invariant is a topological invariant for closed hyperbolic 3-manifolds.

3. The Extended Chern-Simons Invariant

The immediate problem in the computation of the Chern-Simons invariant is to find an explicit section of the orthonormal frame bundle. In general, this is difficult, and to allow ourselves more flexibility we will allow certain types of singularities in our framings.

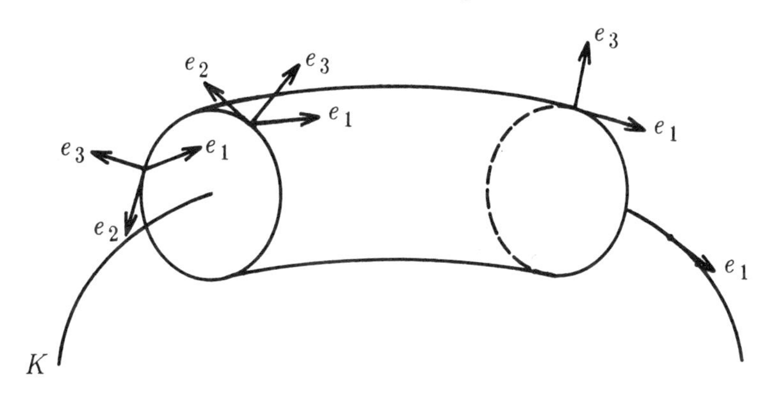

Figure 4.

Definition. A *special singular frame field* on a 3-manifold M is an orthonormal frame field on $M-L$ for some link L which has the following behaviour at each component K in L:

i) (in the limit) e_1 is tangent to K.

ii) e_2 and e_3 determine an index +1 or -1 singularity transverse to K (e.g. see figure 4).

Theorem (Torsion Formula). $(1/8\pi^2) \int_{s(M-L)} Q \dot{+} (1/4\pi) \sum_{K \in L} \tau(K)$ (modulo *half) is an invariant of the hyperbolic 3-manifold* M, *where* $s{:}(M-L) \rightarrow F(M)$ *is a special singular frame field on* M *(singular at the link* L *), and* $\tau(K)$ *is the torsion of the singular curve* K *in* L.

Proof. See [Meyerhoff, 1981, Meyerhoff, 1986] or [Yoshida, 1986 Corollary 1.1] (Yoshida generalises the torsion formula to the case of the eta invariant).

□

Thus, $CS(M)$ can be computed by using a special singular framing. The usefulness of this notion has been brought out in the theorem in section 1, and the results in [Meyerhoff, 1981] and [Yoshida, 1986].

The torsion formula can be used to extend the definition of the Chern-Simons invariant to include cusped hyperbolic 3-manifolds. First, we need to fix the frame fields at the cusps.

Definition. Let the cusp correspond to the point at infinity in the upper-half-space model of hyperbolic 3-space. Then a *linear frame field on a cusp* is of the form $e_1 = t(\partial/\partial t), e_2 = t(\partial/\partial x)$ and $e_3 = t(\partial/\partial y)$.

Given a link L' in a 3-manifold M and a complete hyperbolic structure on $M-L'$, it is quite possible that the linear framing at L' does not extend to an orthonormal framing of M. However, there exists a link L'' in M (disjoint from L') such that M has a special singular framing s which is singular along L'' and linear along L'; for notational convenience we will say that L' *is a part of the special singular set* $L = L' \cup L''$. L can be specified a priori, for example if M is the 3-sphere, and L has an even number of components, each having odd linking number with the union of the others, then a frame field with special singularities at L exists. This sort of construction is carried out in [Meyerhoff, 1981, Meyerhoff, 1986] and [Yoshida, 1986 proposition 3.1].

Definition. The *Chern-Simons invariant* of the cusped hyperbolic 3-manifold M is defined to be

$$(1/8\pi^2) \int_{s(M-L)} Q + (1/4\pi) \sum_{K \in L''} \tau(K) \pmod{\tfrac{1}{2}}.$$

The fact that this definition is well-defined, independent of the choices made, is shown in [Meyerhoff, 1981 section 3.3] or [Meyerhoff, 1986 section 4.2]. For notational convenience, we could define $\tau(K)$ to be 0 for K in L', the cusp locus, in which case the extended Chern-Simons invariant would have the same form as the torsion formula. Also, the Chern-simons invariant is defined modulo 1, but the proofs of the torsion formula lose a factor of 2; thus the extended Chern-Simons invariant holds modulo half.

Acknowledgement: I thank Bill Dunbar and Bill Thurston for helpful conversations.

References

Chern-Simons, 1974.
S. S. Chern and J. Simons, "Characteristic forms and geometric invariants," *Annals of Math.*, **99**, pp. 48-69, 1974.

Meyerhoff, 1981.
R. Meyerhoff, *The Chern-Simons Invariant for hyperbolic 3-manifolds*, 1981. Princeton University Ph.D. Thesis

Meyerhoff, 1986.
R. Meyerhoff, *Density of the Chern-Simons invariant for hyperbolic 3-manifolds*, 1986. This volume

Riley, 1970.
R. Riley, "An Elliptical Path from Parabolic Representations to Hyperbolic Structures," *Lecture Notes in Mathematics*, **722**, pp. 99-133, 1970.

Rolfsen, 1976.
Rolfsen, D., *Knots and links,* Publish or Perish, Berkeley, 1976.

Steenrod, 1951.
N. Steenrod, *The Topology of Fibre Bundles,* Princeton University Press, Princeton, 1951.

Thurston, 1979.
W. P. Thurston, *The Geometry and Topology of 3-Manifolds,* Princeton University Mathematics Department, 1979. Part of this material – plus additional material – will be published in book form by Princeton University Press

Yoshida, 1986.
T. Yoshida, "The eta invariant of hyperbolic 3-manifolds," *Inventiones Mathematicae*, 1986. (to appear)

R.Meyerhoff,
Boston University,
Boston,
Mass. 02215,
USA.

Density of the Chern-Simons Invariant for Hyperbolic 3-Manifolds

Robert Meyerhoff

1. Introduction

The Chern-Simons invariant grew out of attempts by Chern and Simons to develop a purely combinatorial formula for the first Pontrjagin number of a 4-manifold. The Chern-Simons invariant was an intractable boundary term which got in the way of the combinatorial formula. This invariant — for a closed, oriented Riemannian 3-manifold M — is obtained by integrating a certain 3-form Q over a section s of the positively-oriented orthonormal frame bundle, $F(M)$, of M (i.e., s represents a framing of M). Because $dQ = 0$ it can be shown that this integral $(1/8\pi^2) \int_{s(M)} Q \pmod 1$ is independent of the section s — thus it is an invariant of the Riemannian manifold M. It turns out to be a conformal invariant, and by Mostow's theorem, it is a topological invariant for closed hyperbolic 3-manifolds (see Section 2.2 for more details).

In this paper we prove

Theorem. *The hyperbolic 3-manifolds obtained from Dehn surgery on any link with hyperbolic complement take on a dense set of Chern-Simons values in the circle* $\mathbb{R}/\tfrac{1}{2}\mathbb{Z}$.

The method of proof of this theorem is outlined in the next three paragraphs.

The main obstacle to computing $CS(M)$ is in finding a "good" frame field to integrate Q over. This problem is partially alleviated by allowing frame fields with "special" singularities (see Section 3.1), and generalising the Chern-Simons invariant to allow such types of frame fields. In particular,

Theorem.

$$(1/8\pi^2) \int_{s(M)} Q + (1/4\pi) \sum_{c} \tau_c \quad (\text{mod half})$$

is equal to the Chern-Simons invariant (mod *half*) *of the manifold* M.

(here s is a "section" which may have the above type singularities, and τ_c is the total torsion of the singular curve c).

The torsion formula enables us to extend the definition of the Chern-Simons invariant to include hyperbolic 3-manifolds with cusps (see Section 4.2). What properties does the new (extended) definition have? For example, the volume has the following property: Consider $M_{(p_i,q_i)}$ the hyperbolic manifold obtained by performing (p_i,q_i) Dehn surgery on some (hyperbolic) knot in the 3-sphere; as $p_i^2 + q_i^2 \to \infty$ the volume of $M_{(p_i,q_i)}$ approaches the volume of the knot complement, M. Does the Chern-Simons invariant have this property? The answer is no, and in fact $CS(M_{(p_i,q_i)})$ takes on a dense set of values in $\mathbb{R}/½\mathbb{Z}$ as $p_i^2 + q_i^2$ approaches infinity along all possible routes.

An explanation of this fact (which completes the proof) is as follows. To study $CS(M_{(p_i,q_i)})$ a frame field is needed. The easiest way to get one is by "perturbing" the frame field on M. This leads to "good" singularities in the framing along the core (surgered) geodesic. Then, according to the torsion formula, $CS(M_{(p_i,q_i)})$ is the sum of the integral of Q over the singular section plus the torsion of the core geodesic. As $p_i^2 + q_i^2 \to \infty$, the first term in the sum approaches $CS(M)$. But the torsion takes on a dense set of values in $\mathbb{R}/½\mathbb{Z}$. Thus the theorem is proved.

The torsion formula is the backbone of this paper. The importance of the torsion formula is also borne out by the results of T. Yoshida (see [Yoshida, 1986]). Yoshida extends the torsion formula to the eta invariant (see the introduction to [Atiyah-Patodi-Singer, 1975a, Atiyah-Patodi-Singer, 1975b] for the definition of the eta invariant), and uses it to prove a conjecture of Neumann and Zagier (see the introduction to [Neumann-Zagier, 1986]) relating volume and Chern-Simons invariants for hyperbolic 3-manifolds. Further, Yoshida comes up with a formula for the eta invariant of hyperbolic 3-manifolds obtained by performing Dehn surgery on the figure-eight knot in the 3-sphere. This formula should be implementable by computer. Since the eta invariant is a real-valued generalisation of the Chern-Simons invariant modulo half (see Proposition 4.19 of [Atiyah-Patodi-Singer, 1975b]), Yoshida's torsion formula proof gives an alternative to the torsion formula proof presented here.

This paper contains results from my Princeton University Ph.D. thesis ([Meyerhoff, 1981]). I would like to thank Bill Dunbar and Bill Thurston for helpful conversations.

2.1 Hyperbolic Preliminaries

In this paper we will be dealing with complete hyperbolic 3-manifolds (Chapters 1 through 6 of [Thurston, 1979] is a good reference). A *hyperbolic manifold* is a differentiable manifold with a Riemannian metric of constant sectional curvature -1. *Complete* (i.e., complete as a metric space) implies that the manifold M can be thought of as H/Γ where H is hyperbolic 3-space and Γ is a discrete subgroup of the group of orientation-preserving isometries of H. All hyperbolic 3-manifolds we consider will be assumed complete.

An important model for hyperbolic 3-space is the upper-half-space model, UHS.

$$\text{UHS} = \{(x,y,t): x,y,\ t \text{ are real numbers and } t \text{ is positive}\},$$

the Riemannian (hyperbolic) metric is $ds^2 = (dx^2 + dy^2 + dt^2)/t^2$, and the volume form is $dV = (dx\ dy\ dt)/t^3$. The boundary of UHS is the Riemann sphere $\{z = x + iy\} \cup \{\infty\}$, and the orientation-preserving isometries of UHS correspond to elements of

$$\text{PSL}(2,\mathbb{C}) = 2\times 2 \text{ complex matrices of determinant } 1/\pm\text{Id}.$$

These matrices act on the bounding Riemann sphere by

$$z \mapsto (az + b)/(cz + d),$$

i.e., they are Moebius transformations (which take circles to circles), and this action can be extended to UHS (taking hemispheres to hemispheres) by using a hemisphere associated to a circle in the bounding complex plane.

Mostow's theorem [Thurston, 1979, Section 5.7] implies that two complete hyperbolic 3-manifolds of finite volume which are homeomorphic are necessarily isometric. That is, if a topological 3-manifold has a complete hyperbolic structure of finite volume then the structure is unique. Therefore, the volume is a topological invariant for complete hyperbolic 3-manifolds.

Most link complements in the 3-sphere have a complete hyperbolic structure of finite volume (see the introduction to [Riley, 1970]). The end determined by a component of the link is topologically the Cartesian product of a torus with a half-open interval, i.e., $T^2 \times [0,\infty)$; while geometrically it must have the following form (see [Thurston, 1979, Proposition 5.4.4]): It is the quotient of UHS (above a certain height t = constant) by the translations $z \to z+1$, and $z \to z+w$, where w is a complex number with non-zero imaginary part. Such an end is referred to as a *cusp*. We will sometimes perform *hyperbolic Dehn surgery* on these manifolds with cusps (see [Thurston, 1979, Chapter 4]). The result of such surgery is the same (homeomorphic) as that of regular Dehn surgery, but the resultant manifold has a natural hyperbolic structure. To make this a closed manifold the cusp had to be completed by adding a geodesic, which will be referred to as the *core geodesic*.

$\mathbb{H}$, the space of all isometry classes of complete hyperbolic 3-manifolds of finite volume, possesses a *geometric topology* (see [Thurston, 1979, Section 5.11]). In this topology a cusped manifold is the limit of the closed manifolds obtained by performing higher and higher order Dehn surgeries on the cusp. The volume is a continuous function with respect to the geometric topology ([Thurston, 1979], Section 5.11).

2.2 The Chern-Simons Invariant

Given a section, s, of the positively-oriented orthonormal frame bundle, $F(M)$, of a closed, oriented Riemannian 3-manifold M, the *Chern-Simons integral* is defined as $(1/8\pi^2)\int Q$ (mod 1) where the integration is performed over $s(M)$ and Q, the *Chern-Simons form*, is the 3-form

$$\omega_{12} \wedge \omega_{13} \wedge \omega_{23} + \omega_{12} \wedge \Omega_{12} + \omega_{13} \wedge \Omega_{13} + \omega_{23} \wedge \Omega_{23}.$$

Here, the ω_{ij}'s are the connection forms and the Ω_{ij}'s are the curvature forms. Such a section always exists for a closed 3-manifold (see [Steenrod, 1951, Section 39.9]).

It can be shown (see [Chern-Simons, 1974]) that the Chern-Simons integral is independent of the section, s. The fiber $SO(3)$ of $F(M)$ is normalized to have volume $8\pi^2$. Thus, the Chern-Simons integral is actually an invariant of M, called the *Chern-Simons invariant*, $CS(M)$. Chern and Simons also show that $CS(M)$ is a conformal invariant of the Riemannian 3-manifold M. The 3-manifolds we will be concerned with will have constant sectional curvature -1, i.e., they will be hyperbolic 3-manifolds. By Mostow's theorem (see Section 2.1) closed hyperbolic 3-manifolds have unique hyperbolic structures and therefore the Chern-Simons invariant is a topological invariant for closed hyperbolic 3-manifolds.

A couple of useful properties of the Chern-Simons invariant which follow from the definition are

1) reversing the orientation of the manifold changes the sign of the Chern-Simons invariant, and

2) taking a geometric n-fold cover of a manifold multiplies the Chern-Simons invariant by a factor of n.

3.1 The Torsion Formula

The immediate problem in the computation of the Chern-Simons invariant is to find an explicit section of the orthonormal frame bundle. In general, this is difficult, and for more flexibility we will allow certain types of singularities in our framings.

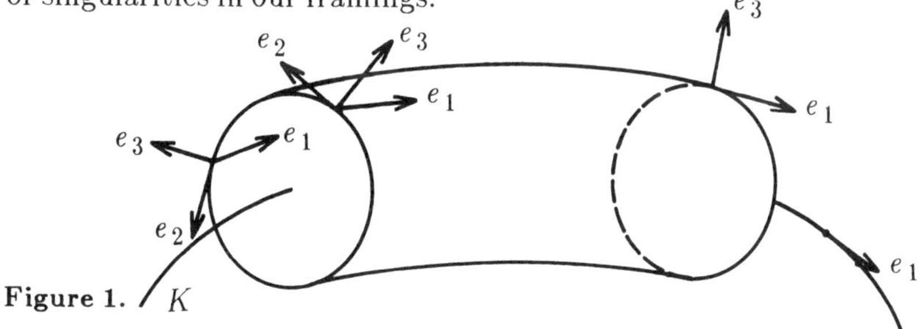

Figure 1.

Definition. A *special singular frame field* on a 3-manifold M is an orthonormal frame field on $M-L$ for some link L which has the following behaviour at each component K in L:

i) (In the limit) e_1 is tangent to K.

ii) e_2 and e_3 determine an index $+1$ or -1 singularity transverse to K (e.g., see Figure 1).

Theorem (Torsion Formula).

$$(1/8\pi^2) \int_{s(M-L)} Q + (1/4\pi) \sum_{K \in L} \tau(K) \ (\text{modulo } 1/2)$$

is an invariant of the hyperbolic 3-manifold M, where $s:(M-L) \to F(M)$ is a special singular frame field on M(singular at the link L), and $\tau(K)$ is the torsion of the singular curve K in L.

Proof. The proof will be given in sections 3.2, 3.3, and 3.4.

A pair of "moves" which will allow us to eliminate the singularities in the framing will be studied in Sections 3.2 and 3.3. In particular, the effect of the moves on the Chern-Simons integral will be computed. In Section 3.4, a method for using these moves to get from a special singular framing to a non-singular framing will be described, and the total effect on $CS(M)$ computed. This will prove the torsion formula.

Remark. *The torsion formula implies that the Chern-Simons invariant (modulo half) can be computed by using a special singular frame field.*

3.2 Homotopy of a Special Singular frame field

We are given a hyperbolic 3-manifold M with a special singular framing. The associated singular curves form a link in M and we now assume that the frame field on M is homotoped in such a way that one of the singular curves, L_0, moves to a new singular curve, L_1, and all others are fixed. In this section we will prove that the change in the Chern-Simons integral under this homotopy is $(1/4\pi)(\tau_1 - \tau_0)$, where τ_0 and τ_1 are the (total) torsions of L_0 and L_1, respectively. The *total torsion* of L is defined as $\int_c \omega_{23}$ mod 2π, where c represents a framing of L with e_1 tangent to L. The total torsion of L is independent (mod 2π) of the choice of c. Here, we are assuming that the index of the vector field formed by the e_2's transverse to the singular curve L_0 has index 1.

Let H be the (possibly degenerate) 4-manifold in $F(M)$ swept out by the homotopy from the 3-manifold G_0 in $F(M)$ to the 3-manifold G_1 in $F(M)$, where G_0 and G_1 represent the appropriate special singular frame fields on M. Then, by Stokes' theorem (using the fact that $dQ = 0$), we have that $0 = \int_H dQ = \int_{\partial H} Q = \int_{G_1} Q - \int_{G_0} Q + \int_N Q$, where N is the 3-manifold in $F(M)$ swept out by the boundary tori of G_0 under the homotopy, and Q is the Chern-Simons form (see Section 2.2). So, the change in the Chern-Simons integral is

$$\Delta(CS) = \frac{1}{8\pi^2}\int_{G_1} Q - \frac{1}{8\pi^2}\int_{G_0} Q = -\frac{1}{8\pi^2}\int_N Q.$$

Here it should be pointed out that the orientation on N is induced from the orientation on H given by the 4 vectors h, e_1, e_2, e_3 (h is "homotopy direction" and the e_i are horizontal lifts to $F(M)$ of the e_i in M). Thus, the orientation for N is given by h, e_1, and f where $f \in T_x$ (fiber) is clockwise e_2e_3 spin about e_1. It will come into play that $\omega_{23}(f) = -1$. We can use the frame vectors h, e_1 and f to integrate Q over N. Since $\omega_{12}(f)$, $\omega_{13}(f)$, $\Omega_{12}(f, \cdot)$, and $\Omega_{13}(f, \cdot)$ all equal 0, Q reduces to

$$\omega_{12} \wedge \omega_{13} \wedge \omega_{23} + \omega_{23} \wedge \Omega_{23} = \omega_{23} \wedge d\omega_{23}.$$

$d\omega_{23}(f, \cdot) = 0$ and integrating (along the "fiber" corresponding to f) yields

$$\int_N \omega_{23} \wedge d\omega_{23} = \int_f \omega_{23} \cdot \int_{c_0 \times [0,1]} d\omega_{23} = (-2\pi)\left[\int_{c_1} \omega_{23} - \int_{c_0} \omega_{23}\right].$$

Here, N is homeomorphic to $T^2 \times [0,1]$, c_0 is a closed curve in $F(M)$ lying above L_0, c_1 is the closed curve which is the homotopic image of c_0, and thus, c_0 and c_1 induce framings on L_0 and L_1 with e_1 tangent to the singular curve (the choice of c_0 is arbitrary, e.g., it could be the Darboux framing). Now, $\int_{c_1} \omega_{23} = \tau_1$, $\int_{c_0} \omega_{23} = \tau_0 \pmod{2\pi}$ and we have $\Delta(CS) = (1/4\pi)(\tau_1 - \tau_0)$ modulo half. The ambiguity one half comes from the 2π ambiguity in the total torsion.

Remarks.

1) If we homotope a special singular frame field to another special singular frame field such that $L_0 = L_1$, then $\Delta(CS) = 0$ (mod half). Here we need to have the e_1 vectors point in the same direction at the end of the homotopy as at the beginning.

2) A special singular curve with $+1$ index can be homotoped to a -1 index (rotate each point around the normal to the osculating plane to L). At times t not equal to 0 or 1, e_1 does not point along L. However, the $\Delta(CS)$ calculation is essentially the same and $\Delta(CS) = (1/4\pi)(\tau_1 - \tau_0)$ modulo half. Here, $\tau_1 = -\tau_0$.

3.3 The Crossing Move

The second move is described by Figure 2, where L, L_1 and L_2 are singular curves for special singular framings.

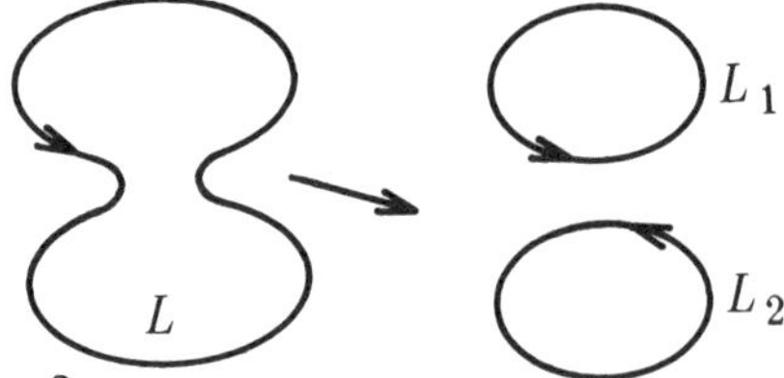

Figure 2.

By the result of section 3.2 we can assume that all singular curves have +1 transverse index, and that the move takes place in as small a ball as we like, i.e., locally the two singular arcs in Figure 3a are taken to be the two singular arcs in Figure 3b.

Figure 3. 3a 3b

Thus, we will be doing an essentially Euclidean calculation and the only non-trivial term of Q is $\omega_{12} \wedge \omega_{13} \wedge \omega_{23}$, i.e., the volume form for $SO(3)$. For computational convenience, we assume that the four arcs lie on a totally geodesic surface.

We want to measure the change in the Chern-Simons invariant gotten by replacing Figure 3a by Figure 3b where the frame field on the boundary of the ball is fixed. We would like to use the Stokes' theorem argument of Section 3.2. To do this, we consider the manifold N (with boundary) in $F(M)$ sitting above the first small ball, and bore out a small solid tube S connecting the two cylinders sitting above the singular arcs.
The local picture in $F(M)$ is given in Figure 4, and by the assumption in the first paragraph we can assume that this is the local picture in $SO(3)$. Now, the Chern-Simons integral for the first small ball is

$$(1/8\pi^2) \int_{N-S} Q + (1/8\pi^2) \int_S Q$$

and we perform the homotopy argument of Section 3.2 on $N-S$, where the homotopy takes the boundary described in Figure 5a to the boundary described in Figure 5b (this is in $SO(3)$).

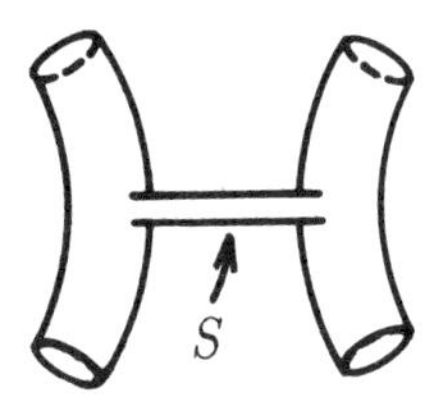

Figure 4.

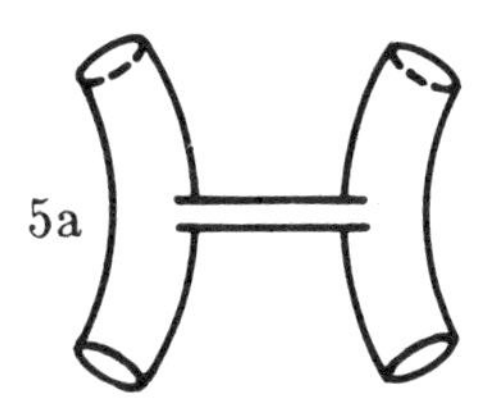

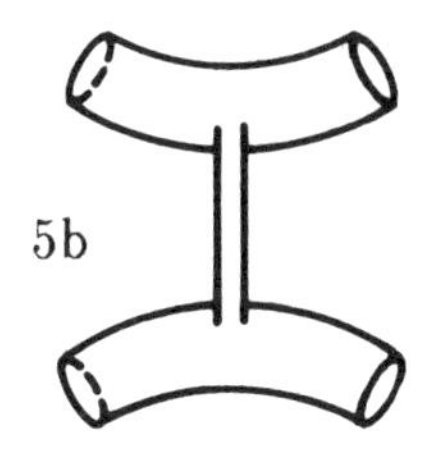

Figure 5.

The change in the Chern-Simons integral will be the volume of the object in $SO(3)$ swept out by the homotopy connecting these two boundaries. For computational ease we will consider $SO(3)$ as the unit-tangent bundle of S^2, where points in S^2 correspond to directions of e_1 vectors. Thus, the fiber over a point in S^2 represents 2π worth of e_2's. So, the integral of $\omega_{12} \wedge \omega_{13} \wedge \omega_{23} = Q$ over the swept out region in $SO(3)$ is approximately equal to 2π times the integral of $\omega_{12} \wedge \omega_{13}$ over the area in S^2 swept out by the e_1 vectors. Here we are ignoring the term associated with the small solid tube which, as we will see, is negligible.

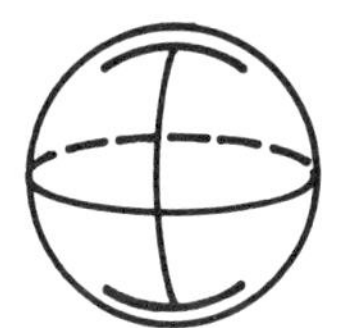

Figure 6.

In S^2 the e_1 vectors from Figure 4 map out the region shown in Figure 6, where the heavy line represents the small connecting tube. (Here we may have had to homotope our original frame field to get the solid tube in the desired position, but this would have no effect on the Chern-Simons integral because the singular curves would have stayed fixed.) We can homotope this as follows. (See Figure 7.) So,

$$\int \omega_{12} \wedge \omega_{13} = 1/2 \ \text{area}(S^2) + \epsilon;$$

where we get half the area of S^2 because our original set-up had the singular arcs lying on a totally geodesic surface, and ϵ comes from the arbitrarily

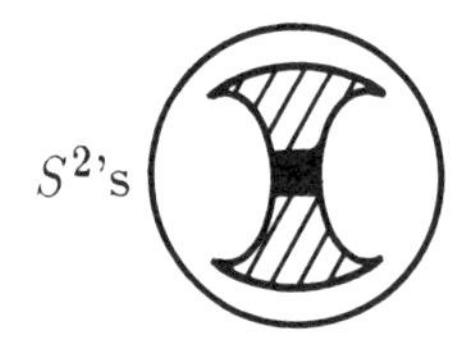

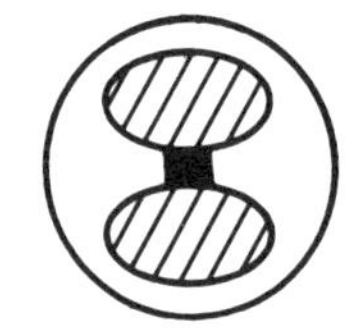
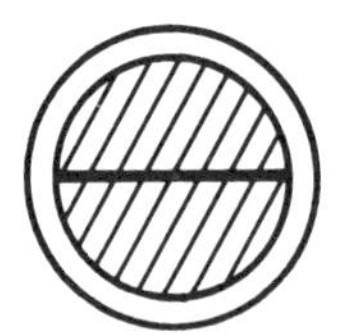

Figure 7.

small area swept out by the small solid tube. Thus $\Delta(CS) = (1/16\pi^2)(\text{volume}(SO(3)))$ which is zero modulo half. It remains to show that the small solid tube in Figure 5b is homotopically trivial, so that we

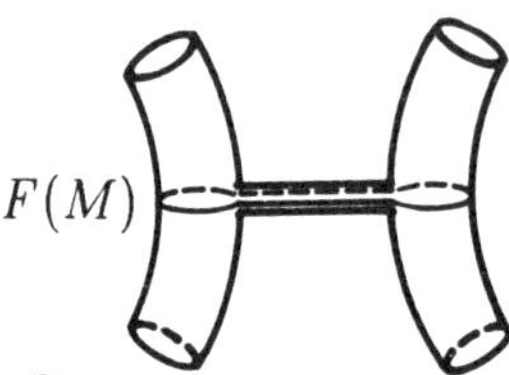

Figure 8.

can fill it in (with negligible volume). This can be seen by tracing out the path in $SO(3)$ determined by the curve in Figure 8.

Summing up, we see that the beginning and ending small solid tubes can be ignored (just take the beginning tube as small as necessary, and note that the region swept out under the homotopy is as small as we want) and the change in the Chern-Simons integral is zero modulo half for the crossing move.

3.4 Combining the Moves

The link of singular curves associated with a special singular framing forms a 1-cycle Z in the 3-manifold M. Z represents the obstruction to extending the framing on the 1-skeleton of M to the 2-skeleton. Since there exists a framing on M, we have (using Poincaré duality) that Z represents the trivial element in $H_1(M; \mathbb{Z}/2\mathbb{Z})$. Thus there is an embedded surface S in M with boundary Z.

We can define a non-degenerate Morse function h (minimal at the boundary) on S in such a way that the level sets of h begin with Z and end with a contractible singular curve Z' (this is a standard result in Morse theory). The procession of level sets describes a passage from Z to Z' by the elimination moves of Sections 3.2 and 3.3. In particular, critical points correspond to the crossing move of Section 3.3.

The results of Sections 3.2 and 3.3 show that $(1/8\pi^2)\int_{s(M)} Q + (1/4\pi)\sum\tau$ (mod half) is unaffected by the moves described by h. Thus, to prove the torsion formula we need only prove that eliminating the null-homotopic singular curve Z' has no effect on the Chern-Simons integral.

As in Section 3.3, we can assume we are working in a sufficiently small ball so our calculation reduces to determining the volume in $SO(3)$ of

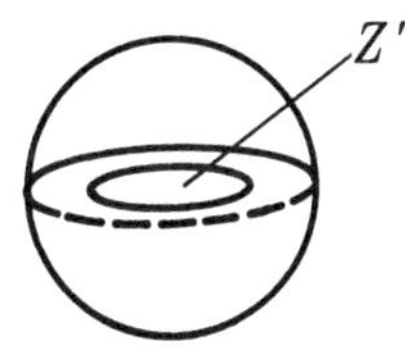

Figure 9.

the frames in the small ball containing Z' (see Fig. 9). Since $\pi_2(SO(3)) = 0$ we can assume the boundary of the ball has constant frame field in the "ball coordinate chart" i.e., the frames on the boundary of the ball map to a single point in $SO(3)$. Further, we can use the result in Section 3.2 and assume that Z' lies in a totally geodesic surface. Explicitly mapping out the frames in $SO(3)$ yields an object with boundary two copies of a torus in $SO(3)$. The fact that Z' lies in a totally geodesic surface results in the object having volume equal to one-half the volume of $SO(3)$. Thus, the contribution of the small ball containing Z' to the total Chern-Simons integral of M is the same as the contribution of the small ball with constant framing would be. So, we can carry out this switch with no effect on the Chern-Simons integral, and the invariance of the torsion formula is proved.

We note here (though it will not be used later) that the torsion formula's invariance holds for closed, oriented Riemannian 3-manifolds with constant sectional curvature. In fact, the proof is step-by-step the same as the one given in Sections 3.2 to 3.4.

4.1 A Linear Frame Field on a Cusp

The torsion formula can be used to extend the definition of the Chern-Simons invariant to include cusped hyperbolic 3-manifolds. This extension is carried out in Sections 4.1, 4.2, and 4.3. First, we need to fix the frame fields at the cusps.

Definition. Let the cusp correspond to a point at infinity in UHS. Then a *linear frame field on a cusp* is of the form

$$e_2 = t[\cos(r)\,\partial/\partial x + \sin(r)\,\partial/\partial y],$$

$$e_3 = t[-sin(r)\,\partial/\partial x + \cos(r)\,\partial/\partial y],$$

$$e_1 = t\,\partial/\partial t$$

where r is a constant.

We now evaluate the Chern-Simons form Q over the section, s, associated with a linear framing. Rather than do the computation in the 6-dimensional orthonormal frame bundle, we will use s to pull the computation back to the cusp in the manifold itself. This is Cartan's method of moving frames. Thus, the dual 1-forms on the cusp are

$$\theta_2 = (\cos(r)dx + \sin(r)dy)/t,$$

$$\theta_3 = (-sin(r)dx + \cos(r)dy)/t,$$

$$\theta_1 = dt/t$$

Claim.

$$Q = \omega_{12} \wedge \omega_{13} \wedge \omega_{23} + \omega_{12} \wedge \Omega_{12} + \omega_{13} \wedge \Omega_{13} + \omega_{23} \wedge \Omega_{23}$$

is pointwise zero.

To show this we compute the ω_{ij} (note that for constant curvature 3-manifolds M, $\Omega_{ij} = k\theta_i \wedge \theta_j$, where k is the curvature) by solving the structural equation $d\theta = -\omega \wedge \theta$; which, expanded in local coordinates, is

$$d\theta_1 = -\omega_{12} \wedge \theta_2 - \omega_{13} \wedge \theta_3$$

$$d\theta_2 = \omega_{12} \wedge \theta_1 - \omega_{23} \wedge \theta_3$$

$$d\theta_3 = \omega_{13} \wedge \theta_1 + \omega_{23} \wedge \theta_2.$$

We get $\omega_{12} = \theta_2$, $\omega_{13} = \theta_3$, $\omega_{23} = 0$. Thus,

$$Q = \theta_2 \wedge \theta_3 \wedge 0 + \theta_2 \wedge (k\theta_1 \wedge \theta_2) + \theta_3 \wedge (k\theta_1 \wedge \theta_3) + 0 \wedge (k\theta_2 \wedge \theta_3) = 0.$$

So, linear frame fields on cusps yield Chern-Simons forms Q which are pointwise zero. It should be pointed out that this does not mean that all frame fields put on the cusp have Chern-Simons invariant zero. The point is that the cusp has a boundary, and manifolds with boundary do not have the independence-of-the-section property mentioned in Section 1.1.

However, if the frame field at the boundary is fixed, then the independence-of-the-section property holds by essentially the same proof as that given in [Chern-Simons, 1974] for the no-boundary case.

What is the effect of changing from one linear frame field to another on a cusp?

Claim. *If, in the definition of linear frame field, r is a function of x, y, and t, then Q is still pointwise zero.*

This follows by essentially the same computation as above, although in solving the structural equations we find that $\omega_{23} = -(\partial r/\partial x\, dx + \partial r/\partial y\, dy + \partial r/\partial u\, du)$. Thus, we can change from one linear framing to another on a cusp.

4.2 The extended definition of $CS(M)$

In Section 4.3 we will prove that for complete hyperbolic 3-manifolds of finite volume with an even number of cusps satisfying a certain linking property, linear framings on the cusps can be extended over the entire manifold M. We define the Chern-Simons invariant of such a manifold M to be $(1/8\pi^2) \int_{s(M)} Q \pmod 1$ where s is such a framing. By the computations in Section 4.1 this is well-defined.

We generalise this as follows. Consider a complete hyperbolic 3-manifold M of finite volume, with cusps. By Theorem 5.11.2 of [Thurston, 1979] there is a hyperbolic link L in S^3 such that M is obtained by performing Dehn surgery on some components of the link. We can bore out additional closed curves in S^3 in such a way that the new link L' has an even number of components each having odd linking number with the union of the others. Now we can use Theorem 4.3 to put a frame field on M which has linear frame fields on the cusps and special singularities at the other components of L'. Here, some of the components of L may have had Dehn surgery performed on them (to produce M), but this is compatible with their being singular curves. We now define the Chern-Simons invariant of M as

$$(1/8\pi^2) \int_{s(M)} Q + (1/4\pi) \sum_{L''} \tau \pmod{\text{half}}$$

where $L'' = L' -$ cusps, and we need only show that this definition is independent of the choices made, i.e., independent of L''. But L''

represents the obstruction to extending the linear framings on the cusps to all of M, and as such, is homologous to any other such obstruction. We now apply the Morse theory argument of Section 3.4 to get independence. Thus, our extended definition of $CS(M)$ is well-defined. It should be mentioned that we used the independence-of-the-section property referred to in Section 4.1. Also, note the remark at the end of Section 4.3.

Actually, a frame field (with special singularities) can be put on M, a 3-manifold with torus boundary components, extending linear frame fields on the tori without resorting to Theorem 4.3. Although this approach is less concrete, it does show that special singularities arise naturally. We give it step-by-step as follows.

1) Put an e_1 vector field on M such that e_1 points out at the torus boundary components. This determines a 2-plane at each point of M.

2) Choose a section s of the 2-plane bundle of M (this corresponds to a choice of e_2 vectors) such that the section is linear on the boundary of M, and such that the section is transverse to the 0-section.

3) The 0-section and s are 3-dimensional manifolds sitting in the 5-dimensional 2-plane bundle, and as such, they intersect generically in a 1-dimensional manifold. Hence, projecting to M, the singularities of s occur along curves.

4) Homotope the resultant frame field so that the e_1 vectors are tangent to the singular curves.

5) The fact that s is transverse to the 0-section implies that the first derivative of the e_2 vector field is non-singular, and thus, locally the e_2 vector field is linear.

6) The fact that the e_3 vector field is locally linear transverse to the singular curve implies that it has a standard form — either an index $+1$ or -1 singularity.

So, special singularities are natural, and $CS(M)$ can be defined as

$$(1/8\pi^2) \int_{s(M)} Q + (1/4\pi) \sum_{\substack{\text{singular} \\ \text{curves}}} \tau \quad (\text{mod half}),$$

where s is a special singular section which is linear on the cusps.

4.3 Frame Fields Linear Near the Boundary

Which complete hyperbolic 3-manifolds of finite volume have frame fields which are linear on the cusps? This is essentially a topological question, i.e., if we consider the manifold with torus boundary components obtained by chopping off the open ends of cusps, we want to know whether we can extend linear framings on the boundary tori to a framing of the entire manifold.

Theorem 4.3. *Let M be a 3-manifold with torus boundary components. There exists a frame field on M which is linear near the boundary, if and only if any cycle on the boundary of M which bounds a surface in M has an even number of components.*

Proof. This is an exercise in obstruction theory (see [Steenrod, 1951, Sections 29.4, 33.1, and 34.2]). Assume we have a 3-manifold M with torus boundary components and linear frame fields on these tori. Given a triangulation of $(M, \partial M)$ we can put a framing (3 orthonormal vectors) on the 1-skeleton which agrees with the boundary framing. This induces a map f from the 1-skeleton of M to $F(M)$. If we can extend f to the 2-skeleton then we would have the desired framing, because $\pi_2(SO(3)) = 0$ implies there is no obstruction to extending the framing from the 2-skeleton to the 3-skeleton. On the other hand, $\pi_1(SO(3)) = \mathbb{Z}/2\mathbb{Z}$, and we can extend f from the 1-skeleton to the 2-skeleton if and only if the obstruction co-cycle $c(f)$ determined by f and the 1-skeleton is zero in $H^2(M,\partial M;\mathbb{Z}/2\mathbb{Z})$. The co-cycle $c(f)$ assigns 0 or 1 to a 2-cell in the triangulation depending on whether the framing on the boundary of the 2-cell yields a trivial or a non-trivial loop in $SO(3)$

We will show that $[c(f)] = 0$ if and only if M satisfies the hypothesis of the theorem. There is a perfect pairing between $H^2(M,\partial M;\mathbb{Z}/2\mathbb{Z})$ and $H_2(M,\partial M;\mathbb{Z}/2\mathbb{Z})$, so showing $[c(f)] = 0$ amounts to showing that any 2-cycle paired with $c(f)$ yields 0. The 2-cycle is made up of 2-cells in the 2-skeleton and by cutting and pasting a smooth surface S representing its homology class can be found. Since we can rotate a linear frame field on the boundary to any other linear frame field on the boundary, we can assume that e_2 vectors on the boundary of M are tangent to the boundary of S (this will come into play shortly). Here we are ruling out the possibility of S having trivial boundary components in the boundary of M. In computing $\langle c(f), S\rangle$ we consider 2 cases;

... **Case 1.** *The surface S is orientable.*

Put a frame field on S such that e_3 is normal to S and e_1, e_2 are tangent to S with e_1 pointing out at the boundary components of S. The e_1-vectors define a (possibly singular) vector field on S and the index (mod 2) of this vector field is $\langle c(f), S\rangle$, the evaluation of the 2-co-cycle $c(f)$ on the 2-cycle S. But the index of the vector field (mod 2) is the Euler characteristic (mod 2), which is zero if and only if S has an even number of boundary components. So, $[c(f)] = 0$ in $H^2(M,\partial M;\mathbb{Z}/2\mathbb{Z})$ if and only if the hypothesis of the theorem is satisfied.

... **Case 2.** *S is non-orientable.*

We want to put a vector field on S as in Case 1, but first we have to cut S up into an oriented surface.

There are 2 types of cutting curves to consider.

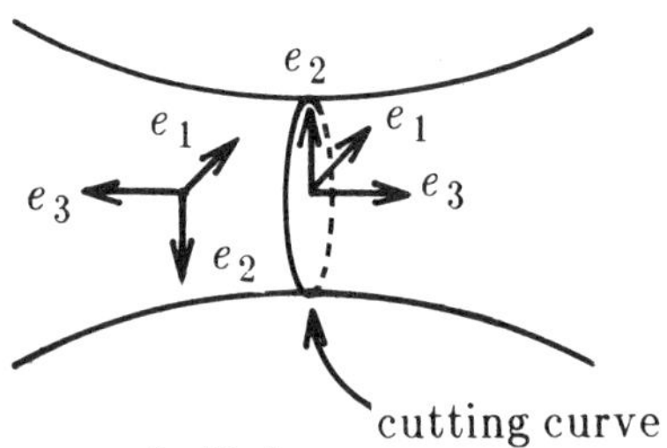

Figure 10. *2-sided*

2-sided (Figure 10): we rotate 180 degrees about e_1 across the cutting curve. This contributes 0 to the index of the vector field (mod 2) on S.

1-sided (Figure 11): the contribution along the cutting curve to the index (mod 2) is 1. This can be seen by looking at the rotation needed to go from one frame to the other across the cutting curve.

So, $\langle c(f), S\rangle$ equals the Euler characteristic of the oriented surface obtained by cutting, plus the number of 1-sided (on S) cutting curves (mod 2). But the Euler characteristic of the oriented surface is equal to the number of boundary components of the oriented surface (mod 2). Adding in one boundary component for each one-sided cutting curve and two boundary components for each 2-sided cutting curve yields the number of boundary components of the original surface (mod 2). Thus, $\langle c(f), S\rangle$ is equal to the number of boundary components of the original surface (mod 2).

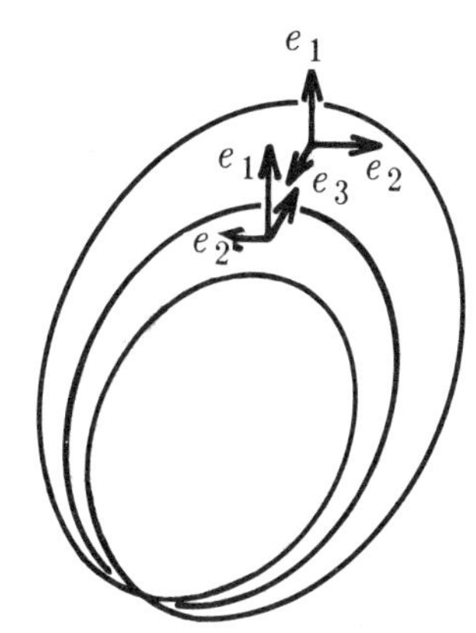

Figure 11. *1-sided*

This proves that the obstruction to extending the frame field vanishes if and only if the boundary cycle hypothesis of the theorem is satisfied.

Theorem 4.3

Corollary. *If* $M = S^3 - L$, *where* L *is a link with an even number of components each having odd linking number with the union of the others, then a linear frame field on the boundary components can be extended to the entire manifold* M.

Proof. Think of $\langle c(f), S \rangle$ as a homomorphism from $H_2(M,\partial M;\mathbb{Z}/2\mathbb{Z})$ to $\mathbb{Z}/2\mathbb{Z}$. To prove the corollary we need only show that this map is zero on the generating set for $H_2(M,\partial M;\mathbb{Z}/2\mathbb{Z})$. A set of generators is given by spanning surfaces for each component of the link L. By the definition of linking number (see [Rolfsen, 1976 page 133]), this number is zero for odd linking numbers, because $\langle c(f), S \rangle$ simply counts ∂-components (mod 2). □

Note. The proof of the corollary did not use the fact that L has an even number of components. However, this condition is necessary (a Seifert surface for an odd component link would lead to a contradiction in Theorem 1.4). Thus, odd linking numbers implies even number of components. This could also be proved by a combinatorial analysis.

Remark. Topologically, linear frame fields on cusps are the same as special singular frame fields on solid torus neighbourhoods of special singular curves. Thus, Theorem 4.3 and the corollary can be extended to links L made up of cusps and special singular curves.

5.1 How Special Singularities arise from Dehn Surgery

The work of section 4.2 raises the question of how natural the definition of the Chern-Simons invariant for hyperbolic 3-manifolds (possibly) with cusps is. The most natural result we could hope for is that the Chern-Simons invariant is continuous with respect to the geometric topology on $\mathbb{H}$, the space of all isometry classes of complete hyperbolic 3-manifolds of finite volume (see Section 2.1). That this is not the case will be proven in Sections 5.1, 5.2, and 5.3. In particular, we will prove that near a point in $\mathbb{H}$ representing a hyperbolic 3-manifold with at least one cusp, the Chern-Simons invariant (modulo half) takes on a dense set of values in the circle $\mathbb{R}/\frac{1}{2}\mathbb{Z}$. The proof will involve a connection between hyperbolic Dehn surgery (see Section 2.1) and special singularities. In this section we will show that performing hyperbolic Dehn surgery on a cusp produces a special singularity with singular curve the core geodesic upon which Dehn surgery was performed.

We will generally think in terms of Thurston's description of hyperbolic Dehn surgery, where the core geodesic will correspond to the t-axis in UHS and the nested tori around this curve will lie on cones (hyperbolically) equidistant from the t-axis (see [Thurston, 1979] page 4.13). We put the following frame field on the neighbourhood of the core geodesic (t-axis):

$$e_1 = t\,\partial/\partial t, \quad e_2 = (t)r\,\partial/\partial\theta, \quad e_3 = (-t)\partial/\partial r;$$

the "cylindrical co-ordinates frame field". In the limit, as we approach a cusped manifold in $\mathbb{H}$, this becomes a linear frame field (see Section 4.1) on a neighbourhood of the cusp.

We now show that this cylindrical frame field induces a special singularity at the core geodesic. As the tori nest closer and closer to the core geodesic, the cones in UHS become closer and closer to the t-axis. That is, these cones become nearly vertical and in the limit e_1 points in the direction of the core geodesic (the t-axis). Taking a slice in the direction transverse to the t-axis, we can "see" what the core geodesic "sees" and because we have a radial vector field in the transverse direction the core geodesic "sees" an index one singularity. So, the above cylindrical frame field has a special singularity at the core geodesic.

5.2 An Attempt to Prove that the Chern-Simons Invariant is Continuous with Respect to the Geometric Topology on $\mathbb{H}$

In this section we will set up a method for checking the continuity of the Chern-Simons invariant with respect to the geometric topology on $\mathbb{H}$. In Section 5.3 we will exploit this set-up to show that we do not have continuity, i.e., the Chern-Simons invariant of M, a cusped hyperbolic 3-manifold, is not the limit of the Chern-Simons invariants of all sequences of hyperbolic 3-manifolds M_i obtained by performing Dehn surgeries of type (p_i, q_i) on the cusp, where $p_i^2 + q_i^2$ approaches infinity.

We will split the Chern-Simons invariant for a Dehn surgered manifold M' into a sum of 3 terms:

1) a term coming from the torsion of the core geodesic,

2) a term coming from the thickened torus neighbourhood of the core geodesic (this neighbourhood corresponds to the chimney above a portion of the cone in Thurston's model of hyperbolic Dehn surgery), and

3) a term for the remaining portion of the manifold.

Continuity will hold for the second and third terms, but will fail for the first (torsion) term.

Dealing with the second term first, we consider a thickened torus neighbourhood of the core geodesic. We will show that the framing of Section 5.1 yields a Chern-Simons form Q which is pointwise zero. We compute $\theta_1 = (1/t)dt$, $\theta_2 = (1/t)r d\theta$, and $\theta_3 = (-1/t)dr$, so that $d\theta_1 = 0$,

$$d\theta_2 = (1/t)^2 r\, d\theta \wedge dt + (1/t) dr \wedge d\theta = \theta_2 \wedge \theta_1 - \theta_3 \wedge d\theta,$$

and

$$d\theta_3 = (1/t^2) dr \wedge dt = \theta_3 \wedge \theta_1.$$

Now solve the structural equations to find $\omega_{12} = \theta_2$, $\omega_{13} = \theta_3$, and $\omega_{23} = -d\theta$. Thus, the Chern-Simons form Q is equal to

$$\begin{aligned} &\omega_{12} \wedge \omega_{13} \wedge \omega_{23} + \omega_{12} \wedge \Omega_{12} + \omega_{13} \wedge \Omega_{13} + \omega_{23} \wedge \Omega_{23} \\ &= -\theta_2 \wedge \theta_3 \wedge d\theta - \theta_2 \wedge \theta_1 \wedge \theta_2 - \theta_3 \wedge \theta_1 \wedge \theta_3 + d\theta \wedge \theta_2 \wedge \theta_3 \\ &= 0. \end{aligned}$$

We have proved that the thickened torus neighbourhood contribution is 0, which continuously approaches the cusp neighbourhood contribution of 0 (see Section 4.1).

We now show that the term associated with the portion of M' outside of a solid torus neighbourhood of the core geodesic is continuous with respect to the geometric topology. Consider the thick and thin parts of M' separately (see [Thurston, 1979 Section 5.10]). By the above calculation plus the calculation of Section 4.1, we can set up the framing on our Dehn surgered manifolds M' so that the thin parts contribute nothing to $CS(M')$. This leaves us with compact thick parts which are approximately isometric (see [Thurston, 1979 page 5.59]) which gives us the desired continuity of the torsion term away from the core geodesic.

It remains to consider the continuity of the torsion term near the

5.3 The Chern-Simons Invariant Takes on a Dense Set of Values Near a Cusped Manifold in $\mathbb{H}$

We will now study the relationship of the Dehn surgery performed to the torsion of the core geodesic, g (which is represented by the t-axis). Recall that the total torsion is defined as $\int_c \omega_{23}$ (mod 2π, where c represents a framing of g with e_1 tangent to g. That is, the total torsion is the amount of $e_2 e_3$ spin as we travel along g.

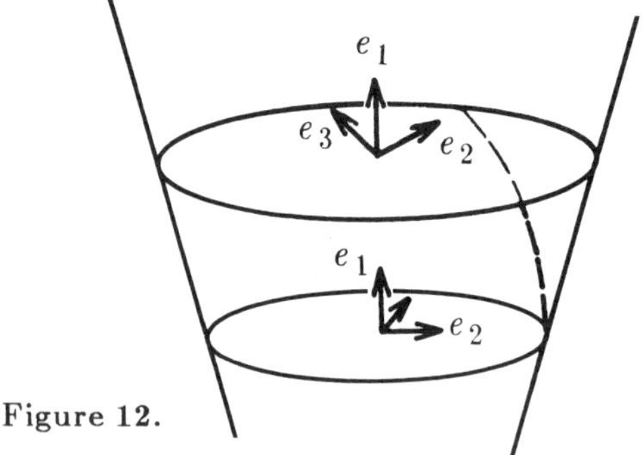

Figure 12.

From Figure 12 we see that the torsion is, in Thurston's notation (see [Thurston, 1979] Section 4.4), the imaginary part of $\log(H'(\alpha_2))$. Here, $\alpha_2 \in \pi_1$ (link of the ideal vertex at ∞) is such that $\|H'(\alpha_2)\|$ generates the image of the map $\alpha \to \|H'(\alpha)\|$; and $H'(\ )$ is the derivative of the holonomy of the link of the vertex. A generator for the kernel of this map is denoted α_1, which for Dehn surgery must have angle 2π. Thus, the torsion of g is the amount of spin of $H'(\alpha_2)$.

Let a be the meridian associated with g, and let b be a longitude. Performing hyperbolic Dehn surgery of type (p,q) corresponds to solving the equation

$$\log H'(\alpha_1) = p \log(H'(a)) + q \log(H'(b)) = 2\pi i.$$

This equation implies that $p\theta(a) + q\theta(b) = 2\pi$ where $\theta(a)$ is the argument of $\log(H'(a))$. Further, we can express $H'(\alpha_2)$ in terms of $H'(a)$ and $H'(b)$: $\log H'(\alpha_2) = p' \log H'(a) + q' \log H'(b)$ where p', q' are integers satisfying $p'q - q'p = 1$. Thus, the torsion of the core geodesic is $\theta(\alpha_2) = p'\theta(a) + q'\theta(b)$.

Computing, we see that the torsion satisfies

$$\begin{aligned} p'\theta(a) + q'\theta(b) &= ((1+pq')/q)\theta(a) + ((q'q)/q)\theta(b) \\ &= \theta(a)/q + (p\theta(a) + q\theta(b))(q'/q) \\ &= \theta(a)/q + 2\pi q'/q. \end{aligned}$$

As we approach the cusp, i.e., as $p^2 + q^2 \to \infty$, either $q \to \infty$ in which case $\theta(a)/q \to 0$ and the torsion of g approaches $2\pi q'/q$; or q is bounded, in which case $p \to \infty$ and $\theta(a) \to 0$ (since $p\theta(a) + q\theta(b) = 2\pi$), so that the torsion again approaches $2\pi q'/q$.

By judicious choice of q' and q (essentially forcing p' and p) we can achieve a dense set of torsion values for g between 0 and 2π. Section 5.2 implies that the Chern-Simons invariant (modulo ½) for high order Dehn surgeries is made up of a fixed term (in the limit) plus the torsion. This fixed term is equal to $CS(M)$, where M is the original cusped manifold, and, for $p^2 + q^2$ close to infinity, we have

$$CS(M_{(p,q)}) \approx CS(M) + (2\pi/4\pi)(q'/q) \text{ modulo } 1/2.$$

This takes on a dense set of values in the circle $\mathbb{R}/(\frac{1}{2}\mathbb{Z})$, and we have proved

Theorem 5.3. *Consider a complete hyperbolic 3-manifold M of finite volume with at least one cusp. In a neighbourhood of M in the geometric topology on $\mathbb{H}$ the Chern-Simons invariant (modulo ½) takes on a dense set of values in the circle $\mathbb{R}/\frac{1}{2}\mathbb{Z}$.*

Thus, the Chern-Simons invariant is not continuous with respect to the geometric topology on $\mathbb{H}$.

Remark The ambiguity in the choice of (p',q') does not affect q'/q modulo 1. Also, a different choice of longitude has no effect on q' and q.

References

Atiyah-Patodi-Singer, 1975a.
M.F. Atiyah, V.K. Patodi, and I.M. Singer, "Spectral Asymmetry and Riemannian Geometry I," *Math.Proc.Camb.Phil.Soc.*, **77**, pp. 43-69, 1975.

Atiyah-Patodi-Singer, 1975b.
M.F. Atiyah, V.K. Patodi, and I.M. Singer, "Spectral Asymmetry and Riemannian Geometry II," *Math.Proc.Camb.Phil.Soc.*, **78**, pp. 405-432, 1975.

Chern-Simons, 1974.
S. S. Chern and J. Simons, "Characteristic forms and geometric invariants," *Annals of Math.*, **99**, pp. 48-69, 1974.

Meyerhoff, 1981.
R. Meyerhoff, *The Chern-Simons Invariant for hyperbolic 3-manifolds*, 1981. Princeton University Ph.D. Thesis

Neumann-Zagier, 1986.
W. Neumann and D. Zagier, "Volumes of Hyperbolic 3-Manifolds," *Topology*, 1986.

Riley, 1970.
R. Riley, "An Elliptical Path from Parabolic Representations to Hyperbolic Structures," *Lecture Notes in Mathematics*, **722**, pp. 99-133, 1970.

Rolfsen, 1976.
Rolfsen, D., *Knots and links,* Publish or Perish, Berkeley, 1976.

Steenrod, 1951.
N. Steenrod, *The Topology of Fibre Bundles,* Princeton University Press, Princeton, 1951.

Thurston, 1979.
W. P. Thurston, *The Geometry and Topology of 3-Manifolds,* Princeton University Mathematics Department, 1979. Part of this material – plus additional material – will be published in book form by Princeton University Press

Yoshida, 1986.
T. Yoshida, "The eta invariant of hyperbolic 3-manifolds," *Inventiones Mathematicae*, 1986. (to appear)

R.Meyerhoff,
Boston University,
Boston,
Mass. 02215,
USA.

Note on a result of Boileau-Zieschang

*José María Montesinos**

1. Introduction

Various concepts of *genus* of a closed, orientable 3-manifold M can be defined. There is a well known *Heegaard genus*, $Hg(M)$, which is the smallest possible g among all Heegaard splittings (M,F_g) where F_g is a closed, orientable 2-manifold of genus g, embedded in M, such that the closures of the two components of $M-F$ are handlebodies V,W. Craggs proposes the concept of *the extended Nielson genus of* M (see [Craggs]), $EN(M)$, which is the smallest possible m among all possible presentations $P = | y_1, \ldots, y_m : r_1, \ldots, r_m |$ of $\pi_1(M)$ which come, by extended Nielsen operations, from any presentation of $\pi_1(M)$ associated to any Heegaard diagram $(F_g;\partial v,\partial q)$ of M. Remember that $(F_g;\partial v,\partial w)$, where v and w are complete systems of meridians of V and W, resp., defines a presentation for $\pi_1(M)$ with generators $v = (v_1, \ldots, v_g)$ and relations $w = (w_1, \ldots, w_g)$, where w_i is a word which records the intersections of ∂w_i with the (oriented) curves of ∂v (cfr. [Waldhausen, 1978]). It is also known that any two of these presentations are related by extending Nielsen operations [Singer, 1933].

Finally, I propose to call *big genus of* M, $Bg(M)$, the smallest possible integer r, such that, if $M_\star$ is M minus the interior of a 3-ball, $M_\star \times I$ has a handle decomposition $H_0 \cup rH_1 \cup rH_2$, with just one 0-handle and the same number r of 1- and 2-handles. We also have the *rank of* M, $rk(M)$ i.e. the minimum of elements of $\pi_1(M)$ which suffice to generate $\pi_1(M)$. We then have

$$rk(M) \leq EN(M), \; Bg(M) \leq Hg(M).$$

* Supported by CAICYT.

$$rk(M) \leq Hg(M)$$

From the recent and important work of Boileau-Zieschang [Boileau-Zieschang, 1984] we know that, in some cases, $rk(M) < Hg(M)$. In fact, they have exhibited some Seifert manifolds of rank 2 and Heegaard genus 3. Craggs remarks that that implies the existence of some M with $EN(M) < Hg(M)$ [Craggs]. In this note we will exhibit a concrete M satisfying this. The Andrews-Curtis conjecture is just the statement that if $rk(M) = 0$ then $EN(M) = 0$.

All this is related to Poincaré Conjecture. Waldhausen asked [Haken, 1970] if $rk(M) = Hg(M)$ always holds. This implies Poincaré Conjecture. Though the answer to his question is now known to be negative [Boileau-Zieschang, 1984], one still might ask if $rk(M) = Bg(M)$ always holds. This, again, implies Poincaré Conjecture, because if $Bg(M) = 0$ (i.e. $M^3_\star \times I$ is the 4-ball) then $S^3 = M \# M$, and M is S^3. Accordingly we state the problem

Problem. *Is $rk(M) = Bg(M)$, for every M?*

This problem can be broken into two problems.

1) Is $rk(M) = EN(M)$?

2) Is $EN(M) = Bg(M)$?

The first problem generalizes the Andrew-Curtis conjecture. Craggs has been studying if $EN(M) \geq Bg(M)$ in [Craggs], but since there is a difficulty with Lemma 5.3 of [Craggs], the question is still an open problem. We can test this second problem with the example M which has $EN(M) = 2$ and $Hg(M) = 3$ mentioned above. We will obtain a handle presentation $H_0 \cup 3H_1 \cup 3H_2$, for $M_\star \times I$, realizing $P = [y_1, y_2, y_3 : r_1, r_2, r_3]$ such that reduces to one with two generators and two relations by generalized Nielsen operations. Then we will realize these Nielsen operations by geometric moves ("band moves") on the handle presentation, to obtain a simpler handle presentation whose associated group presentation is of the form $[y_1, y_2, y_3 : r_1, r_2, y_3 = y_3(y_1, y_2)]$. This can be easily done, and in an infinite number of ways, but the so called Mazur phenomenon makes the geometric cancellation of the handles corresponding to y_3 and $y_3 = y_3(y_1, y_2)$ impossible. We have done this in many different ways, with negative result. One wonders if there is some sort of obstruction measuring this experimental impossibility, and one gets the strong feeling that, at least for this particular example, $Bg(M) = 3$.

2. Heegaard diagrams

To define $M(a, b)$ we are using a modification of Seifert notation [Seifert, 1980 Satz 5; pages 3 & 6]: the first two letters O, o will mean that the Seifert manifold is orientable and that the base of the Seifert fibration is orientable; the zero, in the third place, is the genus of the base. We depart from the Seifert notation in using angle brackets instead of parentheses, to indicate that one invariant might not be normalized (in our case b can be negative or bigger than $2a + 1$). When this invariant is normalized, as in the statement of Theorem 3, we will use parentheses.

In [Boileau-Zieschang, 1984] it is proved that the Seifert manifolds $M(a,b) = \langle O\ o\ 0 \mid -1;\ (2,1),\ (2,1),\ (2,1),\ (2a+1,b)\rangle$, where $b \neq -a$, $-(a+1)$, have genus 3 and rank 2.

Proposition 1. *Let* $(F_3;\ v = (v_1,v_2,v_3),\ (w_1,w_2))$ *be the Heegaard diagram of Figure 1. This diagram represents an orientable 3-manifold* N, *bounded by a torus* T^2 *with basis* m, 1. *Moreover the* $180°$ *rotation around the point* O *extends to an involution on* N *which defines a 2-fold covering* $p{:}N \to B^3$ *with the branching set shown in Figure 2. The curves* $(m,1)$ *are liftings of the arcs* $(\overline{m},\overline{1})$ *of Figure 2.*

Proof.

The quotient of N under the rotation of $180°$ around O is the ball B^3 of Figure 3, to which two move 3-balls should be pasted along the arcs pw_1, pw_2. From here, Figure 2 can be easily obtained.

□

Corollary 2. *A Heegaard diagram for the Seifert manifold* $M(a,b) = \langle O\ o\ 0 \mid -1;\ (2,1),\ (2,1),\ (2,1),\ (2a+1,b)\rangle$ *is* $(F_3, v = (v_1, v_2, v_3),\ w = (w_1, w_2, w_3(a,b)))$, *where* $(F_3;\ v, (w_1, w_2))$ *is the diagram of Figure 1, and the curve* $w_3(a,b)$ *is a simple curve in* $F_3 - (v \cup w_1 \cup w_2)$, *homologous to* $(2a+1)m + b\,1$.

Proof. The manifold represented by the Heegaard diagram $(F_3; v, w)$ is clearly the 2-fold covering of S^3 branched over the link obtained by pasting a rational tangle $(2a+1)/b$ to the ball of Figure 2. Using [Montesinos, 1973], that manifold is $M(a,b)$.

□

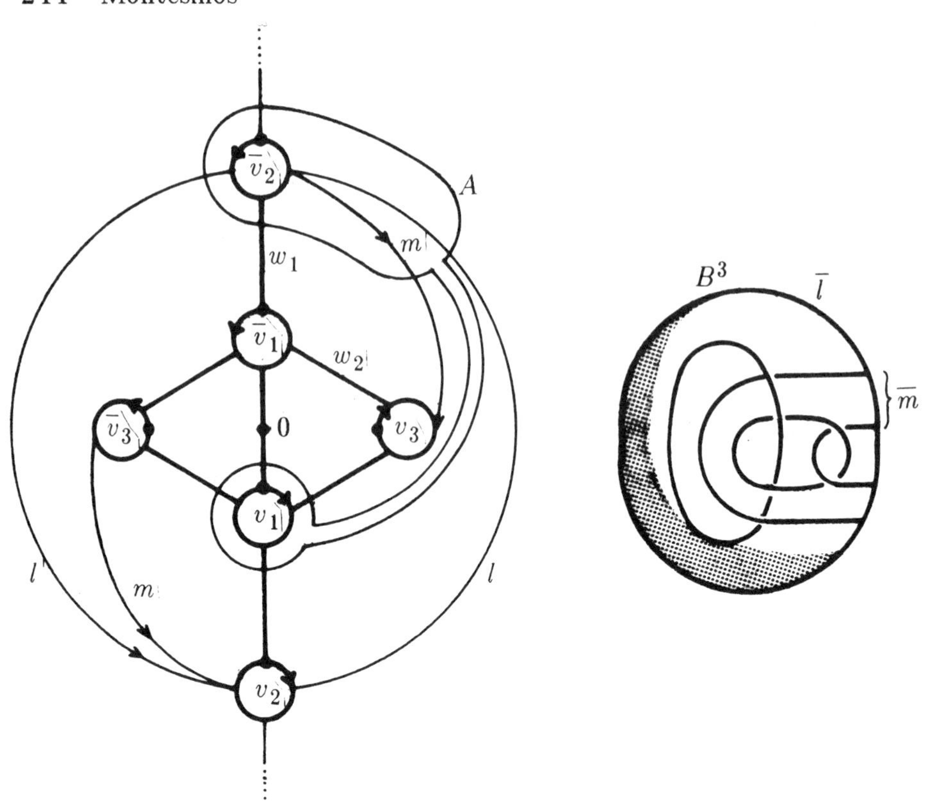

Figure 1 **Figure 2.**

We now change the system (v_1, v_2, v_3) to the system $(A, v_2 = B, v_3 = C)$ where A is shown in Figure 1. We obtain a new Heegaard diagram for $M(a,b)$ which, for the simpler case $(a,b) = (1,1)$, is depicted in Figure 4. We study this simpler case.

The presentation for $\pi_1(M(1,1))$ coming for this diagram is

$$\{ A, B, C : BA^{-1}BA^{-3} = 1, CACA^{-1} = 1 \ (CBA^{-1})^3(BA^{-1})^2 = 1 \} \qquad (1)$$

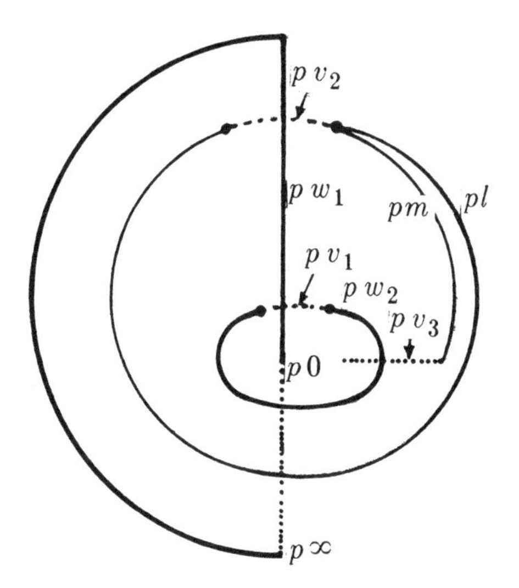

Figure 3.

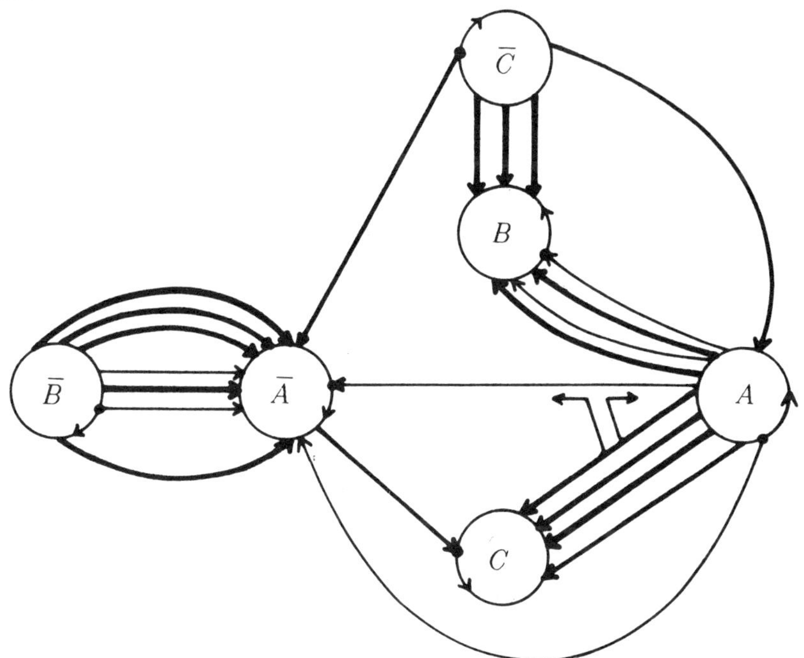

Figure 4.

3. Band moves

We now change the above group presentation by the following sequence of Nielsen operations, by which only the third relation is modified.

$(CBA^{-1})^3(BA^{-1})^2,$

$(CBA^{-1})^3(BA^{-1})^2.\underline{(BA^{-1})^{-2}A^2} = CBA^{-1}CBA^{-1}CBA,$

$CBA^{-1}C.\underline{C^{-1}AC^{-1}A^{-1}}.BA^{-1}CBA = CBC^{-1}A^{-1}BA^{-1}CBA,$

$CBC^{-1}A^{-1}B.\underline{B^{-1}AB^{-1}A^3}.A^{-1}CBA = CBC^{-1}B^{-1}AACBA,$

$CBC^{-1}B^{-1}AAC.\underline{C^{-1}A^{-1}C^{-1}A}.BA = CBC^{-1}B^{-1}AC^{-1}ABA,$

$CBC^{-1}B^{-1}AC^{-1}.\underline{CA^{-1}CA}.ABA = CBC^{-1}B^{-1}CAABA,$

$CBC^{-1}B^{-1}CAA.\underline{AB^{-1}A^3B^{-1}}.BA = CBC^{-1}B^{-1}CAAAB^{-1}AAAA,$

$CBC^{-1}B^{-1}CA^3B^{-1}.\underline{BA^{-3}BA^{-1}}.A^4 = CBC^{-1}B^{-1}CBA^3,$

$CBC^{-1}B^{-1}CBA^2.\underline{A^{-2}BA^{-1}BA^{-1}}.A = CBC^{-1}B^{-1}CB^2A^{-1}B$

Thus the above presentation becomes the presentation

$$\{ A,B,C : BA^{-1}BA^{-3} = 1, BCBC^{-1}B^{-1}CB^2 = A \}.$$

This proves the following theorem:

Theorem 3. *The Seifert manifold*

$$M(1,1) = (Oo0 \mid -1; (2,1),(2,1),(2,1),(3,1))$$

has $Hg(M(1,1)) = 3$ *and* $EN(M(1,1)) = 2$.

Remark It might be that the same statement is true for the other manifolds $M(a,b)$ of Heegaard genus 3, but the above manipulation with the relations becomes very difficult in cases other than $M(1,1)$.

4. Testing $EN(M) = Bg(M)$

We realise the Nielsen operations used to prove Theorem 3 by means of band moves done in a handle presentation $M(1,1)_\star \times I = H_0 \cup 3H_1 \cup 3H_2$. This handle presentation is depicted in Figure 4, which must be interpreted as follows. The picture lies on S^3 with pairs, $A,\bar{A}$, $B,\bar{B}$, $C,\bar{C}$, of 3-balls deleted and pasted together to get $3 \# S^1 \times S^2 = \partial(H_0 \cup 3H_1)$. The attaching spheres of the 2-handles are the curves of the Heegaard diagram and their framings are assumed to be given by parallel curves lying on the paper, (see [Montesinos, 1979]). The geometric counterpart of the Nielsen operations are the band moves explained in [Montesinos, 1979]. We will give a set of pictures starting with Figure 4 and finishing with Figure 12. Each picture comes from the preceding one by a band move followed by some isotopic reduction of the diagram. In each picture is sketched the band move which gives rise to the next picture. The last Figure realizes the group presentation written just before the statement of Theorem 3. We can isotope Figure 12 a little further to eliminate the two obvious loops, but the rest of the picture is so tangled that one convinces oneself easily of the impossibility of the picture.

These pictures have a different interpretation. Projecting them on the (paper) plane we get singular Heegaard diagrams for $M(1,1)$. The numbers are placed to help the interested reader in finding the singularities of the meridian disks. But we advise the reader to manipulate the singular Heegaard diagrams with care: A little mistake (in the region $2,\bar{3},\bar{4}$) caused me to believe incorrectly that the big genus of $M(1,1)$ was 2.

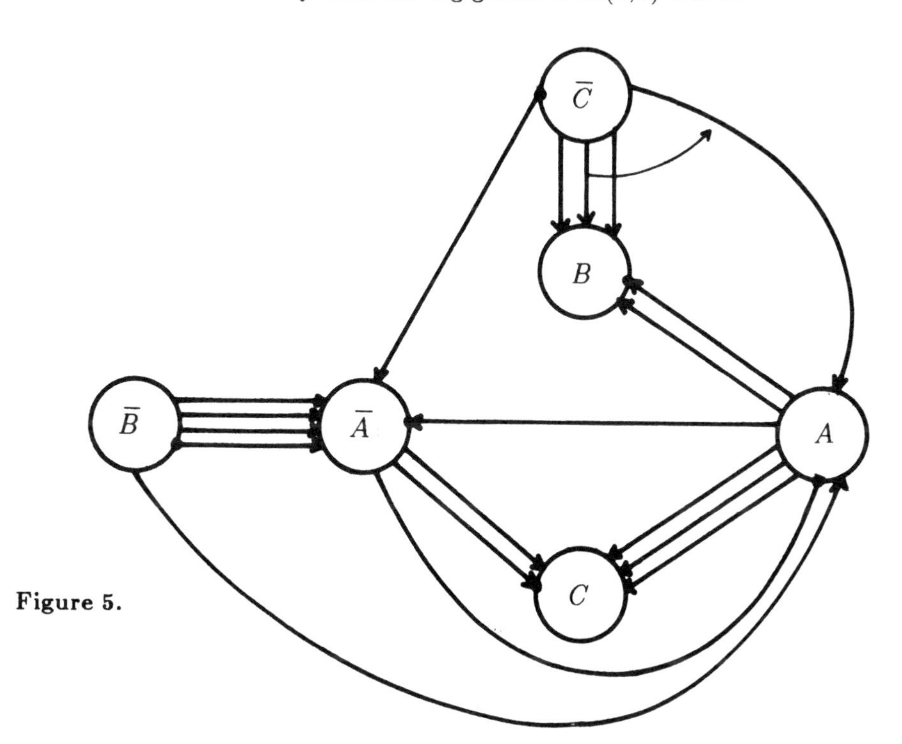

Figure 5.

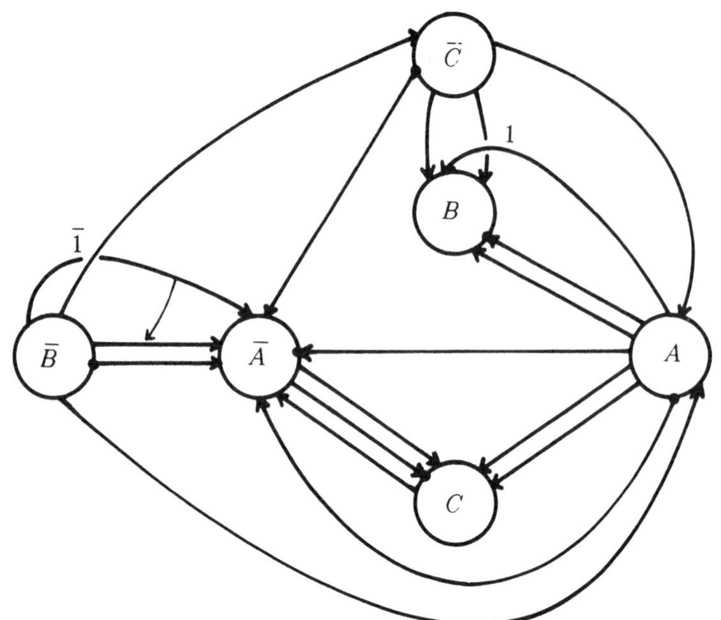

Figure 6.

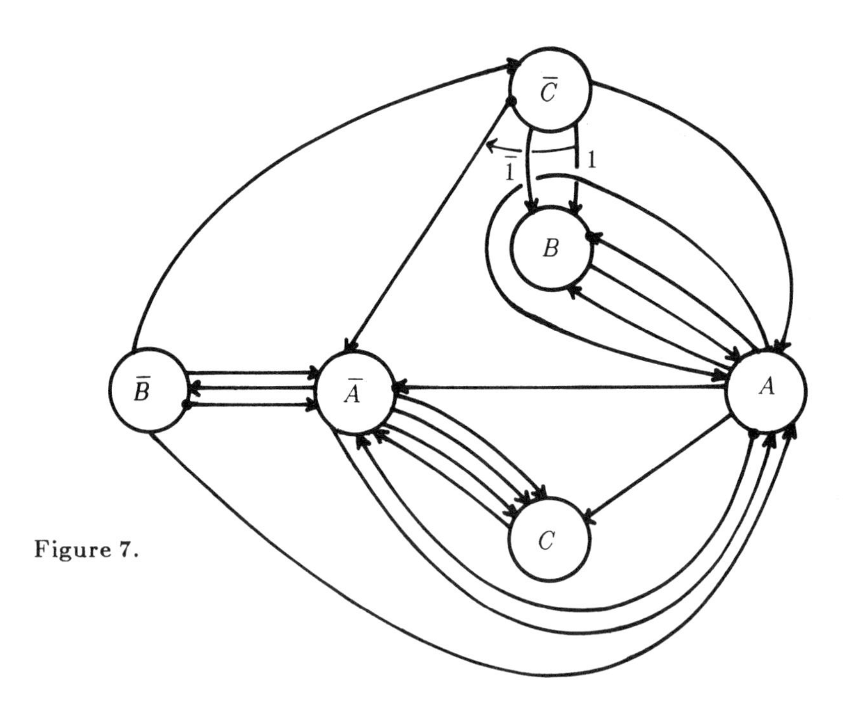

Figure 7.

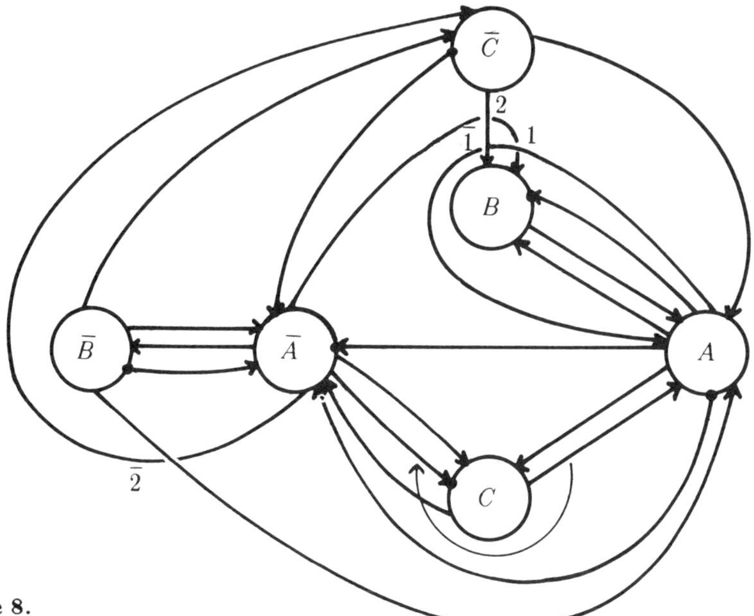

Figure 8.

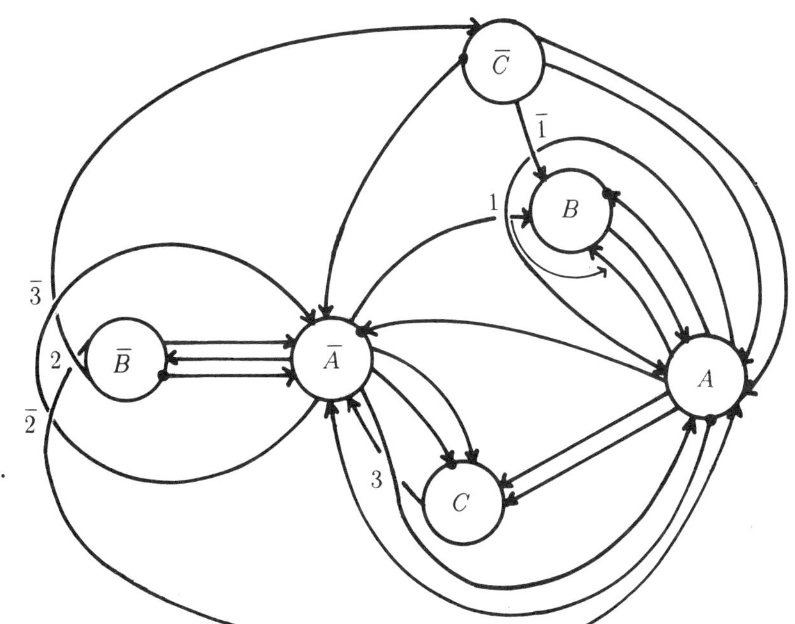

Figure 9.

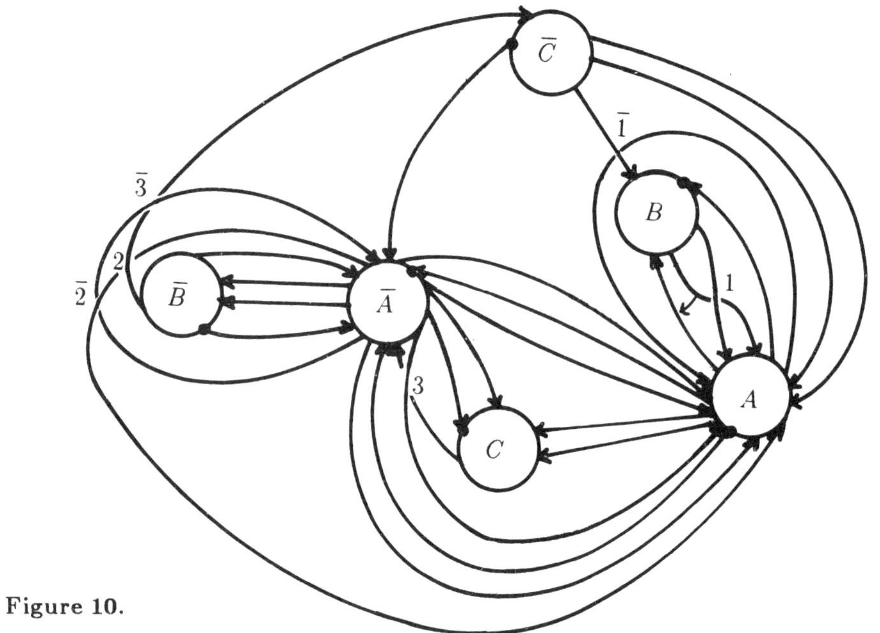

Figure 10.

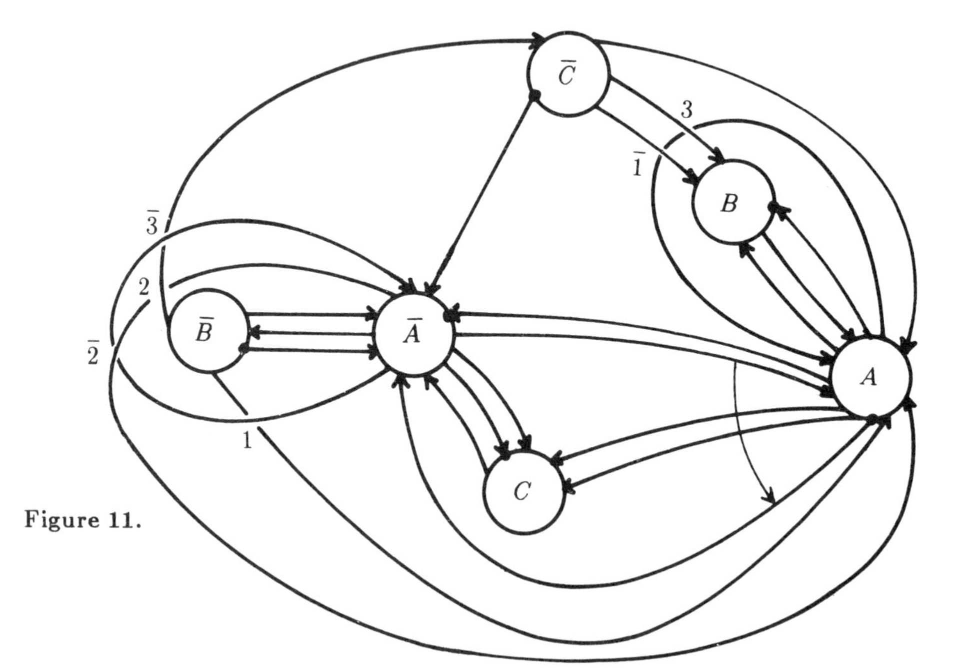

Figure 11.

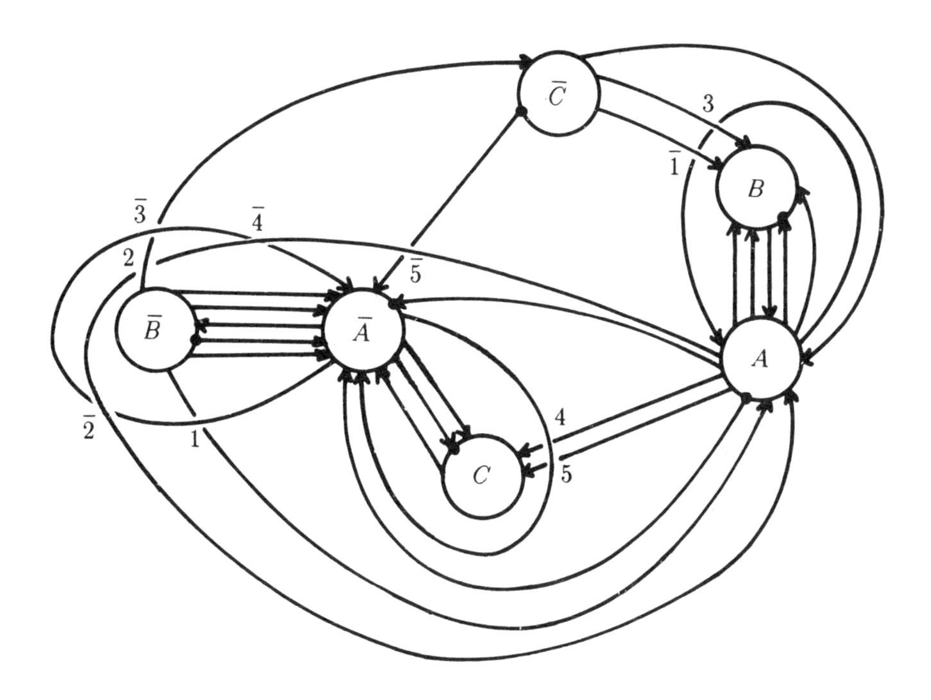

Figure 12.

References

Boileau-Zieschang, 1984.
C. M. Boileau and H. Zieschang, "Heegaard genus of closed orientable Seifert 3-manifolds," *Inventiones Math.*, **76**, pp. 455-468, 1984.

Craggs.
R. Craggs, *The Poincaré conjecture is equivalent to a restricted form of the Andrews-Curtis conjecture.* preprint

Haken, 1970.
W. Haken, "Various aspects of the 3-dimensional Poincaré problem," in *Topology of Manifolds*, pp. 140-152, Markham, Chicago, 1970. Proc. Inst. Univ. of Georgia 1969

Montesinos, 1973.
J. M. Montesinos, "Variedades de Seifert que son recubridores cíclos ramificados de dos hojas," *Boletín Soc. Mat. Mexicana*, pp. 1-32, 1973.

Montesinos, 1979.
J. M. Montesinos, "Heegaard diagrams for closed 4-manifolds," in *Geometric Topology*, ed. J. C. Cantrell, pp. 219-237, Academic Press, New York, 1979. Proc. Inst. Univ. of Georgia 1977

Seifert, 1980.
H. Seifert, "Topologie dreidimensionaler gefaster Raüme," in *Seifert and Threlfall: A Textbook of Topology*, pp. 147-288, Academic Press, London, 1980. English translation of the 1933 paper

Singer, 1933.
J. Singer, "Three dimensional manifolds and their Heegaard diagrams," *Trans. AMS*, **35**, pp. 88-111, 1933.

Waldhausen, 1978.
Waldhausen, F., "Some problems on 3-manifolds," *Proc. of Symposia in Pure Math.*, **32**, pp. 313-322, 1978.

J. Montesinos,
Departamento de Geometria Y Topologia,
Facultad de Ciences,
Zaragoza,
SPAIN.

Homology Branched Surfaces: Thurston's Norm on $H_2(M^3)$

Ulrich Oertel

In his paper "A norm on the homology of 3-manifolds" [Thurston, 1986], W. Thurston introduced the norm which is the subject of this paper. To prove some of the deeper facts about the norm, Thurston used foliations of 3-manifolds. In closely related work David Gabai has constructed foliations with compact leaves which represent homology classes, [Gabai, 1983], [Gabai]. In this paper we used branched surfaces to describe Thurston's norm. Beyond the initial elementary facts, our development will be independent of Thurston's. In addition, we will describe the relationship between the "homology branched surfaces" which arise and the foliations constructed by Gabai.

Needless to say, this paper owes its existence to W. Thurston. I thank Robert Edwards, with whom I worked through Thurston's paper.

1. Definitions and Statement of Results

Let $M = M^3$ be a compact 3-manifold, possibly with boundary. We shall always assume that M is orientable and irreducible. Further, if M has boundary we assume that M is ∂-irreducible, i.e. that ∂M is incompressible in M. We will be concerned with embedded surface representatives of $H_2(M, \partial M) = H_2(M, \partial M; \mathbb{Z})$, classes which appear as integer lattice points in $H_2(M, \partial M; \mathbb{R})$. Most theorems will be stated in this section in full generality and will be proved in later sections, but for simplicity they will be proved only in the case $\partial M = \varnothing$. Thus statements of theorems will involve $H_2(M, \partial M)$ while proofs and discussions deal with $H_2(M)$.

We begin with some definitions from [Thurston, 1986] and [Oertel, 1984]. If F is a surface, $\chi_-(F)$ is defined by $\chi_-(F) = \sum_i |\chi(F_i)|$ where the sum is over components F_i of F with $\chi(F_i) \leq 0$. Assuming $\partial M = \varnothing$, each class $a \in H_2(M)$ is dual to a cohomology class $\phi \in [M, S^1] \cong H^1(M)$. When ϕ is made transverse to a point $\star$ of S, the surface $\phi^{-1}(\star)$ represents the homology class $a \in H_2(M)$. For an integral lattice homology class $a \in H_2(M; \mathbb{R})$ Thurston defines the *norm* of the class a by $x(a) = \min \{\chi_-(F): [F]=a, F \text{ an embedded surface in } M\}$. A surface F which realizes this minimum, so that $x(a) = \chi_-(F)$ and $[F] = a$, is called a *minimal-χ_- representative* of its homology class. W. Thurston begins his paper [Thurston, 1986] on the norm on second homology by proving the following:

Lemma 1 (Thurston).

(a) If $a \in H_2(m, \partial M; \mathbf{Q})$ is an integer lattice point and k is an integer, then $x(ka) = |k|x(a)$.

(b) If a, b $H_2(M, \partial M; \mathbf{Q})$ are integer lattice points then

$$x(a+b) \leq x(a) + x(b).$$

The reader is referred to Thurston, 1986 for the proof of the lemma. Thurston proves (a) by showing that a minimal-χ_- representative of $ka \in H_2(M)$ must be a union of k minimal-χ_- representatives of a. By Lemma 1, the function x defined on the integer lattice points of $H_2(M; \mathbf{Q})$ can be extended to a rational-valued pseudonorm on $H_2(M; \mathbf{Q})$ as follows: for each $a \in H_2(m; \mathbf{Q})$ choose an integer n so that (na) in an integer lattice point and define $x(a) = (1/n)x(na)$. This ensures that x will be linear on rays from the origin.

We will not use any other results from Thurston's paper; instead we shall describe the norm using branched surface techniques.

A *branched surface* B with generic branch locus is a compact space locally modeled on the space shown in Figure 1.1a. We shall also use branched surfaces with non-generic branch-locus; these are branched surfaces whose branch loci need not be self-transverse with at most double-point self-intersections. The *sectors* of B are the closures in B of the components of B-(branch locus of B). An *invariant measure* $\vec{w}$ on a branched surface is an assignment of a non-negative weight w_i to each sector E_i of B such that *branch equations* hold as shown in Figure 1.1a. Figure 1.1b shows the local model of the *fibered neighbourhood* of B which we denote $N(B)$. The figure also shows the portions of $N(B)$ which we call the *horizontal boundary* $\partial_h N(B)$ and the *vertical boundary* $\partial_v N(B)$. There is a projection map π associated to B which collapses fibers of $N(B)$; π can be regarded as a map $\pi: N(B) \to B$ or as a map $\pi: M \to M/\sim$ where $M/\sim$ is the

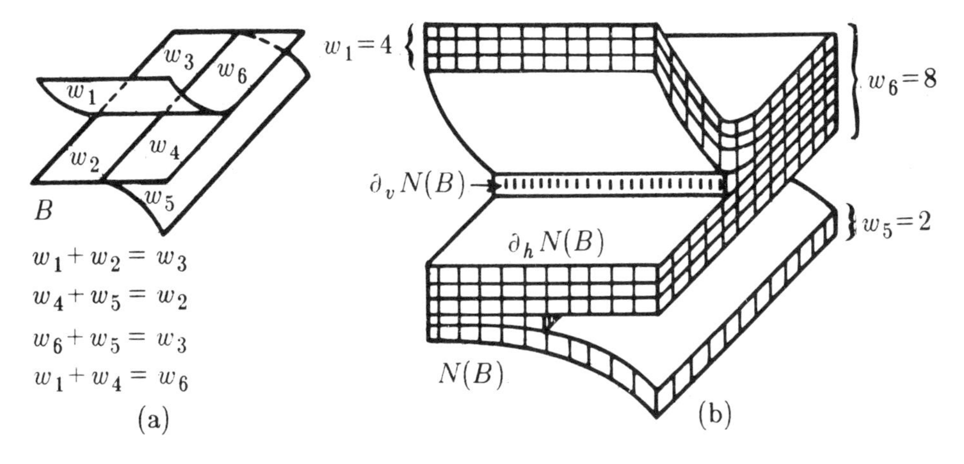

Figure 1.1.

quotient space in which each fiber of $N(B)$ becomes a point of $M/\sim$. The space $M/\sim$ can be identified with M. An integer invariant measure $\vec{w}$ on B determines a surface $B(\vec{w})$ *carried* by B as shown in Figure 1.1b. The surface $B(\vec{w})$ can be embedded in $N(B)$ transverse to fibers, and conversely any surface embedded in $N(B)$ transverse to fibers is carried by B. After possibly replacing $F = B(\vec{w})$ by $\partial N(F)$, we may assume that F is embedded in $N(B)$ with $\partial_h N(B) \subset F$ as shown in Figure 1.1b.

In [Oertel, 1984] conditions are given on a branched surface B embedded in M which ensure that every surface carried by B is injective (∂-injective):

(i) B has no *disc of contact* (or half-disc of contact), i.e. there does not exist a disc (half-disc) D embedded in $N(B)$ transverse to fibers with $\partial D \subset$ int $(\partial_v N(B))$ $(\cup \partial M)$.

(ii) $\partial_h N(B)$ is incompressible (and ∂-incompressible) in $M - \mathring{N}(B)$.

(iii) There is no monogon in $M - \mathring{N}(B)$, i.e. there does not exist a disc D properly embedded in $M - \mathring{N}(B)$ with $\partial D \cap \partial_v N(B)$ a fiber of $\partial_v N(B)$ and with $\partial D \cap \partial_h N(B)$ equal to the complementary arc of ∂D.

(iv) V has no Reeb components, i.e. B does not carry a torus T bounding a solid torus $\overline{T}$ such that for some surface G carried by B with positive weights, $\overline{T} \cap G$ is a collection of compressing discs for T. (B does not carry an annulus A cutting a solid torus $\overline{T}$ from M such that for some surface G carried by B with positive weights, $G \cap \overline{T}$ is a collection of ∂-compressing discs for A.)

We say a branched surface B embedded in M is a RIB, a *"Reebless" incompressible branched surface*, if it satisfies (i) – (iv) and carries some surface with positive weights. The following theorem is the fundamental theorem about incompressible branched surfaces. A weaker version appears in [Floyd-Oertel, 1984]; the version given here is proved in [Oertel, 1984].

Theorem 2 [Oertel, 1984].

(a) Given an orientable, irreducible, ∂-irreducible 3-manifold M, there exists a finite collection of RIB's such that every 2-sided incompressible, ∂-incompressible surface in M is carried by a RIB of the collection.

(b) Any surface F carried by a RIB in M is injective and ∂-injective. (recall that F is injective (∂-injective) if and only if $\partial N(F)$ is incompressible (∂-incompressible).)

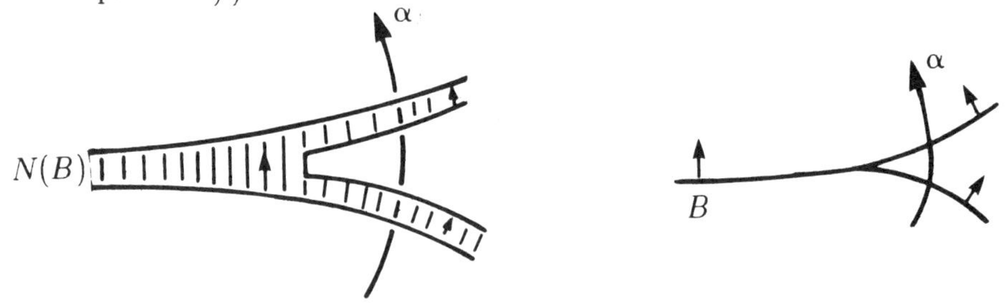

Figure 1.2.

We shall prove analogous results for branched surfaces carrying minimal-χ_- representatives of homology classes. First we identify branched surfaces which carry only surfaces non-trivial in homology. An *oriented branched surface* is a branched surface B with an orientation on the 1-foliation of $N(B)$ whose leaves are the fibers of $N(B)$. A *closed oriented transversal* for an oriented branched surface B is a simple closed curve α in M which intersects $N(B)$ in fibers of $N(B)$ so that the orientation of α coincides with the orientations of the fibers $\alpha \cap N(B)$ of $N(B)$, see Figure 1.2. A *homology branched surface* is an oriented branched surface B embedded in M such that for each point of B there exists a closed oriented transversal of B intersecting B at the given point and possibly other points. Such a branched surface could be called an *oriented transversely recurrent* branched surface. A *homology RIB* is a branched surface which is simultaneously a homology branched surface and a RIB.

Proposition 3.

(a) *If B is a homology branched surface, then any surface carried by B is non-trivial in $H_2(M, \partial M)$.*

(b) *If B is a homology RIB, then any surface carried by B is an incompressible (∂-incompressible) representative of a non-trivial class in $H_2(M, \partial M)$.*

(c) *If B is an oriented branched surface; in particular, if B is a homology RIB, and $B(\vec{w})$ denotes the surface $B(\vec{w})$ with orientation induced by the orientation on B, then*

$$[B(\vec{w}_1)] + [B(\vec{w}_2)] = [B(\vec{w}_1 + \vec{w}_2)]$$

as classes in $H_2(M, \partial M)$. Here $\vec{w}_1$ and $\vec{w}_2$ have non-negative integer entries.

Of course a surface carried by a homology RIB need not be a minimal-χ_- representative of its homology class, but the following theorem shows that homology RIB's can be constructed which carry only minimal-χ_- representatives of homology. If B is an oriented branched surface (homology RIB), which carries only minimal-χ_- representatives of homology, then we say B is a *taut oriented branched surface (taut homology RIB)*

Theorem 4. *Given M, there is a finite collection of taut homology RIB's such that for each class of $H_2(M, \partial M)$ there exists a minimal-χ_- representative of the class carried with positive weights by one of the RIB's of the collection. The taut homology RIB's may be chosen so that for each B of the collection $M - B$ is connected and so that closed oriented transversals can be found through any given point of B which intersect B only at that point.*

The fact that the branched surfaces in the theorem are RIB's deserves mention, but is really not essential in the logical sequence of the paper. For most purposes it is enough to know that they are taut and oriented.

There is a connection between the taut homology RIB's of Theorem 4 and David Gabai's work on foliations of 3-manifolds, see [Gabai, 1983]. For those readers who are familiar with Gabai's work, we describe the relationship. Suppose M is a closed 3-manifold. If B is a homology RIB, then $(M - \mathring{N}(B), \partial_v N(B))$ is a sutured manifold, where $\partial_v N(B)$ is the union of sutures. Gabai proves in [Gabai, 1983] that if a sutured manifold is taut, it has a "taut foliation" which is either C^∞ or has finite depth. We shall see that the sutured manifold $M - \mathring{N}(B)$ is taut if B is a taut homology RIB. If we insert one of Gabai's foliations $\mathcal{F}$ in $M - \mathring{N}(B)$, and then apply π to replace $N(B)$ by B, we get a singular foliation with finitely many singular branched-surface leaves as indicated schematically in Figure 1.3.

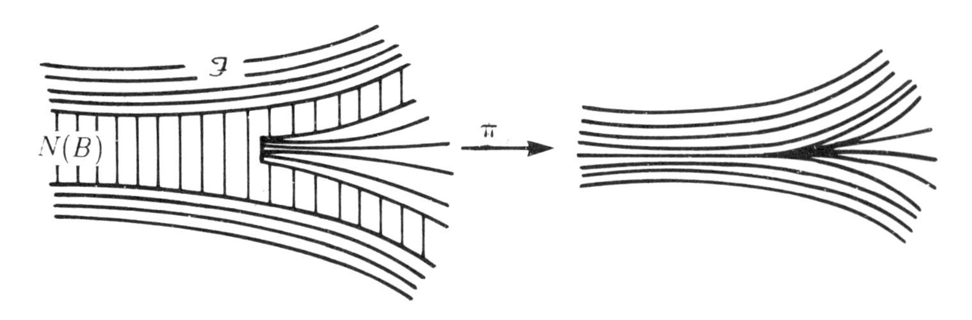

Figure 1.3.

Corollary 5. *Let M be a closed orientable irreducible 3-manifold.*

(a) *If B is a taut homology RIB in M, then $M - \mathring{N}(B)$ is a taut sutured manifold.*

(b) *If B is a taut homology RIB in M then $M - \mathring{N}(B)$ has a taut foliation which is either C^∞ or has finite depth, hence M has a corresponding singular foliation.*

(c) *There is a finite collection of singular foliations (C^∞ or of finite depth), each with finitely many singular leaves which are taut homology RIB's such that a minimal-χ_- representative of each homology class of $H_2(M; \partial M)$ is carried with positive weights by a singular leaf of one of the foliations.*

If M is a simple (atoroidal and anannular) manifold, there is a stronger version of Theorem 4:

Theorem 6. *Suppose M is a simple (orientable, irreducible, and ∂-irreducible) 3-manifold. Then there is a finite collection of taut homology RIB's such that every minimal-χ_- representative of homology is carried with positive weights by a RIB of the collection.*

We return now to Thurston's norm on homology. Theorem 4 yields the following corollary:

Corollary 7 (Thurston).

(a) *The pseudonorm x on $H_2(M, \partial M; \mathbf{Q})$ described earlier has a unique continuous extension to $H_2(M, \partial; \mathbb{R})$.*

(b) *The unit ball $U_x = \{a \in H_2(M, \partial M; \mathbb{R}) : x(a) \leq 1\}$ is a finite, convex, possibly non-compact polyhedron.*

(c) *If M is simple, x is a norm and U_x is a finite convex polyhedron.*

Proposition 8. *Suppose $g, h \in H_2(M, \partial M; \mathbb{R})$ are integer lattice points such that $x(g) \neq 0$ and $x(h) \neq 0$. Then g and h lie in the cone over the same face of ∂U_x if and only if there exists a taut oriented branched surface carrying minimal-χ_- representatives of g and h.*

Theorem 9 (Thurston). *Minimal-χ_- representatives of integer lattice homology classes in $H_2(M, \partial M; \mathbb{R})$ in the cone over an open face of ∂U_x are either all fibers of some fibration of M over S^1, or none are fibers.*

This paper leaves an important question unresolved:

Question. Given a (simple) 3-manifold M, let U_x be the unit ball of the norm x on $H_2(M, \partial M; \mathbb{R})$. For each face of ∂U_x is it possible to find a taut oriented branched surface which carries a minimal-χ_- representative of every projective homology class represented in the face? For each face of ∂U_x is it possible to find a taut oriented branched surface which carries *every* minimal-χ_- representative of integer lattice homology classes in the cone over the face of ∂U_x?

2. Proof of Proposition 3 and Theorem 4

Proof of Proposition 3:

(a) Suppose B is a homology branched surface and F is carried by B. Suppose F is embedded in $N(B)$ transverse to fibers. Since B has a closed oriented transversal passing through any given fiber of $N(B)$, we can find a closed oriented transversal α intersecting F in at least one point. Then F has non-zero algebraic intersection number with α, therefore F is non-trivial in homology.

(b) This part of the proposition follows immediately from Theorem 2.

(c) To prove that $[B(\vec{w}_1)] + [B\vec{w}_2)] = [B(\vec{w}_1 + \vec{w}_2)]$ in $H_2(M)$ we must interpret the addition of invariant measures geometrically. If the surfaces $B(\vec{w}_1)$ and $B(\vec{w}_2)$ are embedded in $N(B)$ transverse to fibers and transverse to each other, then clearly $B(\vec{w}_1 + \vec{w}_2)$ is obtained from $B(\vec{w}_1)$ and $B(\vec{w}_2)$ by *switching* on curves of intersection as shown in Figure 2.1. Since the orientations of $B(\vec{w}_1)$, $B(\vec{w}_2)$, and $B(\vec{w}_1 + \vec{w}_2)$ are all induced by the orientation of B,

$$[B(\vec{w}_1)] + [B(\vec{w}_2)] = [B(\vec{w}_1 + \vec{w}_2)].$$

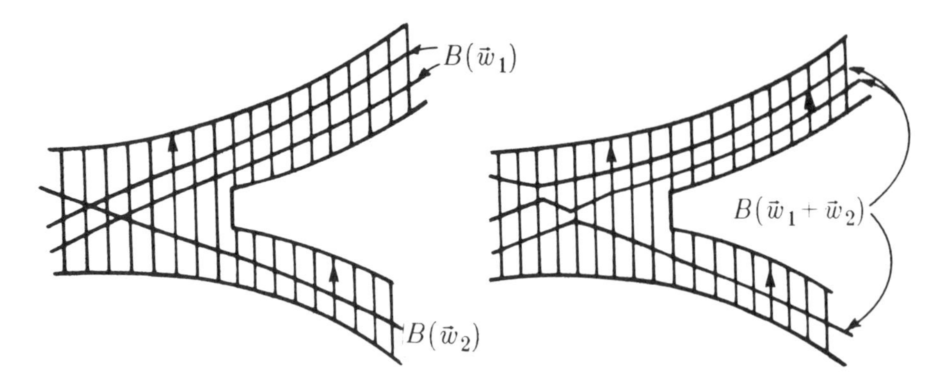

Figure 2.1.

□

The proof of Theorem 4 is based on Haken's normal surface theory, see [Haken, 1961], [Schubert, 1961], which is also described in [Jaco-Oertel, 1984], [Morgan-Shalen], and [Floyd-Oertel, 1984]. The theory of incompressible branched surfaces developed in [Floyd-Oertel, 1984] and [Oertel, 1984] depends heavily on Haken's theory. Here we use normal surface theory to construct homology RIB's. In the following proof we assume that M has a suitable handle-decomposition; for example, a handle-decomposition obtained from a triangulation of M. If F is a normal surface relative to this handle-decomposition, $\gamma(F)$ denotes the complexity of F, where the complexity is defined as the number of discs in which F intersects the union of 2-handles. The complexity is a crude measure of area.

Proof of Theorem 4: The proof will be divided into 5 steps. Given a minimal-χ_- representative of homology, we replace it by a normal surface F, then in step 1 we construct a RIB B_F carrying F with positive weights. In step 2 we show that B_F can be oriented so that the orientation of B_F induces the orientation of F. In step 3 we modify B_F and show that the modified B_F is a homology RIB. In step 4 we show that there are only finitely many possibilities for B_F. Finally, in step 5 we show that the homology RIB's constructed in the previous steps are taut.

Step 1: *Construction of B_F.* Let the oriented surface F be a minimal-χ_- representative of its homology class. Since F is incompressible it may be replaced by an isotopic normal surface. Among minimal-χ_- normal surface representatives of its homology class, we may assume F has minimal complexity.

We construct a branched surface $\overline{B}_F$ from the normal surface F as in [Oertel, 1984]. First we replace F by $\partial N(F)$, two parallel copies of the normal surface F. Then $N(\overline{B}_F) = \overline{N}_F$ is obtained from F as the union of products $D^2 \times I$ where $D^2 \times 0$ and $D^2 \times 1$ are adjacent normally isotopic discs of $F \cap H$; H is a 0, 1, or 2-handle; and $\partial D^2 \times I \subset \partial H$. If we choose the different $D^2 \times I$ product structures so that I-factors agree on handle-boundaries, then $\overline{N}_F$ becomes the fibered neighbourhood of a branched surface $\overline{B}_F$ whose branch locus is non-generic. Since F is incompressible and of minimal complexity in its isotopy class we can proceed as in [Oertel, 1984]. Let

$$\overline{L}_F = (\overline{N}_F \text{ cut on } F - \text{int}_F(\partial_h \overline{N}_F));$$

clearly $\overline{L}_F$ is an I-bundle. We say a component J of $\overline{L}_F$ is *trivial* if each curve in $\partial_h J$ bounds a disc (half-disc) in F. Such a trivial component yields a disc of contact. In [Oertel, 1984] it is shown that each trivial component is a product $E \times I$ where E is a punctured disc. We modify $\overline{N}_F = N(\overline{B}_F)$ by removing the interiors of fibers of trivial components of $\overline{L}_F$ to get $N_F = N(B_F)$. The proof of Theorem 3 in [Oertel, 1984] shows that B_F is a RIB. The fact that B_F is a RIB, however, plays no essential part of this proof; the reader need not concern himself with the details.

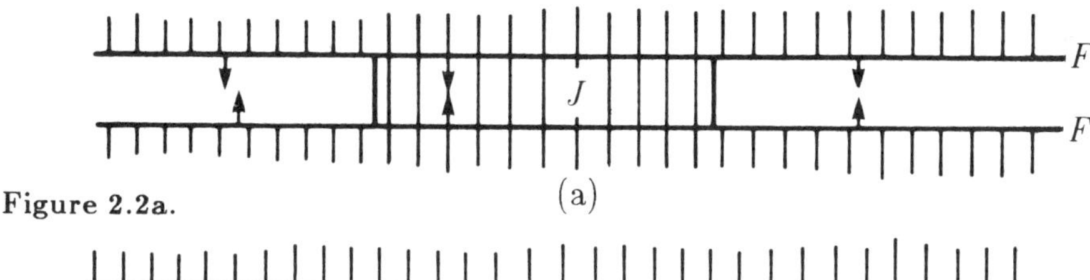

Figure 2.2a.

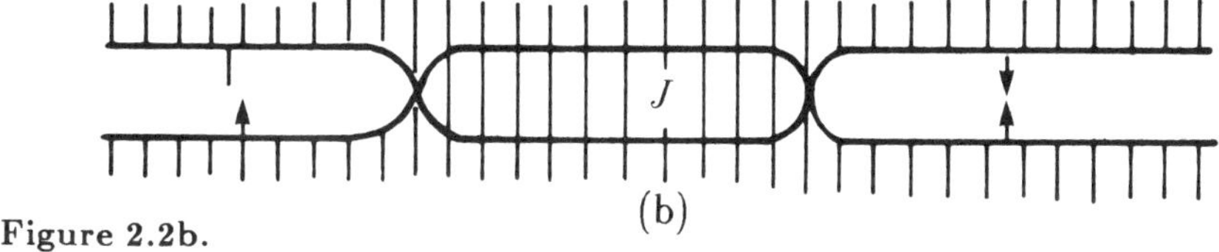

Figure 2.2b.

Step 2: *To show B_F is orientable and that the orientation of B_F can be chosen to induce the orientation of F.* Let $\overline{N}_F$ denote N_F union components of $M - \mathring{N}_F$ of the form $D^2 \times I$ with $\partial D^2 \times I \subset \partial_v N_F$, $D^2 \times \partial I \subset \partial_h N_F$, and with product structures chosen so that the I-fibers of $\partial D^2 \times I$ are contained in fibers of N_F. Thus $\overline{N}_F$ is a 1-foliated submanifold of M. We let $\overline{L}_F$ denote the I-bundle obtained by cutting $\overline{N}_F$ on the closure in F of $F - \partial_h \overline{N}_F$. We show that for each component J of $\overline{L}_F$ the F-orientation on $\partial_h J$ can be extended to the fibers of J. Suppose not, i.e. suppose that the orientation on F is opposite at opposite ends of each fiber

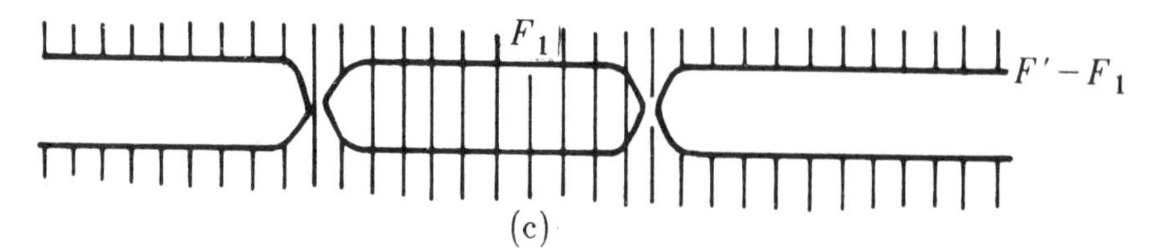

Figure 2.2c.

of J, see Figure 2.2a. We homotope F to an immersed surface as shown in Figure 2.2b. Each point of F on $\partial_h J$ is homotoped to the opposite end of a fiber of J. Finally switch F on its curves of self-intersection as shown in Figure 2.2c to get a normal surface F' homologous to F. One of the components F_1 of F' will be isotopic to ∂J, and therefore F_1 is null-homologous. If J is not a bundle over an annulus or Moebius band, then $\chi(F_1) < 0$, so $\chi_-(F' - F_1) < \chi_-(F)$, which contradicts our choice of F. (Notice that J cannot be a $D^2 \times I$, otherwise J would contain a trivial component of L_F.) If J is a bundle over an annulus or Moebius band, then $F' - F_1$ can be isotoped to a normal surface of complexity smaller than that of F, which again contradicts our choice of F.

Step 3: *To modify B_F and show that the modified B_F has a closed oriented transversal through any given point of B_F which intersects B_F only at that point.* (The proof of this step can be greatly simplified if one only wants to prove that B_F has a closed oriented transversal through each point of B_F, possibly intersecting B_F at other points. An outline of this proof, suggested by W. Thurston, is given below at the end of this proof.) Every component K of $M - \mathring{N}(B_F)$ is a sutured manifold (see [Gabai, 1983]). We can write $\partial K = \partial_+ K \cup \partial_- K \cup \partial_v K$ where $\partial K \cap \partial_h N(B_F) = \partial_+ K \cup \partial_- K$ and $\partial K \cap \partial_v N(B_F) = \partial_v K$ is the union of sutures. The orientations of fibers of $N(B_F)$ point outward (inward) relative to K at points of $\partial_+ K$ ($\partial_- K$). Since the boundary of $\partial_+ K$ ($\partial_- K$) is contained in handle-boundaries, one can define the complexity of $\partial_+ K$ ($\partial_- K$) just as it was defined for normal surfaces. Clearly $\gamma(\partial_+ K) = \gamma(\partial_- K)$, otherwise one could replace F be a homologous surface of smaller complexity, see Figure 2.3, by replacing $\partial_- K \subset F$ by $\partial_+ K$ (or $\partial_+ K$ by $\partial_- K$). Also clearly $\chi(\partial_+ K) = \chi(\partial_- K)$, otherwise one of these replacements shows that F does not have minimal χ_- in its homology class.

We assign an integer "length" to the branch locus of B_F. The handle-decomposition of M induces a cell-decomposition of $\cup \partial H$ where the union is over the handles H of the handle-decomposition. The length of the branch locus is then defined to be the number of 2-cells in which $\partial_v N(B_F)$ intersects the 2-cells of that cell-decomposition.

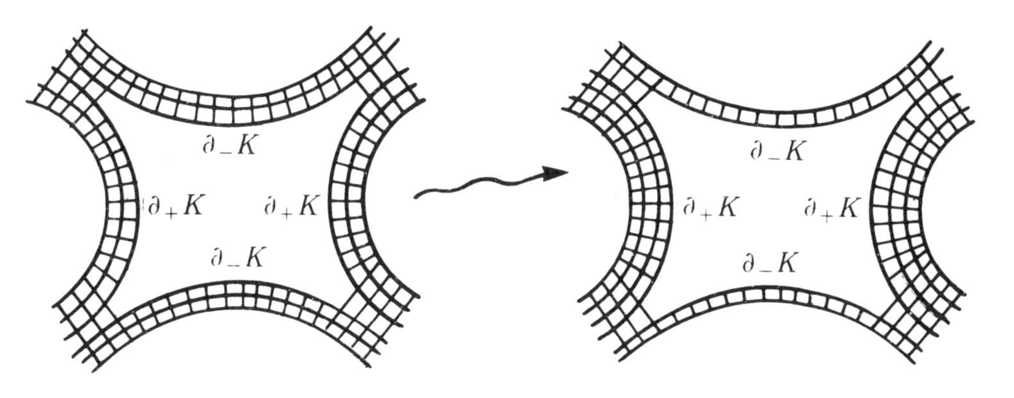

Figure 2.3.

Let $\vec{e}_+$ ($\vec{e}_-$) be the non-invariant measure on B_F corresponding to $\partial_+ K$ ($\partial_- K$). So the i-th entry of $\vec{e}_+$ is the number of intersections of $\partial_+ K$ with the i-th sector of B_F. If $F = B_F(\vec{w})$, then the move shown in Figure 2.3 replaces $B_F(\vec{w})$ by the homologous surface $B_F(\vec{w} + \vec{e}_+ - \vec{e}_-)$. We associate the vector $\vec{k} = \vec{e}_+ - \vec{e}_-$ to K. If $\vec{k} \neq \vec{0}$ for some component K of $M - \mathring{N}(B_F)$, then we replace F by the homologous minimal-complexity minimal-χ_- surface $B_F(\vec{w} + n\vec{k})$ where n is chosen so that $\vec{w} + n\vec{k}$ has at least one null entry and all other entries are non-negative. Then $F_1 = B_F(\vec{w} + n\vec{k})$ is carried with positive weights by a proper branched subsurface B_1 of B_F. This branched subsurface may have discs of contact, but we can remove them by removing trivial components of the I-bundle $L_1 = L_{F_1} = (N(B_1)$ cut on $F_1 - \text{int}(\partial_h N(B_1)))$. (Lemma 4.9 of [Oertel, 1984] shows that as F_1 ranges over minimal complexity incompressible surfaces carried by B_1, removing discs of contact in this way yields finitely many branched surfaces.) We may therefore assume B_1 is again an oriented RIB; the proof is exactly the same as the proof in Steps 1 and 2 that B_F is an oriented RIB. We repeat the process: assuming we have constructed the oriented RIB B_i carrying a surface F_i, and assuming there is a component K_i with associated $\vec{k}_i \neq \vec{0}$, we replace F_i by a homologous surface F_{i+1} carried by a proper branched subsurface B_{i+1} of B_i and eliminate its discs of contact. Since the branch locus of B_{i+1} is shorter than that of B_i, the process must stop with an oriented RIB B_r carrying a minimal-complexity minimal-χ_- surface F_r with positive weights, where F_r is homologous to F, such that every component K of $M - \mathring{N}(B_r)$ has associated $\vec{k} = \vec{0}$.

Let K be a component of $M - \mathring{N}(B_r)$ with associated $\vec{k} = \vec{0}$. We claim $M - \mathring{N}(B_r) = K$, so $M - \mathring{N}(B_r)$ is connected. If a fiber of $N(B_r)$ has one end in $\partial_+ K$, its other end must be in $\partial_- K$. If for some sector E,

$\pi^{-1}(\mathring{E})$ has such a fiber (with ends in ∂K), then every fiber of $\pi^{-1}(\mathring{E})$ is such a fiber. Figure 2.4 shows that if E' is an adjacent sector, then every fiber of $\pi^{-1}(\mathring{E}')$ has one end in $\partial_- K$ and one end in $\partial_+ K$. (Recall that our branched surfaces have non-generic branch locus.) Thus each component with of $N(B)$ which shares some of its boundary with K, shares all of its boundary with K. Therefore $M - \mathring{N}(B_r) = K$ and B_r is connected.

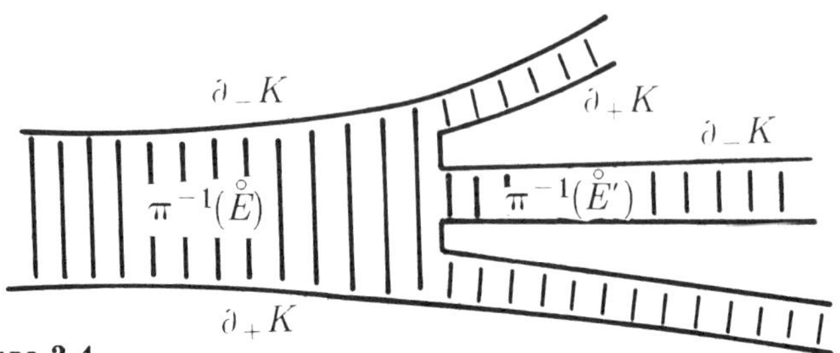

Figure 2.4.

Given a fiber f of $N(B_r)$, we can now easily construct a closed transversal α of $N(B_r)$ such that $\alpha \cap N(B_r) = f$. Simply let $\alpha = f \cup \beta$ where β is an arc in K joining the ends of f.

Step 4: *To show that there are only finitely many possibilities for B_r as F ranges over minimal-complexity minimal-χ_- representatives of homology.* In [Oertel, 1984] it is shown that as F ranges over incompressible normal surfaces of minimal complexity in their own isotopy classes there are just finitely many possibilities for B_F. In our context F ranges over a much smaller class of incompressible normal surfaces, namely minimal-complexity minimal-χ_- representatives of homology, so there are still just finitely many possibilities for B_F. At the i-th modification, changing B_{i-1} to B_i, there are also just finitely many possibilities for B_i. First, there are just finitely many possible branched subsurfaces B_i of B_{i-1}, then by Lemma 4.9 of [Oertel, 1984] there are just finitely many ways to eliminate discs of contact from B_i. Since the positive-integer-valued length of the branch locus of B_i decreases as i increases, the number of modifications is bounded. Thus for a given B_G, as F ranges over minimal-complexity minimal-χ_- representatives carried with positive weights by B_G (so $B_F = B_G$), there are just finitely many possibilities for B_r.

Step 5: *To show that each B_r is taut.* Suppose $F_r = B_r(\vec{w})$ and suppose B_r is not taut. Then there exists a surface $S = B_r(\vec{v})$ which is carried by B_r but is not minimal-χ_- representative of its homology class. Choose n large enough so $nw_i > v_i$ for all i; this is possible since F_r is carried with positive weights by B_r. Then $B_r(n\vec{w}) = B_r((n\vec{w} - \vec{v}) + \vec{v})$ and $B_r(n\vec{w})$

can be obtained from $S = B_r(\vec{v})$ and $F' = B_r(n\vec{w} - \vec{v})$ by embedding S and F' in $N(B_r)$ transverse to the fibers of $N(B_r)$ and transverse to each other, and then switching on curves of $S \cap F'$, see Figure 2.1. Since S is not a minimal-χ_- representative of its homology class, there exists a homologous surface S' which is a minimal-χ_- representative. Similarly, we replace F' by a minimal-χ_- representative of its homology class. Assume S' is transverse to F', eliminate trivial curves of $F' \cap S'$, then switch on curves of intersection to obtain a surface F'' homologous to F with $\chi_-(F'') < \chi_-(F)$. This contradicts the fact that F is a minimal-χ_- representative of its homology class. Therefore B_r is taut.

Theorem 4

We now outline Thurston's simpler argument for constructing a closed oriented transversal for the unmodified B_F constructed in Step 1. We construct a graph Γ whose vertices correspond to the components of $M - B_F$ and whose edges correspond to the sectors of B_F (components of the complement of the branch locus in B_F). We assign an orientation and a weight to each edge of Γ, according to the orientation and complexity of the corresponding sector of B_F. Now, given a component K of $M - \mathring{N}(B_F)$, $\partial_- K$ and $\partial_+ K$ have equal complexity. Therefore there is a weight-preserving flow on Γ. Hence, we can find a closed oriented loop in Γ through any given edge and a corresponding closed oriented transversal through any given point of B_F.

3. Proofs of Subsidiary Results

Proof of Corollary 5:

(a) Suppose B is taut but the sutured manifold $K = M - \mathring{N}(B)$ is not taut. This means that $\partial_+ K$ ($\partial_- K$) is not a minimal-χ_- representative of $H_2(K, \partial_v K)$. It follows that there is a surface $(S, \partial S)$ representing $[\partial_+ K] \in H_2(K, \partial_v K)$ with $\chi_-(S) < \chi_-(\partial_+ K)$. After gluing canceling pairs of components of ∂S, ∂S is isotopic to $\partial(\partial_+ K)$. If F is a surface carried with positive weights by B, F embedded in $N(B)$ transverse to fibers with $\partial_h N(B) \subset F$, then we replace $\partial_+ K \subset F$ by S to get a surface F' homologous to F. Now $\chi_-(F') < \chi_-(F)$ which contradicts the tautness of B.

Parts (b) and (c) of the corollary are immediate consequences of D. Gabai's work in [Gabai, 1983].

□

To prove Theorem 6 we need some more definitions. An *annulus of contact* for a branched surface B is an annulus A embedded in $N(B)$ transverse to fibers with $\partial A \subset \operatorname{int} \partial_v N(B)$. (Both components of ∂A may be in the same component of $\partial_v N(B)$ when A double covers a Moebius band.) A *rectangle of contact* is a disc D embedded in $N(B)$ transverse to fibers with $\partial D \cap \partial M$ a union of two disjoint arcs in ∂D and with

$$\partial D \cap \partial_v N(B) \subset \operatorname{int}(\partial_v N(B))$$

equal to the two complementary arcs of ∂D.

If B is a branched surface with s sectors $\mathcal{C}(B)$ is the subset of points $\vec{w} \in \mathbb{R}^s$ such that $\vec{w}$ is an invariant measure, $\mathcal{C}(B)$ is called the *cone of invariant measures on* B; it is the cone of a convex polyhedron.

Lemma 3.1. *Let B be a RIB in an orientable, irreducible, ∂-irreducible, simple 3-manifold. Then B has just finitely many distinct annuli of contact (rectangles of contact) up to isotopy in $N(B)$ respecting fibers.*

Proof. Suppose we have an infinite sequence (A_j) of annuli of contact for B, where the A_j's are distinct up to isotopy in $N(B)$ respecting fibers. Since each A_j is embedded in $N(B)$ transverse to fibers with $\partial A_j \subset \operatorname{int}(\partial_v N(B))$, A_j determines a non-invariant measure $\vec{a}_j$ in B. The weight a_{ji} on the i-th sector E_i is simply the number of intersections of A_j with a fiber of $\pi^{-1}(\mathring{E}_i)$. Since the A_j's are distinct up to isotopy in $N(B)$ respecting fibers of $N(B)$, the $\vec{a}_j$'s are also distinct. By passing to a subsequence we may assume that $\partial A_j = \partial A_k$ for all j and k, again up to isotopy respecting fibers of $\operatorname{int}(\partial_v N(B))$, and that $a_{(j+1)i} \geq a_{ji}$ for each i. Clearly, for any j, $\vec{w} = \vec{a}_{j+1} - \vec{a}_j$ is an invariant integer measure on B and clearly $\vec{w} \neq \vec{0}$ since $\vec{a}_{j+1}$ and $\vec{a}_j$ are distinct measures. We claim $\chi(B(\vec{w})) = 0$. This is true because $A_{j+1} = B(\vec{a}_{j+1})$ can be obtained from $A_j = B(\vec{a}_j)$ and $T = B(\vec{w})$ by embedding T and A in $N(B)$ transverse to fibers and transverse to each other, then switching on curves of intersection. Thus $\chi(T) + \chi(A_j) = \chi(A_{j+1})$, and since $\chi(A_j) = \chi(A_{j+1}) = 0$ it follows that $\chi(T) = 0$. Therefore T or $\partial N(T)$ is an incompressible torus, a contradiction.

□

Proof of Theorem 6: The proof differs only slightly from the proof of Theorem 4. We indicate the differences in each step of the proof.

Step 1: *Construction of B_F.* Given a minimal-χ_- representative F of homology we only assume F is a normal surface of minimal complexity in its isotopy class, not in its homology class. Then we construct $\bar{B}_F$ and B_F as before. We make one modification: for every component J of $\bar{L}_F$ which is a bundle over an annulus or Moebius band, we remove the interiors of fibres of $J \cap L_F$ from M_F to obtain the fibred neighbourhood of another branched surface, a modified B_F. Since for each J of this type $\partial_h J$ yields one or two annuli of contact for B, Lemma 3.1 ensures that for a fixed B_G as F ranges over incompressible normal surfaces such that $B_F = B_G$, there are only finitely many ways to modify B_F. In the following steps we refer to the modified B_F as B_F.

Step 2: *To show that B_F is orientable.* As before, suppose J is a component of $\bar{L}_F$ such that the orientation of F cannot be extended to J, see Figure 2.2. The modification performed above ensures that J is not an I-bundle over a Moebius band or annulus. We can replace F by a homologous surface $F' - F_1$ (Figure 2.2) of smaller χ_- than F, which contradicts our choice of F.

Step 3: *To show that there is a closed orientable transversal through every point of B_F.* Recall that in the proof of Theorem 4 we modified B_F so it had just one complementary component. We cannot use those modifications here because they involve replacing F by a homologous but non-isotopic surface. We shall find a closed orientable transversal for B_F itself. We still do this, however, by examining branched subsurfaces of B_F which carry surfaces homologous to F.

Let $a_0 \in \mathring{E}_0$ be a given point of B_F, where E_0 is a sector of B_F. Let f_0 be the fiber $\pi^{-1}(a_0)$, and let K_0 be the component of $M - \mathring{N}(B_F)$ on the positive side of $\pi^1(E_0)$, see Figure 3.2, so that $\partial_- K_0$ is on the positive end of fiber f_0. Finally, let $\vec{k}_0$ be the vector associated to K_0. If $\vec{k}_0 = \vec{0}$, $M - \mathring{N}(B_F) = K_0$ and we can easily find a closed transversal as in the proof of Theorem 4. If $\vec{k}_0 \neq \vec{0}$, let $F = B_F(\vec{w})$, then the change from $B_F(\vec{w})$ to $B_F(\vec{w} - \vec{k}_0)$ replaces $\partial_+ K_0 \subset F$ by $\partial_- K_0$ and yields a surface homologous to F. There is a positive integer n_0 such that $\vec{w} - n_0\vec{k}_0$ is a non-negative invariant measure with at least one zero weight. The weight assigned by $\vec{w}_1 = \vec{w} - n_0\vec{k}_0$ to the sector E_0 cannot be zero because the change from $\vec{w}$ to $\vec{w} - n_0\vec{k}_0$ can decrease weights only on sectors corresponding to $\partial_+ K_0$. Thus $F_1 = B_F(\vec{w}_1)$ is carried with positive weights by a branched subsurface B_1 of B_F, where $E_0 \subset B_1$. Our goal is to repeat this process to obtain a sequence of branched surfaces (B_i), $i=0, \ldots, r$, where B_{i+1} is a

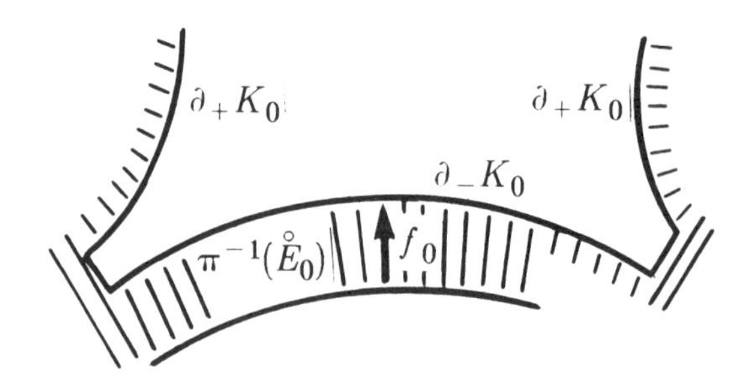

Figure 3.2.

branched subsurface of B_i, $B_0 = B_F$, and B_r has the property that $M - \mathring{N}(B_r) = K_r$ is connected with $\vec{k}_r = \vec{0}$. Assuming we have constructed B_i, Let E_i be the sector of B_i containing E_0 and the point a_0. Let K_i be the component of $M - \mathring{N}(B_i)$ on the positive side of $\pi^{-1}(\mathring{E}_i)$. (π denotes the projection map associated to the branched surface B_i.) If $\vec{k}_i$, the vector associated to K_i, is zero, we are done; we let $i = r$. The branched surface B_i carries a surface $F_i = B_i(\vec{w}_i)$ homologous to F. If $\vec{k}_i \neq 0$, we choose $n_i > 0$ so that $F_{i+1} = B_i(\vec{w}_i - n_i \vec{K}_i)$ is carried with positive weights by a proper branched subsurface B_{i+1} of B_i. The sequence of B_i's must be finite since the branch locus of B_{i+1} is shorter than that of B_i. We now construct the closed transversal by examining the sequence of B_i's in reverse order, see Figure 3.3. The branched surface B_{r-1} is obtained from B_r by adding a collection of sectors to B_r. The added sectors lie in $\pi_{r-1}(\partial_+ K_{r-1})$. Hence there is an arc α_r in $\pi_r(K_r) - \operatorname{int} \pi_{r-1}(K_{r-1})$ from the point a_0 to a point a_{r-1} in $\pi_{r-1}(\partial_+ K_{r-1})$, see Figure 3.3. In the same way one can find an arc α_i from a_i to $a_{i-1} \in \pi_{i-1}(\partial_+ K_{i-1})$, $i = r-1, \ldots, 1$. Then $\alpha = \bigcup_{j=1}^{r} \alpha_i$ is the desired oriented closed transversal, when it is oriented to have positive intersection with $B_F = B_0$.

Step 4: *To show that there are just finitely many possibilities for B_F.* We constructed $\overline{B}_F$ as before. First one shows as in [Oertel, 1984] that the modification eliminating discs of contact yields finitely many possibilities for B_F. Then Lemma 3.1 shows that the further modifications done in Step 1 still yield just finitely many possibilities for B_F.

Step 5: *To show that each B_F is taut.* The proof of this step is the same as before.

Theorem 6

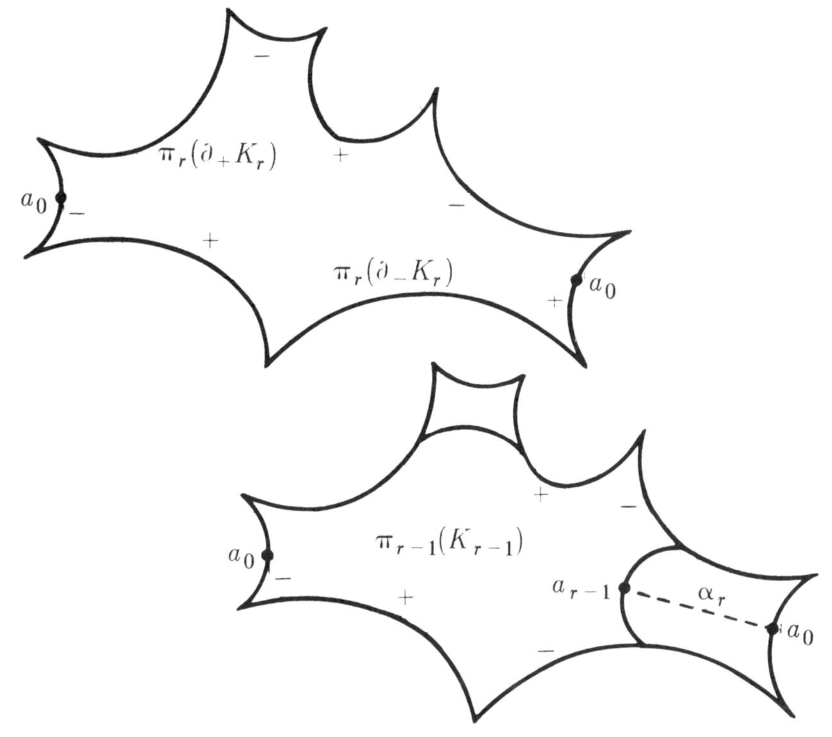

Figure 3.3.

Proof of Corollary 7:

(a) Let $\{B_i\}$ be the finite collection of homology RIB's constructed in Theorem 4. Suppose B_i has s_i sectors. For each i the map h_i: $(\mathcal{C}(B_i)) \cap \mathbf{Q}^{s_i} \to H_2(M; \mathbf{Q})$ defined by $h_i(\vec{w}) = [B_i(\vec{w})]$ is clearly convex-linear. In the convex cone $h_i(\mathcal{C}(B_i))$, for lattice points $h_i(\vec{v})$ and $h_i(\vec{w})$, we have $x([h_i(\vec{v})]) + x([h_i(\vec{w})]) = x([h_i(\vec{v} + \vec{w})])$. This is true because $B_i(\vec{v})$, $B_i(\vec{w})$, and $B_i(\vec{v} + \vec{w})$ are minimal-χ_- representatives of homology and because

$$\chi(B_i(\vec{v} + \vec{w})) = \chi(B_i(\vec{v})) + \chi(B_i(\vec{w})).$$

Recall that χ is additive in this way because $B_i(\vec{v} + \vec{w})$ is obtained from $B_i(\vec{v})$ and $B_i(\vec{w})$ by cut-and-paste operations. It follows that x is linear on $h_i(\mathcal{C}(B_i) \cap \mathbf{Q}^{s_i}) \subset H_2(M; \mathbf{Q})$. It also follows that h_i has a unique linear extension to $\mathcal{C}(B_i)$, which we still call h_i. Finally, we conclude that x has a unique linear extension to $h_i(\mathcal{C}(B_i)) \subset H_2(M; \mathbb{R})$. Since $H_2(M; \mathbb{R}) = \bigcup_i h_i(\mathcal{C}(B_i))$, x has a unique continuous extension to $H_2(M; \mathbb{R})$. The convexity of the extended x follows from the convexity of the x defined on $H_2(M; \mathbf{Q})$, so x is a pseudonorm.

(b) The unit ball U_x is a finite convex polyhedron because in each convex cone $h_i(\mathcal{C}(B_i))$, x is linear.

(c) If M is simple, then x is strictly positive on non-trivial homology classes, so x is a norm.

□

Proof of Proposition 8: If $g, h \in H_2(M)$ have minimal-χ_- representatives G and H carried by a taut oriented branched surface B, then we may assume that $G=B(\vec{v})$ and $H=B(\vec{w})$. Then $B(t\vec{v}+(1-t)\vec{w})$ represents $tg+(1-t)h \in H_2(M;\mathbf{R})$, $0 \le t \le 1$, $t \in \mathbf{Q}$, and $x(tg+(1-t)h) = tx(g)+(1-t)x(h)$ by the proof of Corollary 7. Therefore G and H lie on the same face of ∂U_x.

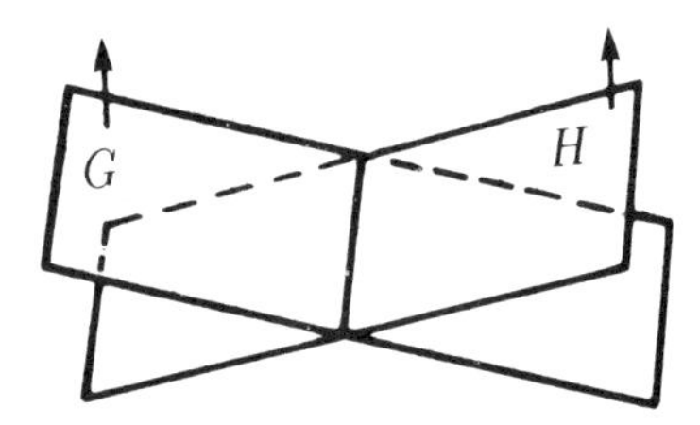

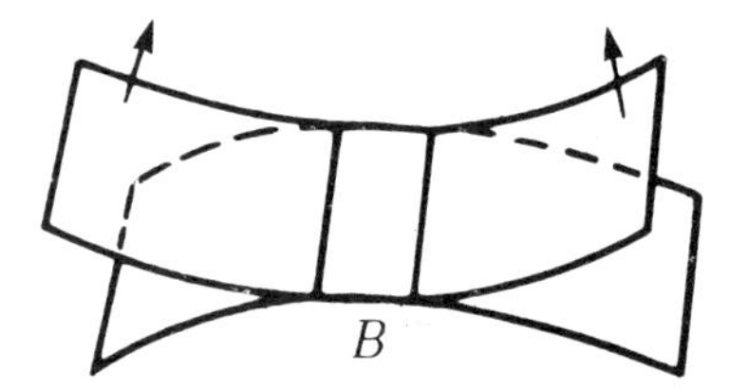

Figure 3.4.

Conversely, suppose g and h lie in the cone over a face of ∂U_x. Let us represent g and h by oriented minimal-χ_- representatives G and H respectively. Assume G and H are transverse to each other, then eliminate trivial curves of $G \cap H$ by isotopy of G and H. Finally construct a branched surface B from $G \cup H$ by changing each curve of $G \cap H$ to an "annulus or Moebius band of contact", see Figure 3.4. It is easy to show that B is a taut oriented branched surface.

□

Proof of Theorem 9: As Thurston observes in [Thurston, 1986], if a homology class in $H_2(M)$ has a representative which is a fiber of some fibration of M over S^1, then every incompressible surface representative of that class is isotopic to a fiber.

Let $a \in H_2(M)$ be a homology class which can be represented by a fiber and suppose a is in the cone over $\mathring{C}$, where C is a face of ∂U_x. We will show that any other lattice point in $H_2(M; \mathbf{R})$ in the cone over $\mathring{C}$ can be represented by a fiber. We are really concerned with rational projective homology classes. Clearly $a/x(a)$ is a rational point in $\mathring{C}$. If b is another integer lattice homology class in the cone over $\mathring{C}$, then $b/x(b)$ is also in $\mathring{C}$.

The line through $a/x(a)$ and $b/x(b)$ intersects ∂C in rational points which can be written $g/x(g)$ and $h/x(h)$ where g and h are integer lattice points of $H_2(M; \mathbb{R})$. Let G and H be minimal-χ_- surface representatives of g and h respectively. As in the proof of Proposition 8, construct a taut oriented branched surface B from $G \cup H$. Then a positive multiple of a is represented by a surface F carried by B with positive weights, and a multiple of b is represented by a surface S carried by B with positive weights. We may assume that F (and S) can be embedded in $N(B)$ so $\partial_h N(B) \subset F$ (or $\partial_h N(B) \subset S$). Since F is a fiber, M cut on F is a product K. The product $(K, \partial_h K)$ can be isotoped so the I-fibres of K agree with those of L_F just as in the proof of Theorem 1 in [Oertel, 1984]. It follows that $M - \mathring{N}(B)$ is a product, therefore S is a union of fibers.

$\square$

References

Floyd-Oertel, 1984.
W.J. Floyd and U. Oertel, "Incompressible surfaces via branched surfaces," *Topology*, **23**, pp. 117-125, 1984.

Gabai.
Gabai, D., *Foliations and genera of links.*

Gabai, 1983.
D. Gabai, "Foliations and the topology of 3-manifolds," *Journal of Differential Geometry*, pp. 445-503, 1983.

Haken, 1961.
Haken, W., "Theorie der Normalflaechen," *Acta. Math.*, **105**, pp. 245-375, 1961.

Jaco-Oertel, 1984.
Jaco, W. and Oertel, U., "An algorithm to determine if a 3-manifold is a Haken manifold," *Topology*, 1984.

Morgan-Shalen.
J. Morgan and P. Shalen, *Degenerations of hyperbolic structures.* preprint — two vols

Oertel, 1984.
Oertel, U., "Incompressible branched surfaces," *Inventiones Math.*, 1984.

Schubert, 1961.
Schubert, H., "Bestimmung der Primfaktorzerlegung von Verkettungen," *Math. Z.*, **76**, pp. 116-148, 1961.

Thurston, 1986.
W. P. Thurston, "A Norm on the Homology of 3-Manifolds," *Memoirs of the AMS*, **339**, 1986.

U. Oertel,
University of Oklahoma,
Norman,
Oklahoma 73019,
U.S.A.

n-relator 3-manifolds with incompressible boundary.

Józef H. Przytycki

ABSTRACT

This paper extends the results of Jaco and the author to the case of manifolds with many curves in the boundary. In particular we find the conditions for a family of two or three curves in the boundary of a handlebody which assure that after adding 2-handles along the curves we get a manifold with incompressible boundary.

0. Introduction

This paper extends results of [Przytycki, 1983] and [Jaco, 1984]. Namely the following fact was proven in [Przytycki, 1983] (see Part 1 for necessary definitions):

Theorem 0.1: [Przytycki, 1983]. *Let H_n $(n > 0)$ be a handlebody of genus n and γ a 2-sided, simple closed curve in ∂H_n. If $\partial H_n - \gamma$ is incompressible, then the manifold $(H_n)_\gamma$ obtained from H_n by adding a 2-handle along γ has incompressible boundary, or $n=1$, H_1 is orientable and γ is a longitude of ∂H_1.*

The proof in [Przytycki, 1983] is algebraic. In [Jaco, 1984], Theorem 0.1 is generalized using a geometric approach.

Theorem 0.2: [Jaco, 1984]. *Let M be a 3-manifold with compressible boundary and γ a 2-sided simple closed curve in ∂M. If $\partial M - \gamma$ is incompressible then ∂M_γ is incompressible or M is a solid torus and γ is its longitude.*

Theorem 0.2 was reproved by different methods by Scharlemann [Scharlemann, 1985] and Johannson [Johannson].

It is shown in [Przytycki, 1983, Examples 1.11-1.13] that a direct generalization of Theorems 0.1 and 0.2 is not possible. Culler suggested that for more than one curve (say $\gamma_1, \ldots, \gamma_n$) in the boundary of the handlebody H_k we should not only assume that $\partial H_k - \bigcup \gamma_i$ is incompressible but also that for each j, $\partial H_k - (\bigcup \gamma_i - \gamma_j)$ is compressible. We will show in Part 2 that for $n = 2$, in fact, $(H_k)_{\{\gamma_1, \gamma_2\}}$ has incompressible boundary (Theorem 2.2) but for $n > 2$ an additional awkward condition is needed (Theorem 2.1). Then we will show that this condition cannot be deleted (Examples 2.5 and 2.6) however, for $n = 3$, some different conditions could be considered instead. In particular we get the following result:

Theorem 0.3. *Let $\{\gamma_1, \gamma_2, \gamma_3\}$ be a family of 2-sided, pairwise disjoint simple closed curves (s.c.c.) in the boundary of a handlebody H_k. Assume that the following three conditions are satisfied:*

(0) $\partial H_k - \bigcup_{i=1}^{3} \gamma_i$ *is incompressible in* H_K,

(1) *For each j, $\partial H_k - (\bigcup_{i=1}^{3} \gamma_i - \gamma_j)$ is compressible in H_k (or, equivalently, the family of elements of $\pi_1(H_k)$ represented by $\{\gamma_i\}_{i=1}^{3} - \gamma_j$ does not bind the free group $\pi_1(H_k)$; see* [Lyon, 1980] *or* [Przytycki, 1983]),

(2) $\gamma_j \in \{\gamma_i\}_{i=1}^{3}$ *does not bind a free factor F_{k-1} of $\pi_1(H_k) = F_{k-1} \times F_1$.*

Then $(H_k)_{\{\gamma_i\}_{i=1}^{3}}$ has incompressible boundary or is equal to D^3.

It is always possible, for a given family of curves, to verify whether the conditions (0)-(2) are satisfied (see [Lyndon-Schupp, 1977] or [Lyon, 1980]). Again, conditions (0)-(2) are not sufficient for $n > 3$ (Example 2.7). We propose a hopefully sufficient set of conditions in Conjecture 2.8.

We end Part 2 by describing one more situation in which incompressibility of a surface is preserved (Theorem 2.10). This theorem is an important tool in [Lozano-Przytycki, 1985].

In Part 3 we show an application of Theorems 2.2 and 2.10 in knot theory.

1. Preliminaries

We work in the PL-category and our terminology is based on that of [Przytycki, 1983]. We recall here some terminology and definitions.

Definition 1.1.

(a) Let M be a 3-manifold and F a surface which is either properly embedded in M or contained in ∂M. We say that F is *compressible* if one of the following conditions is satisfied:

(i) F is a 2-sphere which bounds a 3-cell in M, or

(ii) F is a 2-cell and either $F \subset \partial M$ or there is a 3-cell $X \subset M$ with $\partial X \subset F \cup M$, or

(iii) there is a 2-cell $D \subset M$ with $D \cap F = \partial D$ and with ∂D not contractible in F.

We say that F is *incompressible* if it is not compressible.

(b) Let F be a submanifold of a manifold M. We say that F is π_1 injective in M if the inclusion-induced homomorphism from $\pi_1(F)$ to $\pi_1(M)$ is an injection.

(c) Let F be a surface properly embedded in a compact 3-manifold M. We say that F is ∂-incompressible in M if there is no 2-disk $D \subset M$ such that $D \cap F = \alpha$ is an arc in ∂D, $D \cap \partial M = \beta$ is an arc in ∂D, with $\alpha \cap \beta = \partial\alpha = \partial\beta$ and $\alpha \cup \beta = \partial D$, and α is not parallel to ∂F in F.

Definition 1.2. Let M be a 3-manifold and γ a simple closed, 2-sided curve on ∂M. We define a new 3-manifold M_γ to be M with a 2-handle glued along γ. That is: Let A_γ be a regular neighbourhood of γ in ∂M. Let (D^3,A) be a 3-disk with an annulus on the boundary and ϕ a homeomorphism $A_\gamma \to A$; then $M_\gamma = (M,A_\gamma) \cup_\phi (D^3,A)$. If $\{\gamma_i\}_{i=1}^n$ is a finite collection of pairwise disjoint, simple closed, 2-sided curves on ∂M then

$$M_{\{\gamma_i\}} \overset{\text{def}}{=} (\cdots((M_{\gamma_1})_{\gamma_2})\cdots)_{\gamma_n}.$$

The definition does not depend on the order of the γ_i.

Definition 1.3. Let $(F,\partial F) \to (M,\partial M)$ be a surface properly embedded in a 3-manifold M. We say that F is *unknotted* in M if and only if M − int V_F is a collection of handlebodies, where V_F is a regular neighbourhood of F in M.

Definition 1.4. [Jaco, 1984] Let $\gamma, \gamma_1, \ldots, \gamma_n$ be a family of pairwise disjoint, 2-sided, simple closed curves on ∂M. We say that γ is *coplanar* with $\{\gamma_i\}_{i=1}^n$ if γ cuts out of a disk with holes from ∂M cut open along $\bigcup_{i=1}^n \gamma_i$ (i.e. γ bounds a disk in $M_{\{\gamma_1, \ldots, \gamma_n\}}$).

Definition 1.5. [Lyon, 1980] Let $W \subset F$ be a set of cyclic words in the basis X of a free group F. The incidence graph $J(W)$ is the graph whose vertices are in 1-1 correspondence with the non-trivial words in W, with an edge joining vertices w_1 and w_2 if there exists $x \in X$ such that x or x^{-1} lies in w_1 and x or x^{-1} lies in w_2. W is connected with respect to the basis X if $J(W)$ is connected, and is connected if it is connected with respect to each basis of F. If the set W of cyclic elements is not contained in any proper free factor of F and if W is connected, we say W binds F.

Definition 1.6 We say that a 3-manifold M is a *disk sum* of M_1 and M_2 (i.e. $M = M_1 \Delta M_2$) iff M is obtained from M_1 and M_2 by identifying 2-disks D_1 and D_2 on their boundaries ($D_i \subset M_i$).

2. Adding 2-handles

Theorem 2.1. *Let $\gamma_1, \gamma_2, \ldots, \gamma_n$ be a family of 2-sided, pairwise disjoint, simple closed curves on ∂M. Let the following conditions be satisfied:*

(i) $\partial M - \bigcup_{i=1}^n \gamma_i$ *is incompressible in M,*

(ii) $\partial M - (\bigcup_{i=1}^n \gamma_i - \gamma_j)$ *is compressible in M for each j,*

(iii) *a disk of compression from (ii), say D, can be chosen in such a way that ∂D is not coplanar with $\bigcup_{i=1}^n \gamma_i - \gamma_j$ (or equivalently the kernel of $\pi_1(\partial M - (\bigcup_{i=1}^n \gamma_i - \gamma_j)) \to \pi_1(M)$ is not contained in the normal subgroup generated by $\{\gamma_i\}_{i=1}^n - \gamma_j$).*

Then $M_{\{\gamma_i\}_{i=1}^n}$ has incompressible boundary or is equal to D^3.

Examples 2.5 and 2.6 show that the condition (iii) is essential. The conditions (i) and (ii) are sometimes sufficient, however; for example if $n = 2$ and M is a handlebody.

Theorem 2.2. *Let γ_1 and γ_2 be 2-sided, disjoint, simple closed curves in the boundary of a 3-manifold M and suppose the following three conditions are satisfied:*

(0) $\partial M - (\gamma_1 \cup \gamma_2)$ *is incompressible in M, and*

(1) $\partial M - \gamma_i$ *is compressible in M ($i=1, 2$), and*

(2) *M is not of the form $M_1 \Delta$ (solid torus), where M_1 is a 3-manifold with incompressible boundary and Δ denotes a disk sum.*

Then $M_{\{\gamma_1, \gamma_2\}}$ has incompressible boundary or is equal to D^3.

In order to prove Theorem 2.1, observe that Theorem 0.2 is valid for a non-compact 3-manifold, so as an immediate consequence we get:

Lemma 2.3. *Let $\gamma_1, \ldots, \gamma_n$ ($n > 1$) be a family of 2-sided, pairwise disjoint, simple closed curves on the boundary of a 3-manifold M. Let $\partial M - \bigcup_{i=1}^{n} \gamma_i$ be incompressible in M and $\partial M - \bigcup_{i=1}^{n-1} \gamma_i$ compressible in M. Then $\partial M_{\gamma_n} - \bigcup_{i=1}^{n-1} \gamma_i$ is incompressible in M_{γ_n}.*

Proof. Use Theorem 0.2 for $M' = M - \bigcup_{i=1}^{n-1} \gamma_i$. □

Proof. Theorem 2.1: We prove Theorem 2.1 by induction on n. For $n=1$ this is Theorem 0.2. Assume that for less than n curves ($n > 1$) Theorem 2.1 is true. From Lemma 2.3 we get that $\partial M_{\gamma_n} - \bigcup_{i=1}^{n-1} \gamma_i$ is incompressible. So in order to use the induction assumption we have to prove that for each j ($j \neq n$) $\partial M_{\gamma_n} - \left(\bigcup_{i=1}^{n-1} \gamma_i - \gamma_j\right)$ is compressible in M_{γ_n} and that a disk of compression can be chosen so that its boundary is not coplanar with $\bigcup_{i=1}^{n-1} \gamma_i - \gamma_j$. Let D be a disk of compression of $\partial M - \left(\bigcup_{i=1}^{n-1} \gamma_i - \gamma_j\right)$ in M such that ∂D is an element of the kernel of

$$\pi_1\left(\partial M - \left(\bigcup_{i=1}^{n} \gamma_i - \gamma_j\right)\right) \to \pi_1(M)$$

which is not contained in the normal subgroup generated by $\{\gamma_i\}_1^n - \gamma_j$. Then, adding a 2-handle along γ_n, ∂D is an element of the kernel of

$$\pi_1\left(\partial M\gamma_n - \left(\bigcup_{i=1}^{n-1}\gamma_i - \gamma_j\right)\right) \to \pi_1\left(M\gamma_n\right),$$

which is not contained in the normal subgroup generated by $\{\gamma_i\}_1^{n-1} - \gamma_j$.

Theorem 2.1

Proof. Theorem 2.2: We know, by Lemma 2.3, that $\partial M\gamma_2 - \gamma_1$ is incompressible and we need $\partial M\gamma_2$ to be compressible (then we can use Theorem 0.2). Let D be a disk of compression of $\partial M - \gamma_2$. If ∂D is not coplanar with γ_2 in ∂M then ∂D is not trivial in $\partial M\gamma_2$ so $\partial M\gamma_2$ is compressible in $M\gamma_2$. If ∂D is coplanar with γ_2 then the only nontrivial case can happen when ∂D cuts out from ∂M a punctured torus T and $\gamma_2 \subset T$ does not separate T (Fig. 2.1).

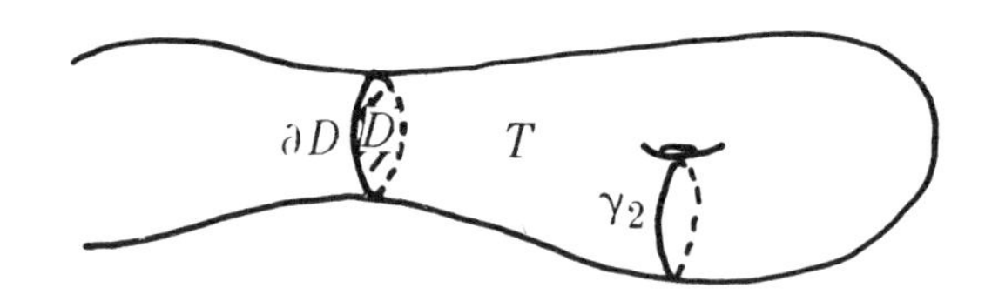

Figure 2.1.

From the assumption (2) we have the following possibilities:

(a) D separates M into genus 1 solid handlebody (with boundary equal to $D \cup T$) and a 3-manifold M_1 with compressible boundary. Then a compressing disk of ∂M_1 is a compressing disk of ∂M too, and the boundary of this disk is not coplanar with γ_2 so $M\gamma_2$ has compressible boundary.

(b) the torus $D \cup T$ is a compressible torus of M cut open along D, then the problem reduces to that of (a).

(c) D separates M into M_1 and M_2. $T \subset \partial M_2$, $D \cup T$ is an incompressible torus in M_2. Then either

(1) ∂M_1 is compressible in M_1 or $\partial M_2 - (D \cup T)$ is compressible in M_2. Then the disk of compression is a disk of compression of $\partial M - \gamma_2$ which is not coplanar with γ_2, or

(2) ∂M_1 and $\partial M_2 - (D \cup T)$ are incompressible; then each compressing disk of ∂M has the boundary isotopic to ∂D (see [Przytycki, 1979], for example). On the other hand $\partial M - \gamma_1 - \gamma_2$ is incompressible

and $\gamma_2 \cap \partial D = \emptyset$ so γ_1 has to cut ∂D nontrivially. Then γ_1 cuts each compressing disk of ∂M, so $\partial M - \gamma_1$ is incompressible, which contradicts the assumption (1) of Theorem 2.2.

(d) D does not separate M. $D \cup T$ is incompressible in M_1, where M_1 is the result of cutting open M along D. Then *either*

(1) $\partial M_1 - (D \cup T)$ is compressible in M_1. Then the disk of compression is a disk of compression of $\partial M - \gamma_2$ which is not coplanar with γ_2, *or*

(2) $\partial M_1 - (D \cup T)$ is incompressible in M_1. Then the proof is the same as that of case (c)(2).

Theorem 2.2

As a corollary of Theorem 2.2, we will now prove Theorem 0.3.

Proof. Theorem 0.3: The idea of the proof is to show that $M\gamma_3$ and $\{\gamma_1, \gamma_2\}$ satisfy the conditions (0)-(2) of Theorem 2.2. For each j, $M - \gamma_j$ and $\{\gamma_i\}_{i=1}^{3} - \gamma_j$ satisfy the conditions (0)-(2) of Theorem 2.2; in particular $M - \gamma_j$ is not of the form $M_1\Delta$ (solid torus) with ∂M_1 incompressible, because if the opposite happens then $M = H_{k-1}\Delta$ (solid torus) and $\gamma_j \subset \partial H_{k-1}$, and $[\gamma_j]$ binds $\pi_1(H_{k-1})$ (see [Lyon, 1980] or [Przytycki, 1983]). But this contradicts the condition (2) of Theorem 0.3. Therefore by Theorem 2.2 and its proof, $M\gamma_i - (\{\gamma_i\}_{i=1}^{3} - \gamma_i)$ is incompressible and $M\gamma_i - \gamma_j$ $(i \neq j)$ is incompressible. So $M\gamma_i$ and $\{\gamma_i\}_{i=1}^{3} - \gamma_i$ satisfy the conditions (0)-(1) of Theorem 2.2. Now, in order to prove Theorem 0.3, it is enough to prove the condition (2) of Theorem 2.2, i.e. to show that $M\gamma_3$ is not of the form $M_1\Delta$ (solid torus) with ∂M_1 incompressible. Assume the opposite. Then γ_3 is a curve in the boundary of an H_{k-1} factor of $H_k = H_{k-1}\Delta$ (solid torus) and γ_3 binds $\pi_1(H_{k-1})$ (compare [Przytycki, 1983]). This contradicts the assumption (3) of Theorem 0.3.

Theorem 0.3

Remark 2.4. If we consider M to be irreducible in Theorems 0.2, 2.1 and 2.2 then the resulting 3-manifold (also in the case of Theorem 0.3) is irreducible, possibly with holes. We use in the proof pre-spheres in place of pre-disks of [Jaco, 1984] (compare also with [Floyd]). (Let $P \subset M$ be a disk with holes properly embedded in M, and $\gamma \subset \partial M$ 2-sided s.c.c. P is a pre-sphere with respect to γ if

(1) $\partial P \subset \partial M - \gamma$ and

(2) each component of ∂P is coplanar with γ in ∂M, and

(3) $P^{\wedge}$ does not bound a 3-cell in $M_{\gamma}^{\wedge}$ where $M_{\gamma}^{\wedge}$ is obtained from M_{γ} by capping off each 2-sphere component of ∂M_{γ} with a 3-cell and $P^{\wedge}$ is a natural extension of P to $M_{\gamma}^{\wedge}$.)

The following examples show that all assumptions of theorems 2.1, 2.2 and 0.3 are necessary.

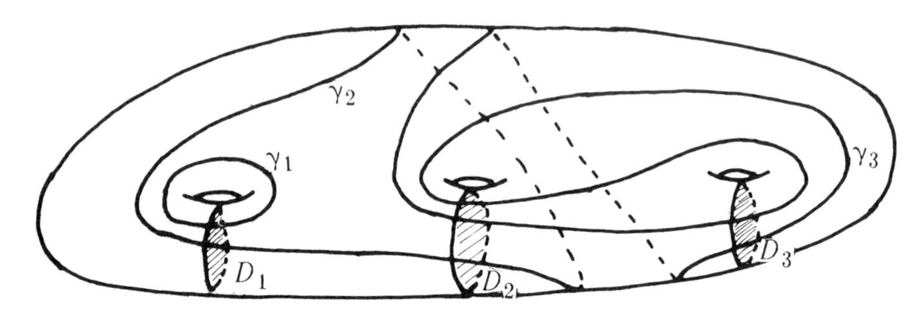

Figure 2.2.

Example 2.5. Consider $\gamma_1, \gamma_2, \gamma_3 \subset \partial H_3$ (H_3 – genus 3 orientable handlebody) as in Figure 2.2: We get, in some standard system of generators of $\pi_1(H_3)$ dual to the disks D_1, D_2, D_3, that:

$$[\gamma_1] = x_1,$$

$$[\gamma_2] = x_1 x_2,$$

$$[\gamma_3] = x_2 x_3 x_2^{-1} x_3^{-1}.$$

Now we conclude that

(0) $\partial H_3 - (\gamma_1 \cup \gamma_2 \cup \gamma_3)$ is incompressible in H_3 (because $\{\gamma_1, \gamma_2, \gamma_3\}$ bind $F_3 = \pi_1(H_3)$; see [Lyon, 1980]),

(1) $\partial H_3 - (\gamma_1 \cup \gamma_2)$, $\partial H_3 - (\gamma_1 \cup \gamma_3)$ and $\partial H_3 - (\gamma_2 \cup \gamma_3)$ are incompressible (see [Lyon, 1980]). But $(H_3)_{\{\gamma_i\}_{i=1}^3}$ is a solid torus (with a hole) so it has compressible boundary. We also observe that $[\gamma_3]$ binds $\pi_1(H_2)$ where $H_3 = H_1 \Delta H_2$.

Example 2.6. Consider $\gamma_1, \gamma_2 \subset \partial M$ where $M = (H_3)_{\gamma_3}$; see Example 2.5. $\partial M - (\gamma_1 \cup \gamma_2)$ is incompressible in M (see Lemma 2.3) and $\partial M - \gamma_1$ and $\partial M - \gamma_2$ are compressible in M. However $M_{\{\gamma_1, \gamma_2\}}$ is a sold torus (with a hole) so it has compressible boundary.

(M is equal to $A \times S^1 \Delta S^1 \times D^2$ where A is an annulus and Δ denotes the disk sum; this presentation was the source of Examples 2.5 and

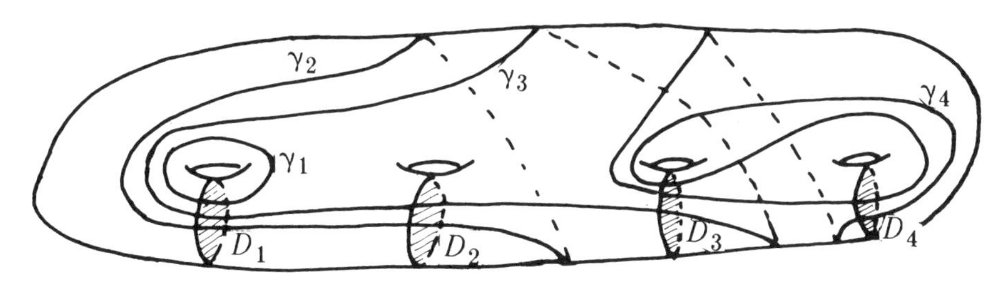

Figure 2.3.

2.6).

Example 2.7. Consider $\{\gamma_i\}_{i=1}^4 \subset \partial H_4$ as in Figure 2.3. In some standard system of generators (dual to disks $D_1, \ldots, D_4$)

$$[\gamma_1] = x_1,$$

$$[\gamma_2] = x_1 x_2,$$

$$[\gamma_3] = x_1 x_2 x_3,$$

$$[\gamma_4] = x_3 x_4 x_3^{-1} x_4^{-1}.$$

Then:

(0) $\partial H_4 - \{\gamma_i\}_{i=1}^4$ is incompressible in H_4,

(1) $\partial H_4 - \{\gamma_i, \gamma_j, \gamma_k\}$ is compressible in H_4 for each triplet of different i, $j, k \in \{1, 2, 3, 4\}$,

(2) For each pair i, j $(i \neq j)$ $\{[\gamma_i], [\gamma_j]\}$ does not bind a free factor F_3 of $\pi_1(H_4) = F_3 \star F_1$ and $(H_4)_{\{\gamma_1, \gamma_2, \gamma_3, \gamma_4\}}$ is equal to a solid torus with a hole so it has the compressible boundary.

We propose the following set of conditions which should guarantee the incompressibility of the boundary of a manifold obtained from a handlebody by adding a family of 2-handles:

Conjecture 2.8. *Let $\{\gamma_1, \ldots, \gamma_n\}$ be a family of 2-sided, pairwise disjoint simple closed curves (s.c.c.) in the boundary of a handlebody H_k. Assume that the following conditions are satisfied:*

(0) $\partial H_k - \bigcup_{i=1}^{n} \gamma_i$ is incompressible in H_K,

(1) for each j $\partial H_k - (\bigcup_{i=1}^{n} \gamma_i - \gamma_j)$ is compressible in H_k (or, equivalently, the family of elements of $\pi_1(H_k)$ represented by $\{\gamma_i\}_{i=1}^{n} - \gamma_j$ does not bind the free group $\pi_1(H_k)$; see [Lyon, 1980] or [Przytycki, 1983]),

(2) for each pair j, s $(j \neq s)$ $\{\gamma_i\}_{i=1}^{n} - \{\gamma_j, \gamma_s\}$ does not bind any free factor F_{k-1} of $\pi_1(H_k) = F_k = F_{k-1} \times F_1$,

...

(p) no $(n-p)$-element subfamily of $\{\gamma_i\}_{i=1}^{n}$ binds a free factor F_{k-p+1} of $\pi_1(H_k) = F_k = F_{k-p+1} \times F_{p-1}$,

...

$(n-1)$no curve $\gamma_j \in \{\gamma_i\}_{i=1}^{n}$ binds a free factor F_{k-n+2} of $\pi_1(H_k) = F_{k-n+2} \times F_{n-2}$.

Then $(H_k)_{\{\gamma_i\}_{i=1}^{n}}$ has incompressible boundary or it is equal to D^3.

All the assumptions of Conjecture 2.8 are necessary as can be shown by examples similar to Examples 2.5 and 2.7.

Theorem 2.2 leads to the following algebraic conjecture:

Conjecture 2.9. *Let $G = \{x_1, x_2, \ldots, x_n : w_1, w_2\}$ be a group with n generators and two relators which satisfies the following conditions:*

(i) $\{w_1, w_2\}$ binds F_n and

(ii) neither w_1 nor w_2 bind F_n.

Then G cannot be decomposed into nontrivial free product.

The following fact is very useful in further considerations.

Theorem 2.10. *Let γ_1, γ_2 be 2-sided, disjoint simple closed curves on the boundary of the 3-manifold M. Let $\partial M - (\gamma_1 \cup \gamma_2)$ be incompressible in M. Denote by A_i a tubular neighbourhood of γ_i in ∂M. Let $f: A_1 \rightarrow A_2$ be a homeomorphism and $\overline{M}$ a manifold obtained from M by identification (using f) of A_1 and A_2. Denote by γ' and γ'' the boundary of A_1. Then $\partial\overline{M} - \gamma'$ and $\partial\overline{M}$ are incompressible in $\overline{M}$.*

Proof. $\partial\overline{M} - (\gamma' \cup \gamma'')$ is incompressible in $\overline{M}$; if not then a disk of compression can be found disjoint from A_1 which would be a disk of compression of $\partial M - (\gamma_1 \cup \gamma_2)$ (γ_1 and γ_2 are nontrivial in M so γ' and γ'' are nontrivial in $\overline{M}$). If $\partial\overline{M} - \gamma'$ has been compressible then, from Lemma 2.3, $\overline{M}_{\gamma''} - \gamma'$ would be incompressible in $\overline{M}_{\gamma''}$. But $\overline{M}_{\gamma''}$ is equal to $M_{\{\gamma_1, \gamma_2\}}$

with a one-handle added and γ' does not cut the handle. This contradicts an incompressibility of $\partial \overline{M}_{\gamma''} - \gamma'$. Therefore $\partial \overline{M} - \gamma'$ is incompressible in $\overline{M}$. Furthermore $\partial \overline{M}_{\gamma'}$ is compressible in $\overline{M}$. So from Lemma 2.3, it follows that $\partial \overline{M}$ is incompressible in $\overline{M}$.

□

3. Application to surfaces in knot complements

From Theorems 2.2 and 2.10 we can get the following result which generalizes Theorem 1.4 of [Przytycki, 1983] to surfaces with four boundary components.

Theorem 3.1. *Let $K \subset S^3$ be a knot in a 3-sphere such that $S^3 - K$ does not contain any closed, incompressible, orientable, non-peripheral surface. Let S be a 2-sided, incompressible, ∂-incompressible, non-peripheral surface with four boundary components properly embedded in $S^3 - \operatorname{int} V_K = M$, where V_K is a regular neighbourhood of K in S^3. Let $M^{\partial S}$ be obtained from M by adding to M (along ∂M) a solid torus with meridian parallel to ∂S, and let $S\hat{}$ be the natural extension of S to $M^{\partial S}$. Then $S\hat{}$ is incompressible in $M^{\partial S}$ or $S\hat{} = S^2$ and $M^{\partial S}$ is a connected sum of lens spaces (allowing S^3 as a factor).*

It should be added that the author does not know any example in which $S\hat{}$ is a compressible 2-sphere in $M^{\partial S}$.

Proof. (suggested by M. Culler [Culler, 1982]). It follows from [Przytycki, 1983; Proposition 1.5] that S disconnects M into two handlebodies H and H'. $H \cap \partial M$ (resp. $H' \cap \partial M$) consists of two annuli A_1, A_2 with cores γ_1, γ_2 (resp. A'_1, A'_2 with cores γ'_1, γ'_2). To prove Theorem 3.1 it is enough to show that H, γ_1, γ_2 and H', γ'_1, γ'_2 satisfy the assumptions of Theorem 2.2. Assume the opposite. Let, for example, $\partial H - \gamma_1$ be incompressible. Then we can form a new surface S' properly embedded in M by a small isotopy of $S \cup A_2$. S' is unknotted (Proposition 1.5 of [Przytycki, 1983]) and has two boundary components, and by the incompressibility of $\partial H - \gamma_1$ and Theorem 2.10, S' is incompressible in M. By Theorem 1.4 of [Przytycki, 1983], $S'\hat{}$ should be incompressible in $M^{\partial S}$ but this is not the case (a disk of compression of S' is $M^{\partial S}$ can be easily found).

Theorem 3.1

References

Culler, 1982.
M. Culler, *Letter to J.H. Przytycki*, June 11th 1982.

Floyd.
W.J. Floyd, *Hyperbolic manifolds obtained by adding a 2-handle to a 3-manifold with compressible boundary.* preprint

Jaco, 1984.
W. Jaco, "Adding a 2-handle to a 3-manifold: an application of Property R," *Proc. A.M.S.*, **92**, pp. 288-292, 1984.

Johannson.
K. Johannson, *On surfaces in one-relator 3-manifolds.* (This volume)

Lozano-Przytycki, 1985.
M.T. Lozano and J.H. Przytycki, "Incompressible surfaces in the exterior of a closed 3-braid," *Math. Proc. Cambridge Phil. Soc.*, **98**, pp. 275-299, 1985.

Lyndon-Schupp, 1977.
R. C. Lyndon and P.E. Schupp, *Combinatorial group theory,* Springer-Verlag, Berlin, 1977.

Lyon, 1980.
H.C. Lyon, "Incompressible surfaces in the boundary of a handlebody- an algorithm," *Canad.J.Math.*, **32**, pp. 590-595, 1980.

Przytycki, 1979.
J. H. Przytycki, "A unique decomposition theorem for 3-manifolds with boundary," *Bull. Ac. Pol: Math.*, **XXVII(2)**, pp. 209-215, 1979.

Przytycki, 1983.
J. H. Przytycki, "Incompressibility of surfaces after Dehn surgery," *Michigan Math. J.*, **30**, pp. 289-308, 1983.

Scharlemann, 1985.
M. Scharlemann, "Outermost forks and a theorem of Jaco," *Contemporary mathematics*, pp. 189-193, AMS, 1985. Proceedings of conference Rochester 1982

J. Przytycki,
Mathematics Department,
University of Warsaw,
PKiN IXp
00901 Warsaw,
POLAND.

Part C:
Kleinian Groups

Hyperelliptic Klein Surfaces With Maximal Symmetry

*E. Bujalance**

*J. J. Etayo**

May proved [May, 1977a] that a compact Klein surface with boundary [Alling-Greenleaf, 1971] of algebraic genus $p \geq 2$ has at most $12(p-1)$ automorphisms. When a surface attains this bound, we say that it has *maximal symmetry* [Greenleaf-May, 1982] and its group of automorphisms is called an $M^{\star}$-group. [May, 1977b]

Families of Klein surfaces with maximal symmetry have been obtained in [Etayo, 1984], [Greenleaf-May, 1982], [May, 1977b], [May, 1980], [May, 1984], and [Singerman, 1985], studying in some of the cases the topological type of the surfaces. The purpose of this paper is to study which hyperelliptic Klein surfaces have maximal symmetry, and determine their corresponding $M^{\star}$-groups. As a consequence of it, we obtain that the subspace of the Teichmüller space corresponding to the hyperelliptic Klein surfaces with maximal symmetry is a manifold, and calculate its dimension.

A Klein surface with boundary is said hyperelliptic if the canonical Riemann surface associated to it is hyperelliptic.

Klein surfaces and their automorphisms are studied via NEC groups. An NEC group is a discrete subgroup of the group of isometries of the hyperbolic plane with compact quotient space. The subgroup Γ^{+} of an NEC group Γ, formed by the orientation preserving elements, is the canonical Fuchsian group associated to the NEC group. NEC groups may be classified according to their signatures, that have the form

$$(g, +-, [m_1, \ldots, m_r], \{(n_{11}, \ldots, n_{1s_1}), \ldots, (n_{k1}, \ldots, n_{ks_k})\})$$

and determine a presentation of the group [Macbeath, 1967], [Wilkie, 1966].

* Partially supported by "Comisión Asesora de Investigación Científica y Técnica".

The numbers m_i are the proper periods, and the brackets $(n_{i1}, \ldots, n_{is_i})$ are the period-cycles of the group.

A Klein surface of topological genus g, with k boundary components and algebraic genus $p \geq 2$, may be represented as D/K, where $D = \{z \in \mathbb{C}, \operatorname{im}(z) \geq 0\}$ and K is an NEC group with signature $(g, \pm, [-], \{(-), \overset{k}{\ldots}, (-)\})$, with sign '+' if X is orientable, and '−' if non-orientable, [Preston, 1975]. The surface D/K^+ is the Riemann surface associated to D/K. A finite group G is a group of automorphisms of D / K if and only if $G = \Gamma/K$, Γ being an NEC group of which K is a normal subgroup [May, 1977b].

Each NEC group Γ has an associated area, $|\Gamma|$, and if $|\Gamma{:}K| = n$, then $|K| = n|\Gamma|$ [Singerman, 1974a].

The next result (*) will be used in the forthcoming theorem: Let Γ be an NEC group with a non-empty period-cycle $(n_1, \ldots, n_s)$ and $c_0, \ldots, c_s$ the reflections that determine this period-cycle [Macbeath, 1967]. Let Γ_0 be a normal subgroup of Γ, with even index N. If $c_i, c_{i+1}, \ldots, c_j \in \Gamma_0, c_{i-1}, c_{j+1} \notin \Gamma_0$, and n is the index of $c_{i-1}c_{j+1} \bmod \Gamma_0$, then n_i, n_{j+1} are even, and among the period-cycles of the signature of Γ_0 there are at least $\frac{N}{2n}$ between those having the forms

$$\left(\frac{n_i}{2}, n_{i+1}, \ldots, n_j, \frac{n_{j+1}}{2}, n_j, \ldots, n_{i+1}, \overset{(n)}{\ldots}, \frac{n_i}{2}, n_{i+1}, \ldots, n_j, \frac{n_{j+1}}{2}, n_j, \ldots, n_{i+1}\right)$$

and

$$\left(\frac{n_{j+1}}{2}, n_j, \ldots, n_{i+1}, \frac{n_i}{2}, n_{i+1}, \ldots, n_j, \overset{(n)}{\ldots}, \frac{n_{j+1}}{2}, n_j, \ldots, n_{i+1}, \frac{n_i}{2}, n_{i+1}, \ldots, n_{i+j}\right),$$

[Bujalance, 1985].

In [Bujalance-Etayo-Gamboa, 1985] we have characterised the hyperelliptic Klein surfaces in terms of NEC groups, obtaining this result: let K be an an NEC group with signature $(g, \pm, [-], \{(-), \overset{k}{\ldots}, (-)\})$.

Then D/K is hyperelliptic if and only if there exists a unique NEC group Γ_1 with $|\Gamma_1{:}K| = 2$, and whose signature is

i) $(0, +, [-], \{(2, \overset{2k}{\ldots}, 2)\})$ if $g = 0$,

ii) $(0, +, [2, \overset{2g+k}{\ldots}, 2], \{(-)\})$ if $g \neq 0$ and K has sign '+',

iii) $(0, +, [\overset{g}{2, \ldots, 2}], \{(\overset{2k}{2, \ldots, 2})\})$ if K has sign '$-$'.

Observe that for each topological type of Klein surfaces with boundary there are infinite hyperelliptic surfaces, excepting the case of sign '+', $g \neq 0$, and $k \geq 3$, in which case there is no hyperelliptic surface.

If D/K is a bordered Klein surface with maximal symmetry and G is its group of automorphisms, then $G = \Gamma/K$, and Γ is an NEC group with signature $(0, +, [-], \{(2, 2, 2, 3)\})$ [Greenleaf-May, 1982].

Theorem. *The unique hyperelliptic Klein surfaces with boundary, with maximal symmetry, are the sphere with three holes and the torus with one hole. In both cases, the group of automorphisms is* D_6.

Proof.

The proof will be divided into three steps, according to the signature of the group Γ_1 above.

Let D/K be a hyperelliptic Klein surface.

Step i: If the signature of K is $(0, +, [-], \{\overset{k}{(-), \ldots, (-)}\})$, then Γ_1 has signature $(0, +, [-], \{(\overset{2k}{2, \ldots, 2})\})$. If D/K has maximal symmetry, there exists an NEC group Γ with signature $(0, +, [-], \{(2, 2, 2, 3)\})$ such that Γ/K is the group of automorphisms of D/K, and hence there is an epimorphism θ from Γ onto Γ/K with kernel K. By [Bujalance-Etayo-Gamboa, 1985], Γ_1/K is central in Γ/K, and so $\Gamma_1 \triangleleft \Gamma$. Thus, there exists an epimorphism θ' from Γ onto $\dfrac{\Gamma/K}{\Gamma_1/K}$ with kernel Γ_1, extendible to θ.

The presentation of Γ is given by generators c_1, c_2, c_3, c_4, and relations $c_i^2 = (c_1c_2)^2 = (c_2c_3)^2 = (c_3c_4)^2 = (c_4c_1)^3 = 1$. As $c_i^2 = 1$ and θ' is an epimorphism, $\theta'(c_i)$ either is 1 or has order two. If all of them were 1, in the signature of Γ_1 there would appear the value 3 in the period-cycles, [Bujalance, 1981a] in contradiction with the known signature of Γ_1; and by (*), for the existence of the value 2 in the period-cycles, there must be at least two consecutive c_i's whose image is 1. If there were three, say $\theta'(c_1) = \theta'(c_2) = \theta'(c_3) = 1$, $\theta'(c_4) = x$, then $\theta'(c_4c_1)^3 = x$, in contradiction with $(c_4c_1)^3 = 1$. (Any other situation with three images equal to 1 is analogous.) So we have just two c_i's with image 1. Let them be c_2, c_3, and $\theta'(c_1) = x$, $\theta'(c_4) = y$. As $(c_4c_1)^3 = 1$, the order of xy is 1 or 3. If $xy = 1$, by [Bujalance, 1981b] there would be proper periods in Γ_1, which is impossible. So xy has order 3.

The group Γ/K has order $12(p-1)$, p being the algebraic genus of D/K. As $p = \alpha g + k - 1$, where $\alpha = 2$ for orientable surfaces and $\alpha = 1$ for non-orientable ones, in this case $p = k - 1$, and the order of Γ/K is $12(k-2)$. So the order of $\dfrac{\Gamma/K}{\Gamma_1/k}$ is $6(k-2)$. Now using (*) we obtain that there exist at least $\dfrac{6(k-2)}{6}$ period-cycles in the signature of Γ_1. As we know that it has only one, then $k = 3$.

In this case D/K is the sphere with three holes, and it was obtained in [Bujalance-Gamboa, 1984] that there exist such surfaces with maximal symmetry, and their group of automorphisms is D_6.

Step ii If the signature of K is $(g, +, [-], \{\overset{k}{(-), \ldots, (-)}\})$ with $g \neq 0$, Γ_1 has signature $(0, +, [\overset{2g+k}{2, \ldots, 2}], \{(-)\})$.

Using the same arguments and notation of case i) the images of the c_i' s either are 1 or have order two, and neither are all 1, nor are all different from 1. If two consecutive c_i's had the same image 1, by (*) there would be non-empty period-cycles in the signature of Γ_1; and if two non-consecutive c_i's had image 1, by [Bujalance, 1981b] there would be no proper period in that signature. So there is a unique c_i with image 1. It may not be c_1 or c_4, because $(c_4c_1)^3 = 1$; if we call x the image of c_4 (resp. c_1), we should have $\theta'(c_4c_1)^3 = x \neq 1$, which is impossible. So, let c_2 have image 1. (The argument holds equally if it were c_3). If we call $\theta'(c_1) = x$, $\theta'(c_3) = y$, $\theta'(c_4) = z$, as $(c_4c_1)^3 = 1$, the order of xz is 1 or 3. If $xz = 1$ we have a proper period 3 in Γ_1 [Bujalance, 1981b]. So xz has order 3. Finally $yz = 1$ for the existence of proper periods in Γ_1, [Bujalance, 1981b] and so $y = z$.

In this case the algebraic genus of D/K is $p = 2g+k-1$, and the order of Γ/K is $12(2g+k-2)$. So $\dfrac{\Gamma/K}{\Gamma_1/K}$ has order $6(2g+k-2)$. As the epimorphism of Γ onto $\dfrac{\Gamma/K}{\Gamma_1/K}$ is given by $\theta'(c_1) = x$, $\theta'(c_2) = 1$, $\theta'(c_3) = \theta'(c_4) = y$, by [Bujalance, 1981b] the number of proper periods in Γ_1 is $3(2g+k-2)$, that must be equal to $2g+k$. As $g \neq 0$, it is $g = 1$, $k = 1$. So the surface is the torus with one hole, and its group of automorphisms is

D_6. [Bujalance-Gamboa, 1984]

Step iii If the signature of K is $(g, -, [-], \{\overbrace{(-), \ldots, (-)}^{k}\})$ Γ_1 has signature $(0, +, [\overbrace{2, \ldots, 2}^{g}], \{(\overbrace{2, \ldots, 2}^{2k})\})$. Using arguments of above paragraphs, two consecutive c_i's must have image 1 for Γ_1 having non-empty period-cycles; and the other two c_i's must have a same image $x \neq 1$ for Γ_1 having proper periods. So necessarily $\theta'(c_1) = \theta'(c_4) = x$, $\theta'(c_2) = \theta'(c_3) = 1$, but then $\theta'(c_4c_1) = 1$, and thus there is a proper period equal to 3 [Bujalance, 1981b]. So there is no non-orientable hyperelliptic Klein surface with maximal symmetry.

□

Given an NEC group K, we denote $R(K, G)$ the set of isomorphisms $r : K \to G$ such that $r(K)$ is discrete and $D/r(K)$ is compact. Two elements r_1, $r_2 \in R(K, G)$ are said to be equivalent if for each $\gamma \in K$ there exists $g \in G$ verifying $r_1(\gamma) = g r_2(\gamma) g^{-1}$. The quotient space $T(K, G)$, the Teichmüller space of K, is homeomorphic to a cell of dimension $d(K)$. When K is a Fuchsian group with signature $(g, +, [m_1, \ldots, m_r])$, it is known that $d(K) = 6(g-1)+2r$, and Singerman proves in [Singerman, 1974b] that if K is a proper NEC group $d(K) = \frac{1}{2}d(K^+)$. The Teichmüller modular group $M(K)$ of K [Macbeath-Singerman, 1975] is the quotient space $\mathrm{Aut}(K)/I(K)$, where $\mathrm{Aut}(K)$ is the full group of automorphisms of K and $I(K)$ denotes the inner automorphisms.

Let K be an NEC group with signature $(0, +, [-], \{(-)(-)(-)\})$ and K' an NEC group with signature $(1, +, [-], \{(-)\})$. Then we call $T_M(K) = \{[r] \in T(K, G) | D/r(K)$ is a sphere with three holes and maximal symmetry$\}$, and $T_M(K') = \{[r] \in T(K', G) | D/r(K')$ is a torus with one hole with maximal symmetry$\}$.

Theorem. *$T_M(K)$ and $T_M(K')$ are submanifolds of $T(K, G)$ and $T(K', G)$, respectively, of dimension 1.*

Proof. By [Harvey, 1971] and [Macbeath-Singerman, 1975]

$$T_M(K) = \bigcup_{\bar{\alpha} \in M(\Gamma)} \bar{\alpha} \left(\bigcup_{i_\phi \in \phi(K, \Gamma, \Gamma/K)} i_\phi^\star(T(\Gamma, G)) \right)$$

where Γ is an NEC group with signature $(0, +, [-], \{(2, 2, 2, 3)\})$ and $\phi(K, \Gamma, \Gamma/K)$ is the family of all equivalence classes of surjections ϕ:

$\Gamma \to \Gamma/K$ with $\mathrm{Ker}\,\phi = K$ (modulo the actions of $\mathrm{Aut}(\Gamma)$ and $\mathrm{Aut}(\Gamma/K)$), and $i_\phi^\star$ is the isometry induced by the inclusion i_ϕ :$\mathrm{Ker}\phi \to \Gamma$. By [Bujalance-Gamboa, 1984] $\phi(K, \Gamma, \Gamma/K)$ has a unique element. Then $T_M(K) = \bigcup_{\bar{\alpha} \in M(K)} \bar{\alpha}(i^\star(T(\Gamma, G)))$. Now, $\bar{\alpha}(i^\star(T(\Gamma, G))) \cap i^\star(T(\Gamma, G)) \neq \varnothing$ if and only if $\bar{\alpha}(i^\star(T(\Gamma, G))) = i^\star(T(\Gamma, G))$. This holds because given the group K, there is a unique group Γ with the stated signature of which K is a normal subgroup, as it is the normalizer of K in G. So the proof of lemma 3 in [Maclachlan, 1971a] may be used in this case, by virtue of the results of. [Macbeath-Singerman, 1975] Hence we have proved that $T_M(K)$ is a disjoint union of cells of dimension $\frac{1}{2}d(\Gamma^+) = 1$.

The same arguments hold for K'.

□

It may be observed that there are no hyperelliptic Riemann surfaces with maximal symmetry. A Riemann surface D/K of genus g has maximal symmetry when its group G of automorphisms has $84(g-1)$ elements, and so there is an epimorphism from a Fuchsian group Γ with signature $(0, +, [2, 3, 7])$ onto G having kernel K.

Proposition. *There is no hyperelliptic Riemann surface with maximal symmetry.*

Proof. D/K is a hyperelliptic Riemann surface of genus g if and only if there exists a unique Fuchsian group Γ_1 such that $|\Gamma_1{:}K| = 2$ and Γ_1 has signature $(0, +, [\overset{2g+2}{2, \ldots, 2}])$ [Maclachlan, 1971a]. If D/K had maximal symmetry, there would be an epimorphism θ from Γ onto $\dfrac{\Gamma/K}{\Gamma_1/K}$ with kernel Γ_1. As Γ has signature $(0, +, [2, 3, 7])$, and so it has a presentation $x_1^2 = x_2^3 = (x_1x_2)^7 = 1$, by [Maclachlan, 1971b] $\theta(x_1) = 1$, and the number of proper periods equal to 2 in Γ_1 should be $42(g-1)$. So $42(g-1) = 2g+2$, and $40g = 44$, impossible.

□

References

Alling-Greenleaf, 1971.
Alling, N.L. and Greenleaf, N., "Foundations of the theory of Klein surfaces," *Lect. Notes in Math.*, **219**, Springer, New York, 1971.

Bujalance, 1981a.
Bujalance, E., "Normal subgroups of NEC groups," *Math. Z.*, **178**, pp. 331-341, 1981.

Bujalance, 1981b.
Bujalance, E., "Proper periods of normal NEC subgroups with even index," *Rev. Mat. Hisp.-Amer. (4) 41*, pp. 121-127, 1981.

Bujalance, 1985.
Bujalance, J.A., *Sobre los órdenes de los automorfismos de las superficies de Klein,* U.N.E.D., 1985. Ph.D. Thesis.

Bujalance-Etayo-Gamboa, 1985.
Bujalance, E., Etayo, J.J., and Gamboa, J.M., "Hyperelliptic Klein surfaces," *Quart. J. Math. Oxford*, **36**, pp. 141-57, 1985.

Bujalance-Gamboa, 1984.
Bujalance, E. and Gamboa, J.M., "Automorphism groups of algebraic curves of R^n of genus 2," *Archiv der Math.*, **42**, pp. 229-237, 1984.

Etayo, 1984.
Etayo, J.J., "Klein surfaces with maximal symmetry and their groups of automorphisms," *Math. Ann.*, **268**, pp. 533-8, 1984.

Greenleaf-May, 1982.
Greenleaf, N. and May, C.L., "Bordered Klein surfaces with maximal symmetry," *Trans. Am. Math. Soc.*, **274**, pp. 265-283, 1982.

Harvey, 1971.
Harvey, W.J., "On branch loci in Teichmüller space," *Trans. Am. Math. Soc.*, **153**, pp. 387-399, 1971.

Macbeath, 1967.
Macbeath, A.M., "The classification of non-euclidean plane crystallographic groups," *Can. J. Math.*, **9**, pp. 1192-1205, 1967.

Macbeath-Singerman, 1975.
Macbeath, A.M. and Singerman, D., "Space of subgroups and Teichmüller space," *Proc. London Math. Soc. (3)*, **31**, pp. 221-256, 1975.

Maclachlan, 1971a.
Maclachlan, C., "Smooth coverings of hyperelliptic surfaces," *Quart. J. Math. Oxford (2)*, **22**, pp. 117-123, 1971.

Maclachlan, 1971b.
Maclachlan, C., "Maximal normal Fuchsian groups," *Illinois J. Math.*, **15**, pp. 104-113, 1971.

May, 1977a.
May, C.L., "A bound for the number of automorphisms of a compact Klein surface with boundary," *Proc. Am. Math. Soc.*, **63**, pp. 273-280, 1977.

May, 1977b.
May, C.L., "Large automorphism groups of compact Klein surfaces with boundary," *Glasgow Math. J.*, **18**, pp. 1-10, 1977.

May, 1980.
May, C.L., "Maximal symmetry and fully wound coverings," *Proc, Am. Math. Soc.*, **79**, pp. 23-31, 1980.

May, 1984.
May, C.L., "The species of bordered Klein surfaces with maximal symmetry of low genus," *Pacific J. Math.*, **111**, pp. 371-394, 1984.

Preston, 1975.
Preston, R., *Projective structures and fundamental domains on compact Klein surfaces Ph. D. thesis. Univ. of Texas*, 1975.

Singerman, 1974a.
Singerman, D., "On the structure of non-euclidean crystallographic groups," *Proc. Cambridge Phil. Soc.*, **76**, pp. 233-240, 1974.

Singerman, 1974b.
Singerman, D., "Symmetries of Riemann surfaces with large automorphism group," *Math. Ann.*, **210**, pp. 17-32, 1974.

Singerman, 1985.
Singerman, D., *Orientable and non-orientable Klein surfaces with maximal symmetry*, 26, pp. 31-4, 1985.

Wilkie, 1966.
Wilkie, M.C., "On non-euclidean crystallographic groups," *Math. Z.*, **91**, pp. 87-102, 1966.

E. Bujalance,
Departamento de Matematica Fundamental,
U.N.E.D.
28040 Madrid,
Spain.

J.J. Etayo Gordejuela,
Departamento de Geometria y Topologia,
Facultad de C. Matematicas,
Universidad Complutense,
28040 Madrid,
Spain.

Right-Angled Coxeter Groups

I.M. Chiswell

We consider the group G presented by a finite set of generators, $x_1, \ldots, x_n$, with relators $x_i^{m_i}$ for $1 \leq i \leq n$, and $[x_i, x_j]$ for $(i, j) \in I$, where I is some subset of $\{1, 2, \ldots, n\} \times \{1, 2, \ldots, n\}$, and $1 < m_i \leq \infty$. If $m_i = \infty$ this means that there is no relator $x_i^{m_i}$ corresponding to this value of i. In the special case that all m_i are equal to 2, the relators $[x_i, x_j]$ may be replaced by $(x_i x_j)^2$, so G is clearly a Coxeter group, and we call such a group a *right-angled* Coxeter group. The main aim is to show that G is *virtually accessible*, that is, has an accessible subgroup of finite index.

In this note the term "accessible group" will not be used with its standard meaning [Wall, 1971 p.146]. The class of accessible groups is defined to be the smallest class of groups $\mathcal{C}$, closed under isomorphism, and with the following properties:

(i) the trivial group is in $\mathcal{C}$

(ii) If H is a free product with amalgamation, $H = A \star_B C$, and A, B, C are all in $\mathcal{C}$, then $H \in \mathcal{C}$

(iii) If H is an HNN-extension $\langle t, A; tBt^{-1} = C \rangle$ and A, B, C are all in $\mathcal{C}$, then $H \in \mathcal{C}$.

Using (i) and (iii), it follows inductively that finitely generated free abelian groups are accessible. The study of this class of groups is suggested by the work of Waldhausen on Whitehead groups [Waldhausen, 1973].

We shall show that any torsion-free subgroup of G of finite index is accessible, and then that G has a torsion-free subgroup of finite index, namely, the kernel of the obvious homomorphism $G \rightarrow T$, where T is the torsion subgroup of $G/[G, G]$. By the same method, it is possible to classify the finite subgroups of G. We finish by investigating the virtual cohomological dimension of G, but have no definitive result. Our arguments are simple, and are based on a decomposition of G as a free product with amalgamation.

We begin with some remarks on graphs of groups, and refer to [Cohen, 1974] for notation concerning graphs of groups. Suppose $(\mathcal{G}, Y)$ is a graph of groups, T is a maximal tree of Y, and $H = \pi(\mathcal{G}, Y)$ is the fundamental group, formed using T (the choice of T is unimportant and is suppressed in the notation for fundamental group). If Y is finite, and all edge and vertex groups are accessible, then H is accessible. For let v be a terminal vertex of T, e the unique unoriented edge of T incident with v, and T' the tree obtained by removing v and e from T. Then

$$\pi(\mathcal{G}|_T, T) = \pi(\mathcal{G}|_{T'}, T')\star_{G_e} G_v$$

and it follows that $\pi(\mathcal{G}|_T, T)$ is accessible by induction on the number of edges of T. Since H is obtained from $\pi(\mathcal{G}|_T, T)$ by making a finite succession of HNN-extensions (one for each unoriented edge not in T), where the associated subgroups are, up to isomorphism, the remaining edge groups, it follows inductively that H is accessible.

We shall use the subgroup theorem for amalgamated free products (see [Karrass-Solitar, 1970]), which is best proved using the Bass-Serre Theorem, and best expressed in terms of this theory (see the proof of Theorem 3 in [Cohen, 1974]). We shall also make use of the following result: if $(\mathcal{G}, Y)$ is a graph of groups and $H = \pi(\mathcal{G}, Y)$, then any finite subgroup of H is contained in a conjugate of a vertex group. This follows from the fact that H acts without inversions on the tree $\tilde{Y}$ (see [Cohen, 1974 Theorem 2]), and the stabilizers of the vertices of $\tilde{Y}$ are the conjugates in H of the vertex groups of $(\mathcal{G}, Y)$. Thus, any finite subgroup of H acts without inversion on $\tilde{Y}$ and therefore has a fixed vertex; this follows from [Serre, 1980 I.4.3, Prop.19]. (See also [Serre, 1980 I.6] in this context.)

Recall that G is defined by

$$G = \langle x_1, \ldots, x_n \,;\, x_i^{m_i}\ (1 \le i \le n), [x_i, x_j], ((i,j) \in I)\rangle \qquad (1)$$

Without loss of generality, we assume that, if $(i, j) \in I$, then $i < j$. Let A be the group obtained from G by deleting x_n from the generators, and all relators in which x_n appears. Then A has a presentation of the same form as G, with one less generator. Now renumber the generators so that $\{i; (i,n) \in I\} = \{1, 2, \ldots, r\}$, and let B be the subgroup of A generated by $x_1, \ldots, x_r$. Then,

$$G = A \star_B C \qquad (2)$$

where $C = B \times \langle x_n; x_n^{m_n}\rangle$. It follows that A is the subgroup of G generated by $x_1, \ldots, x_{n-1}$. It then follows inductively that B is the group obtained from G by deleting $x_{r+1}, \ldots, x_n$ and all the relators in which they appear, so B has a presentation of the same form as G, and therefore so does C. (In the case of a right-angled Coxeter group, this follows from generalities about Coxeter groups – see [Bourbaki, 1968 Ch. 4, 1.8]).

Proposition 1. *Suppose H is a torsion-free subgroup of finite index in the group G defined by (1) above. Then H is accessible.*

Proof. The proof is by induction on n. If $(i, j) \in I$ for all pairs (i, j) with $1 \leq i < j \leq n$, then G is finitely generated abelian, so any torsion free subgroup of finite index is finitely generated free abelian, hence accessible. Otherwise, renumber the generators so that, for some $i < n$, $(i, n) \notin I$. Then in the decomposition (2) above, A, B, C have presentations of the same form as G, but with fewer generators (since $r < n - 1$), so by induction all torsion-free subgroups of finite index in A, B, C, and in their conjugates in G, are accessible.

By the subgroup theorem for amalgamated free products we have $H = \pi(\mathcal{G}, Y)$, where Y is finite, and the vertex and edge groups of $(\mathcal{G}, Y)$ have the form $H \cap gDg^{-1}$, where D is either A, B or C and $g \in G$. Since $H \cap gDg^{-1}$ is torsion-free of finite index in gDg^{-1}, it is accessible, and the result follows.

□

With G defined by equation (1) above, let

$$T_G = \prod_{\substack{1 \leq i \leq n \\ m_i \neq \infty}} \langle x_i ; x_i^{m_i} \rangle$$

and let $K_G = \text{Ker } f$, where $f : G \to T_G$ is the homomorphism defined by $f(x_i) = x_i$ if $m_i \neq \infty$, $f(x_i) = 1$ if $m_i = \infty$.

Proposition 2. *K_G is torsion-free and of finite index in G*

Proof. Since T_G is finite abelian, K_G is of finite index in G. We show K_G is torsion-free by induction on n. Again, if $(i,j) \in I$ for all (i,j) with $i < j$, $K_G = \prod_{\substack{1 \leq i \leq n \\ m_i = \infty}} \langle x_i ; x_i^{m_i} \rangle$ is free abelian.

Otherwise, renumber generators so that $(i,n) \notin I$ for some $i < n$, and use the decomposition $G = A \star_B C$ in (2) above. Applying the subgroup theorem for amalgamated free products, $K_G = \pi(\mathcal{G}, Y)$, where the vertex groups of $(\mathcal{G}, Y)$ have the form $K_G \cap gAg^{-1}$ and $K_G \cap gCg^{-1}$ for certain $g \in G$. Now T_A is a subgroup of T_G in an obvious way, so $K_G \cap gAg^{-1} =$

$g(K_G \cap A)g^{-1} = gK_A g^{-1}$, and likewise $K_G \cap gCg^{-1} = gK_C g^{-1}$. By induction all vertex groups of $(\mathcal{G}, Y)$ are torsion-free, hence so is K_G (using the remark made earlier, that any finite subgroup of $\pi(\mathcal{G}, Y)$ is in a conjugate, in $\pi(\mathcal{G}, Y)$, of a vertex group).

$\square$

Propositions 1 and 2 were proved for right-angled Coxeter groups by B. Levy, using hyperbolic geometry. The proofs given above seem substantially simpler.

Again G denotes the group defined by equation (1) above. Suppose $x_{i_1}, \ldots, x_{i_k}$ are generators, such that, if $i, j \in \{i_1, \ldots, i_k\}$ and $i < j$ then $[x_i, x_j]$ is a defining relator and such that $m_{i_j} < \infty$ for $1 \leq j \leq k$. Then $x_{i_1}, \ldots, x_{i_k}$ generate a finite subgroup of G, mapping isomorphically onto a subgroup of T_G via the homomorphism f defined immediately before Proposition 2. (Recall our assumption that, if $(i, j) \in I$, then $i < j$). We call $\langle x_{i_1}, \ldots, x_{i_k}\rangle$ an *obvious* finite subgroup of G.

Proposition 3. *Any finite subgroup of G is conjugate, in G, to a subgroup of an obvious finite subgroup.*

Proof. Again we use induction on n. If G is itself an obvious subgroup then the result is, so to speak, obvious. Otherwise we can make the decomposition (2), $G = A *_B C$, where A, B and C have fewer generators than G. Note that an obvious finite subgroup of A or C is an obvious finite subgroup of G. Now G is the fundamental group of the graph of groups

$$A \circ\!\!-\!\!\!-\!\!\!-\!\!\!-\!\!\circ\ C \quad (\text{edge } B)$$

so by earlier remarks if F is a finite subgroup of G, F is contained in a conjugate, in G, of either A or C (see [Serre, 1980, Corollary to Theorem 8 in I.4.3]). Thus, for some $g \in G$, $g^{-1}Fg$ is a finite subgroup of either A or C, so by induction, is either conjugate in A to an obvious finite subgroup of A, or is conjugate in C to an obvious finite subgroup of C. In any case, F is conjugate in G to an obvious finite subgroup of G.

$\square$

If on the other hand $x_{i_1}, \ldots, x_{i_k}$ are generators such that for no $i, j \in \{i_1, \ldots, i_k\}$ with $i < j$ is $[x_i, x_j]$ a defining relator, then $x_{i_1}, \ldots, x_{i_k}$ generate, in G, the free product $\langle x_{i_1}\rangle * \ldots * \langle x_{i_k}\rangle$ of cyclic groups of orders

$m_{i_1}, \ldots, m_{i_k}$. This can be seen by considering the quotient obtained by adding relations $x_i = 1$ to G, for all $i \notin \{i_1, \ldots, i_k\}$. Note also that each generator x_i has exact order m_i in G; this can be seen by abelianizing. A further useful remark is the following. Suppose $X_1, \ldots, X_k$ are disjoint subsets of the set of generators of G, such that, if $x \in X_i$ and $y \in X_j$, where $i \neq j$, either $[x, y]$ or $[y, x]$ is a defining relator. let G_i be the subgroup of G generated by X_i, so $[G_i, G_j] = 1$ for $i \neq j$. If $X = X_1 \cup \ldots \cup X_k$, then the subgroup of G generated by X is $G_1 \times \ldots \times G_k$. This follows by considering the quotient of G obtained by setting $x_i = 1$ for all $x_i \notin X$.

We should also note that, in the case of a right-angled Coxeter group, Prop. 3 follows from general results on Coxeter groups. See [Bourbaki, 1968 Ch.5, Theorem 2 in 4.8 and Example 2(d), p.130].

By Propositions 1 and 2, the group G defined by (1) is virtually torsion-free, so has a well-defined virtual cohomological dimension, abbreviated to *vcd* (see [Serre, 1971 p.99]). To investigate $vcd(G)$, we introduce a graph X_G. The vertices are all pairs $\{x_i, x_j\}$ for which $i < j$, both m_i and m_j are finite, and such that $[x_i, x_j]$ is not a defining relator, together with all singletons $\{x_i\}$ for which $m_i = \infty$. Two vertices v and w are joined by an (unoriented) edge if, for each $x_i \in v$ and $x_j \in w$, either $[x_i, x_j]$ or $[x_j, x_i]$ is a defining relator. We define $w(G)$ by

$$w(G) = \max\{\, k; X_G \text{ has a complete subgraph with } k \text{ vertices}\}$$

Proposition 4. $w(G) \leq vcd(G) < \infty$.

Proof. Let $v_1, \ldots, v_k$ be the vertices of a complete subgraph of X_G. If v_i has the form $\{x_s, x_t\}$ where $s < t$, let G_i be the subgroup of G generated by x_s and x_t, so $G_i = \langle x_s \rangle \star \langle x_t \rangle$ is a free product of two finite cyclic groups of orders m_s and m_t, by the remarks after Proposition 3. Thus G_i has an infinite cyclic subgroup, for example $\langle x_s x_t \rangle$. If v_i is of the form $\{x_s\}$ where $m_s = \infty$, let $G_i = \langle x_s \rangle$, an infinite cyclic group. By the remarks after Proposition 3, the subgroup of G generated by $G_1, \ldots, G_k$ is $G_1 \times \ldots \times G_k$. Hence G has a subgroup of the form $C_\infty \times \ldots \times C_\infty$ (k copies of the infinite cyclic group), which has cohomological dimension k. Therefore $vcd(G) \geq k$, using Remarques (b) and (c) on p.99 of, [Serre, 1971] and so $vcd(G) \geq w(G)$.

To see that $vcd(G) < \infty$, we use an inductive argument as in the proof of preceding results. If the defining relators imply that all generators of G commute with each other, then G is finitely generated free abelian by finite, so has finite vcd; otherwise we can use the decomposition (2), together with induction on n and [Serre, 1971 Prop. 15(a)] to see that $vcd(G) < \infty$.

□

Conjecture. $vcd(G) = w(G)$.

Proposition 5. *Assume the conjecture is true in the case that all $m_i < \infty$. Then it is true in general.*

Proof. Again we use induction on n, the number of generators. Suppose G has a generator x_i with $m_i = \infty$, and renumber so that $i = n$. Let A and B be defined as in the decomposition (2) above. Then G is an HNN-extension:

$$G = \langle x_n, A;\, x_n B x_n^{-1} = B \rangle.$$

By induction, $vcd(A) = w(A)$ and $vcd(B) = w(B)$. Moreover, X_A is a subgraph of X_G, so $w(A) \leq w(G)$, and $w(B) < w(G)$, because any complete subgraph of X_B is a subgraph of a complete subgraph of X_G with one more vertex, namely $\{x_n\}$. By the analogue of[Serre, 1971 Prop 15(a)] for HNN-extensions,

$$vcd(G) \leq \max\{vcd(A),\ 1 + vcd(B)\}$$

It follows that $vcd(G) \leq w(G)$, and by Proposition 4, $vcd(G) = w(G)$. (The analogue of [Serre, 1971 Prop.15(a)] for HNN-extensions is proved by the same method as for free products, using the tree on which an HNN-extension acts; see [Cohen, 1974 p.398]).

□

The proof of Proposition 5 shows more than is claimed. For example, the conjecture is true if all $m_i = \infty$. This is related to results of Warren Dicks on polynomial rings. [Dicks, 1981]

In the case that all $m_i < \infty$, the best result we have been able to prove is the following.

Proposition 6. *Suppose G is defined by (1) above, and all $m_i < \infty$. Then $vcd(G) \leq n/2$, with equality if and only if $w(G) = n/2$.*

Proof. Again we use induction on n. If all generators of G commute with each other, then G is finite and X_G is empty, so $vcd(G) = w(G) = 0$. Otherwise, we can make the decomposition (2), $G = A \star_B C$, where $B = \langle x_1, \ldots, x_r \rangle$, $C = B \times \langle x_n \rangle$, and x_{n-1} and x_n do not commute, so $r < n-1$ (as in the proof of Prop. 1). By induction, using [Serre, 1971 Prop. 15(a)],

$$vcd(G) \leq \max\{vcd(A), 1+vcd(B)\} \leq \max\{(n-1)/2, 1+(n-2)/2\} = n/2.$$

(We are using the fact that C is a finite extension of B, so $vcd(C) = vcd(B)$).

Suppose $vcd(G) = n/2$; then the above inequalities imply that $vcd(B) = (n-2)/2$, so by induction, $r = n - 2$ and $w(B) = (n - 2)/2$. Now by definition of X_B, a complete subgraph of X_B with vertices $v_1, \ldots, v_k$ has $v_i \cap v_j = \phi$ for $i \neq j$, and each v_i contains two distinct generators of B. Hence, $k \leq (n - 2)/2$ since B has only $(n - 2)$ generators, and $w(B) = (n - 2)/2$ means that X_B is a complete graph.

Now interchange the rôles of $x_n - 1$ and x_n, to obtain an analogous decomposition: $G = A' \star_{B'} (B' \times \langle x_{n-1} \rangle)$, with $A = \langle x_1, \ldots, x_{n-2}, x_n \rangle$ and $B = \langle x_1, \ldots, x_s \rangle$ where, after renumbering $x_1, \ldots, x_{n-2}$, the generators which commute with x_{n-1} according to the defining relators for G are $x_1, \ldots, x_s$. By the same argument used for B, $s = n-2$. Thus x_n and x_{n-1} both commute with $x_1, \ldots, x_{n-2}$. Hence X_G is the complete graph obtained from X_B by adding the vertex $\{x_{n-1}, x_n\}$ and joining it all to vertices of X_B, so $w(G) = 1 + w(B) = n/2$.

Finally, if $w(G) = n/2$, then $vcd(G) = n/2$ by Proposition 4. □

We emphasise one point from the proof of Proposition 6. If $vcd(G) = n/2$, X_G is a complete graph, and this means that, after suitably renumbering the generators,

$$G = (\langle x_1 \rangle \star \langle x_2 \rangle \times (\langle x_3 \rangle \star \langle x_4 \rangle) \times \ldots \times (\langle x_{n-1} \rangle \star \langle x_n \rangle).$$

References

Bourbaki, 1968.
N. Bourbaki, *Groupes et algèbres de Lie,* Hermann, Paris, 1968. Chapters 4-6

Cohen, 1974.
D. E. Cohen, "Subgroups of HNN- groups," *J. Austral. Math. Soc.*, **17**, pp. 394-405, 1974.

Dicks, 1981.
W. Dicks, "An exact sequence for rings of polynomials in partly commuting indeterminates," *J. Pure Appl. Alg.*, **22**, pp. 215-228, 1981.

Karrass-Solitar, 1970.
A. Karrass and D. Solitar, "The subgroups of a free product of two groups with an amalgamed subgroup," *Trans. Amer. Math. Soc.*, **149**, pp. 227-255, 1970.

Serre, 1971.
J-P. Serre, "Cohomologie des groupes discrets," in *Prospects in Mathematics*, Annals of Mathematics Studies, **70**, pp. 77-169, Princeton University Press, 1971.

Serre, 1980.
J-P. Serre, *Trees,* Springer-Verlag, New York, 1980.

Waldhausen, 1973.
Waldhausen, F., "Whitehead groups of generalised free products," *Lecture Notes in Mathematics*, **342**, pp. 155-179, Springer-Verlag, New York, 1973.

Wall, 1971.
C.T.C. Wall, "Pairs of relative cohomological dimension one," *J. Pure Appl. Alg.*, **1**, pp. 141-154, 1971.

I. Chiswell,
Queen Mary College,
LONDON E1 4NS.

Fuchsian subgroups of the groups $\mathrm{PSL}_2(O_d)$

C. Maclachlan

Let O_d denote the ring of integers in the totally imaginary quadratic number field $\mathbb{Q}(\sqrt{-d})$. The group $\mathrm{PSL}_2(O_d)$ acts by linear fractional transformation on the complex plane and a non-elementary Fuchsian subgroup will fix a circle or straight line in the plane. For small values of d and in particular for $d = 1$, the structure of such subgroups have been considered e.g. [Fine, 1976], [Harding, 1985]. Here, for all values of d, Fuchsian subgroups are shown to exist and may be either cocompact or non-cocompact. Indeed they are all arithmetic Fuchsian groups and for each d are distributed in infinitely many commensurability classes.

1. The elements of $\mathrm{PSL}_2(\mathbb{C})$ act via linear fractional transformation on the complex plane $\mathbb{C}$ and this action can be extended to upper half 3-space regarded as the subset of the classical quaternions

$$H^3 = \{x + yi + zj \mid z > 0\}$$

with the group acting by

$$\begin{pmatrix} a & b \\ c & d \end{pmatrix} : (x + yi + zj) \mapsto (a(x + yi + zj) + b)(c(x + yi + zj) + d)^{-1}$$

where the operations on the right hand side are carried out in the quaternions. If H^3 is endowed with the hyperbolic metric, $ds^2 = \frac{1}{z^2}(dx^2 + dy^2 + dz^2)$, then $\mathrm{PSL}_2(\mathbb{C})$ is the full group of orientation-preserving isometries of H^3. The groups $\mathrm{PSL}_2(O_d)$ are obviously discrete subgroups of $\mathrm{PSL}_2(\mathbb{C})$.

Two subgroups Γ_1, Γ_2 of a group $\mathcal{G}$ are *commensurable* if their intersection is of finite index in both Γ_1 and Γ_2. They are said to be commensurable *in the wide sense* if Γ_1 and some conjugate of Γ_2 are commensurable.

Let ℓ denote a circle or straight line in $\mathbb{C}$. Then $\mathrm{Stab}(\ell, \mathrm{PSL}_2(\mathbb{C})) = \{\gamma \in \mathrm{PSL}_2(\mathbb{C}) \mid \gamma(\ell) = \ell$ and γ maps each component of $\mathbb{C} \setminus \ell$ to itself$\}$ is conjugate in $\mathrm{PSL}_2(\mathbb{C})$ to $\mathrm{PSL}_2(\mathbb{R})$. A *Fuchsian group* is a discrete subgroup of some $\mathrm{Stab}(\ell, \mathrm{PSL}_2(\mathbb{C}))$, and it is *elementary* if its limit set on ℓ has less than 3 points. In this paper, we shall only be concerned with non-elementary Fuchsian groups which necessarily contain infinitely many non-commuting hyperbolic elements (see e.g. [Beardon, 1983, Chapter 5]).

Arithmetic Fuchsian groups are obtained as follows [Borel, 1981], [Vignéras, 1980]: Let k be a totally real number field and A a quaternion algebra over k which is ramified at all but one archimedean place. Then a Fuchsian group Γ is *arithmetic* if there exists a representation σ of A into $M_2(\mathbb{C})$ and an order O in A such that Γ is commensurable with $P\sigma(O^1)$, where O^1 is the group of elements in O of norm 1. If the arithmetic group Γ is a subgroup of $\mathrm{Stab}(\ell, \mathrm{PSL}_2(\mathbb{C}))$, then H^2/Γ has finite volume where H^2 is the hemisphere in $\mathbb{H}^3$ on ℓ, being a hyperbolic plane under the restriction of the hyperbolic metric in $\mathbb{H}^3$. The arithmetic Fuchsian groups encountered in this paper are all defined over $\mathbb{Q}$, i.e. $k = \mathbb{Q}$.

2. Let Γ be a non-elementary Fuchsian subgroup of $\mathrm{PSL}_2(O_d)$, with Γ stabilising the circle or straight line ℓ with equation

$$a\,|z|^2 + Bz + \overline{Bz} + c = 0 \tag{1}$$

where $a, c \in \mathbb{R}$, $B \in \mathbb{C}$.

Suppose $a = 0$, so that ℓ is a straight line. Then I claim that there exists $\gamma \in \mathrm{PSL}_2(O_d)$ so that $\gamma(\ell)$ is a circle. We can assume that $\Gamma = \mathrm{Stab}(\ell, \mathrm{PSL}_2(O_d))$. Let $\gamma \in \mathrm{PSL}_2(O_d)\backslash\Gamma$, so that for every $\tau \in \Gamma$, $\tau\gamma \in \mathrm{PSL}_2(O_d)$. Assume, by way of contradiction, that $\tau\gamma(\ell)$ is a straight line for each τ. Thus there exists a point $x_\tau \in \ell \cup \{\infty\}$ such that $\tau\gamma(x_\tau) = \infty$. So

$$\gamma(x_\tau) \in \tau^{-1}(\infty) \subset \ell.$$

Since Γ is non-elementary, it follows that $\gamma(\ell) = \ell$, which is a contradiction. Thus Γ is conjugate in $\mathrm{PSL}_2(O_d)$ to a non-elementary Fuchsian subgroup whose equation at (1) has $a = 1$.

Since Γ is non-elementary, it contains two hyperbolic elements with distinct fixed points which will lie on ℓ. If one such hyperbolic element is represented by $\begin{pmatrix}\alpha & \beta\\ \gamma & \delta\end{pmatrix}$, then its fixed points are $\dfrac{\alpha-\delta\pm\lambda}{2\gamma}$, where $\lambda^2 = (\alpha + \delta)^2 - 4 > 0$. Note that $\gamma \neq 0$. The perpendicular bisector of the line joining these fixed points has equation

$$\gamma z + \overline{\gamma z} = \gamma\mu + \overline{\gamma\mu}$$

where $\mu = \dfrac{\alpha-\delta}{2\gamma}$. Thus the centre of ℓ is the intersection of two such lines and so $B \in \mathbf{Q}(\sqrt{-d})$. Since (1) can be written in the form

$$|z + \bar{B}|^2 = |B|^2 - c$$

we have that

$$|B|^2 - c = \left|\mu + \frac{\lambda}{2\gamma} + \bar{B}\right|^2 = \left|\mu - \frac{\lambda}{2\gamma} + \bar{B}\right|^2$$

and so $2(|B|^2 - c) = 2(|\mu|^2 + |B|^2 + \mu B + \bar{\mu}\bar{B} + \dfrac{\lambda^2}{4|\gamma|^2})$. Thus $c \in \mathbf{Q}(\sqrt{-d}) \cap \mathbb{R} = \mathbf{Q}$. Thus, it can be assumed that the equation of ℓ at (1) has $a, c \in Z, B \in O_d$, with $a > 0$.

If $T = \begin{pmatrix} \sqrt{a} & \bar{B}/\sqrt{a} \\ 0 & 1/\sqrt{a} \end{pmatrix}$ then $T(\ell) = \ell_D$ where ℓ_D is the circle with centre the origin given by

$$|z|^2 = D \tag{1}$$

where $D \in \mathbb{Z}$. Now the principle congruence subgroup of level a in $\mathrm{PSL}_2(O_d)$ is conjugated by T to a subgroup of $\mathrm{PSL}_2(O_d)$ which contains the principle congruence subgroup of level a^2. Thus $\mathrm{PSL}_2(O_d)$ and $T\mathrm{PSL}_2(O_d)T^{-1}$ are commensurable subgroups of $\mathrm{PSL}_2(\mathbb{C})$. Thus if $\Gamma = \mathrm{Stab}(\ell, \mathrm{PSL}_2(O_d))$ then $T\Gamma T^{-1}$ is commensurable with $\mathrm{Stab}(\ell_D, \mathrm{PSL}_2(O_d))$.

Theorem 1. *Let Γ be a maximal non-elementary Fuchsian subgroup of $\mathrm{PSL}_2(O_d)$. Then Γ is conjugate in $\mathrm{PSL}_2(\mathbb{C})$ to a Fuchsian group commensurable with Stab$(\ell_D, \mathrm{PSL}_2(O_d))$ for some $D \in Z$.*

3. Up to commensurability in the wide sense, it suffices to consider the groups Stab $(\ell_D, \mathrm{PSL}_2(O_d))$. In this section, these groups are shown to be arithmetic and the related quaternion algebras are determined. Since each quaternion algebra determines a wide commensurability class of groups, it will follow that each maximal non-elementary Fuchsian subgroup of $\mathrm{PSL}_2(O_d)$ is arithmetic (see [Vignéras, 1980] for material on quaternions).

Now

$$\text{Now } \mathrm{Stab}(\ell_D, \mathrm{PSL}_2(\mathbb{C}) = P\left\{\begin{pmatrix} \alpha & \sqrt{D}\,\beta \\ \dfrac{1}{\sqrt{D}}\bar{\beta} & \bar{\alpha} \end{pmatrix} \in SL_2(\mathbb{C})\right\}.$$

Also if $\hat{\ell}_D$ denotes the hemisphere on ℓ_D with the restriction of the hyperbolic metric on H^3, then $\text{Stab}(\ell_D, \text{PSL}_2(\mathbb{C}))$ = Group of orientation-preserving isometries of $\hat{\ell}_D$. Let $\mathfrak{D}$ be the hyperboloid model of the hyperbolic plane given by

$$x^2 + y^2 - Dz^2 = -1 \qquad z > 0$$

with metric $dx^2 + dy^2 - Ddz^2$. An isometry ϕ from $\hat{\ell}_D$ to $\mathfrak{D}$ is obtained as follows: project vertically from $\hat{\ell}_D$ to the plane disc in $\mathbb{C}$, take that disc as a cross-section in the projective model given by

$$x^2 + y^2 - Dz^2 < -1 \qquad z > 0$$

and project along rays onto $\mathfrak{D}$. The map ϕ is then given by

$$\phi(x, y, z) = \left(\frac{x}{z}, \frac{y}{z}, \frac{1}{z}\right).$$

Finally, if $\mathfrak{D}'$ is the hyperboloid model of the hyperbolic plane

$$dx^2 + y^2 - Dz^2 = -1$$

with corresponding metric, then $f(x, y, z) = (\sqrt{d}x, y, \sqrt{d}z)$ is an isometry from $\mathfrak{D}$ to $\mathfrak{D}'$. Thus $\psi = f \circ \phi$ induces an isomorphism

$$\psi^*: \text{Stab}(\ell_D, \text{PSL}_2(\mathbb{C})) \to \text{Group of isometries of } \mathfrak{D}'$$

which is the group $SO(F, \mathbb{R}) = \{X \in SL_3(\mathbb{R}) \mid X^t FX = F\}$ where

$$F = \begin{pmatrix} d & 0 & 0 \\ 0 & 0 & 0 \\ 0 & 0 & -dD \end{pmatrix}$$

and furthermore, a tedious calculation shows that ψ^* carries $\text{Stab}(\ell_D, \text{PSL}_2(O_d))$ into $SO(F, Z)$. Now $SO(F, \mathbb{R}) = SO(-DF, \mathbb{R})$ and $-DF$ is the norm form of the quaternion algebra $\left(\frac{dD, D}{\mathbb{Q}}\right) \cong \left(\frac{-d, D}{\mathbb{Q}}\right)$ restricted to the pure quaternions.

We are thus led to consider the quaternion algebra $A = \left(\frac{-d, D}{\mathbb{Q}}\right)$ which clearly splits over $\mathbb{Q}(\sqrt{-d})$ giving a representation $\sigma: A \to M_2(\mathbb{Q}(\sqrt{-d}))$ where

$$\sigma(a_0 + a_1 i + a_2 j + a_3 ij) = \begin{pmatrix} a_0 + a_1\sqrt{-d} & D(a_2 + a_3\sqrt{-d}) \\ a_2 - a_3\sqrt{-d} & a_0 - a_1\sqrt{-d} \end{pmatrix}$$

in terms of the standard basis 1, i, j, ij where $i^2 = -d$, $j^2 = D$, $ij = -ji$. Thus if O is the order $Z[1, i, j, ij]$ in A, then clearly

$$P\sigma(O^1) \subset \mathrm{Stab}(\ell_D\ \mathrm{PSL}_2(O_d))..EN.Now\$P\sigma(O^1)$$

$(\ell_D\ \mathrm{PSL}_2(\mathbb{C}))$. Thus Stab $(\ell_D, \mathrm{PSL}_2(O_d))$ has finite covolume and

$$[\mathrm{Stab}(\ell_D, \mathrm{PSL}_2(O_d)) : P\sigma(O^1)] < 0.$$

Theorem 2. *Every maximal non-elementary Fuchsian subgroup of* $\mathrm{PSL}_2(O_d)$ *is an arithmetic Fuchsian group arising from some quaternion algebra* $\left(\frac{-d, D}{\mathbb{Q}}\right)$. *In particular, they have finite covolume.*

4. Now consider the wide commensurability classes of these Fuchsian subgroups in $\mathrm{PSL}_2(\mathbb{C})$. Let $\Gamma_i = \mathrm{Stab}\ (\ell_{D_i}, \mathrm{PSL}_2(O_{d_i}))$, $i = 1, 2$, and suppose that $t\Gamma_1 t^{-1}$ and Γ_2 are commensurable in $\mathrm{PSL}_2(\mathbb{C})$. Since $t\Gamma_1 t^{-1} \subset$ Stab $(\ell_{D_2}, \mathrm{PSL}_2(\mathbb{C}))$, $t(\ell_{D_1}) = \ell_{D_2}$. But arithmetic subgroups of Stab (ℓ_{D_2}) are commensurable if and only if the corresponding quaternion algebras are isomorphic [Takeuchi, 1977]. Since any two representations of a quaternion algebra in $M_2(\mathbb{C})$ are conjugate, it follows that Γ_1, Γ_2 are commensurable in the wide sense if and only if $\left(\frac{-d_1, D_1}{\mathbb{Q}}\right) \cong \left(\frac{-d_2, D_2}{\mathbb{Q}}\right)$. The isomorphism class of a quaternion algebra is determined by its ramification and in these cases of indefinite algebras over $\mathbb{Q}$, by its ramification at the rational primes.

Theorem 3. *Let A be an indefinite quaternion algebra over $\mathbb{Q}$. Then there exists an arithmetic Fuchsian subgroup of* $\mathrm{PSL}_2(O_d)$ *whose associated quaternion algebra is A if and only if the primes p at which A is ramified are either ramified or inert in* $\mathbb{Q}(\sqrt{-d})$.

Proof. Suppose A gives rise to a Fuchsian subgroup of $\mathrm{PSL}_2(O_d)$ then $A \cong \left(\frac{-d, D}{\mathbb{Q}}\right)$ for some D. Let p be an odd prime which is not ramified in $\mathbb{Q}(\sqrt{-d})$. Then $\left(\frac{-d, D}{\mathbb{Q}}\right)$ is ramified at p if $p \mid D$ and $\left(\frac{-d}{p}\right) = 1$. Thus p is inert in $\mathbb{Q}(\sqrt{-d})$ so $-d \equiv 1 \pmod 4$. Then $\left(\frac{-d, D}{\mathbb{Q}}\right)$ is ramified at 2 if $2 \mid D$ and $-d \equiv 5 \pmod 8$. Thus 2 is inert in $\mathbb{Q}(\sqrt{-d})$.

Now suppose that A is ramified at primes $p_1, p_2, \ldots, p_r$ which are inert in $\mathbf{Q}(\sqrt{-d})$ (and at $r_1, r_2, \ldots, r_s$ which are ramified in $\mathbf{Q}(\sqrt{-d})$), so that $r + s$ is even. Now choose an odd prime q such that

1. $\left(\frac{q}{r_i}\right) = -\left(\frac{p_1 p_2 \ldots p_r}{r_i}\right)$ for each odd r_i and $q \equiv 3p_1p_2\ldots p_r \pmod 8$ if $r_i = 2$.

2. $\left(\frac{q}{t}\right) = \left(\frac{p_1 p_2 \ldots p_r}{t}\right)$ for each odd prime t, $t \mid d$, $t \neq r_i$, 1, 2,...,s and $q \equiv p_1p_2\ldots p_r \pmod 8$ if 2 is ramified in $\mathbf{Q}(\sqrt{-d})$ and $2 \neq r_i$.

It then follows that the quaternion algebra $\left(\frac{-d, p_1p_2\ldots p_r q}{\mathbf{Q}}\right)$ is isomorphic to A.

Theorem 3

Corollary 1. *Each* $\mathrm{PSL}_2(O_d)$ *contains infinitely many wide commensurability classes of maximal non-elementary Fuchsian subgroups.*

Proof. This follows from the remarks preceding Theorem 3 and the fact that there are infinitely many primes which are inert in $\mathbf{Q}(\sqrt{-d})$

Corollary 2. *Let* Γ *be an arithmetic Fuchsian group whose associated quaternion algebra* A *is defined over* $\mathbf{Q}$. *Then there exists a group* $\mathrm{PSL}_2(O_d)$ *containing a non-elementary Fuchsian subgroup commensurable with* Γ.

Proof. If A is ramified at $q_1, q_2, \ldots, q_r$, choose d such that $\left(\frac{-d}{q_i}\right) = -1$ if q_i is odd and $-d \equiv 5 \pmod 8$ if $q_i = 2$, so that $q_1, q_2, \ldots, q_r$ are inert in $\mathbf{Q}(\sqrt{-d})$.

□

Corollary 3. *For each pair of square-free integers* d_1, d_2, *there exist infinitely any wide commensurability classes of arithmetic Fuchsian groups with representatives in both* $\mathrm{PSL}_2(O_{d_1})$ *and* $\mathrm{PSL}_2(O_{d_2})$.

Proof. There are infinitely many primes p such that

$$\left(\frac{-d_1}{p}\right) = \left(\frac{-d_2}{p}\right) = -1.$$

References

Beardon, 1983.
A. F. Beardon, *The Geometry of Discrete Groups,* Springer, New York, 1983.

Borel, 1981.
A. Borel, "Commensurability classes and volumes of hyperbolic 3-manifolds," *Annali Scuola Normale Superiore - Pisa*, **8**, pp. 1-33, 1981.

Fine, 1976.
B. Fine, "Fuchsian subgroups of the Picard group," *Can. J. Math*, **28**, pp. 481-485, 1976.

Harding, 1985.
S. Harding, *Thesis*, Southampton, 1985.

Takeuchi, 1977.
Takeuchi, K., "Commensurability classes of arithmetic triangle groups," *J. Fac. Sci. Univ. Tokyo Sect.*, **1A 24**, pp. 201-222, 1977.

Vignéras, 1980.
M.-F. Vignéras, "Arithmétique des Algèbras de Quaternions," *L.N.M.*, **800**, Springer-Verlag, Berlin, 1980.

C. MacLachlan,
Department of Mathematics,
University of Aberdeen,
Aberdeen AB9 2TY,
Scotland.

Polyhedral Orbifold Groups

Norbert J. Wielenberg

ABSTRACT

The fundamental group of a hyperbolic orbifold whose singular set consists of the vertices and edges of a polyhedron has a nice presentation in terms of amalgamations of vertex stabilizers. The vertices are allowed to lie in hyperbolic space, at infinity, or outside infinity. The presentation makes it possible to compute, in principle, all torsion-free subgroups of finite index. We also give an infinite sequence of polyhedral groups which are not arithmetic.

1. Introduction

An *orbifold* is a topological space which is locally modelled on $\mathbb{R}^n/\Gamma$ where Γ is a finite group acting on $\mathbb{R}^n$. An *orbifold with boundary* is locally modelled on $\mathbb{R}^n/\Gamma$ or $\mathbb{R}^n_+/\Gamma$. The *singular set* of the orbifold is the set of those points with non-trivial isotropy group in a local model $\mathbb{R}^n/\Gamma$. See papers of Thurston [Thurston, 1979] for exact definitions.

We will use the term *polyhedral orbifold* for the quotient of a geometric space under the action of the orientation-preserving subgroup G of a discrete subgroup generated by reflections in the faces of a polyhedron P. There are eight relevant three-dimensional geometries; however, we are concerned only with hyperbolic 3-space $\mathbb{H}^3$.

The polyhedron necessarily has dihedral angles of the form π/m, m an integer greater than or equal to 2. The polyhedra to be considered include all such polyhedra of finite volume. In addition, we allow finite-sided polyhedra with some vertices outside infinity, in a sense to be described below. To each vertex of P is associated a sequence $(m_1, m_2, \ldots, m_l)$, $l \geq 3$, of integers corresponding to the dihedral angles

π/m_i enumerated in order around the vertex. This sequence also determines a 2-orbifold which is a sphere with l elliptic points of order $m_1, \ldots, m_l$. The vertex group for the sequence $(m_1, \ldots, m_l)$ has a presentation $\langle a_1, \ldots, a_l; a_1 \ldots a_l = a_i^{m_i} = 1, 1 \leq i \leq l \rangle$.

A vertex is classified as spherical, euclidean, or hyperbolic according as the Euler number χ of the 2-orbifold is > 0, $= 0$, or < 0. The spherical vertices are $(2\ 2\ n)$, $n \geq 2$, $(2\ 3\ 3)$, $(2\ 3\ 4)$, and $(2\ 3\ 5)$. The corresponding (abstract) vertex groups are the dihedral groups D_n, the alternating group A_4, the symmetric group S_4, and the alternating group A_5. These are finite groups, the vertices are ordinary vertices in $\mathbb{H}^3$. The euclidean vertices are $(2\ 4\ 4)$, $(2\ 3\ 6)$, $(3\ 3\ 3)$ and $(2\ 2\ 2\ 2)$. These vertex groups are infinite and the vertices are ideal vertices lying on the boundary of $\mathbb{H}^3$. The remaining cases are hyperbolic; here the faces of P "incident" to this vertex do not actually meet at a vertex. Rather, we require the existence of a totally geodesic plane $\mathbf{S}$ in $\mathbb{H}^3$ which meets each of these faces orthogonally. It follows that the vertex group is a Fuchsian group which stabilizes $\mathbf{S}$.

We will discuss the existence of these orbifolds, the algebraic structure of the groups, a method of calculating torsion-free subgroups of finite index, and some examples. The examples show, in particular, that not all polyhedral orbifold groups are arithmetic.

I would like to thank Andrew Brunner and Youn Lee for assistance on this project and John Wilker for helpful conversations on these topics.

2. Polyhedral Orbifolds

In the case where all vertices are of spherical type, the orbifold is homeomorphic to S^3 with the singular set being the graph consisting of the set of vertices and edges of P. More generally, each euclidean vertex is deleted from S^3 and at each hyperbolic vertex an open ball is removed from S^3. A neighbourhood of a deleted vertex is the Cartesian product of the euclidean 2-orbifold with an open interval. At each hyperbolic vertex, the orbifold has an S^2 boundary component which is parallel to a hyperbolic 2-orbifold. After excising the infinite volume between the hyperbolic 2-orbifold and the boundary component, at each hyperbolic vertex, the orbifold has finite volume. (In the upper half-space model for $\mathbb{H}^3$, the boundary component is the quotient of a disk in the complex plane $\mathbb{C}$ and the hyperbolic 2-orbifold is the quotient of the totally geodesic hemisphere with the same boundary circle as the disk.)

Consider now an abstract polyhedron, also called P, with an integer ≥ 2 assigned to each edge and where each face of P has at least three edges. We can decide whether there exists a hyperbolic polyhedral orbifold of this type as follows. Each vertex v of P has a neighbourhood consisting of a cone from v to a polygon and its interior, where the vertices of the polygon lie on the edges of P incident to v. The set of these cones can be taken to have disjoint closures. At each hyperbolic vertex, remove the cone and label each new edge with the integer 2. Call the new polyhedron Q. All vertices of Q are of spherical or euclidean type. The theorems of Thurston [Thurston, 1979], [Thurston] or of Andreev [Andreev, 1970] apply to give a criterion for the existence of a hyperbolic structure for Q. Specifically, the cases where Q is a tetrahedron are classified as spherical, euclidean, or hyperbolic by their Coxeter diagrams. When Q is not a tetrahedron, let O be the orbifold consisting of S^3 with the vertices and edges of Q as singular set with the given branching indices, and with the euclidean vertices deleted. Then Q has a hyperbolic structure if and only if O is irreducible and atoroidal. In other words, the criterion is that O must have no bad 2-suborbifolds, each incompressible spherical 2-suborbifold must bound a cone from a vertex to that suborbifold, and each incompressible euclidean 2-suborbifold must isotope into a product neighbourhood of a deleted vertex.

3. Algebraic Structure

Let G be the orientation-preserving subgroup of the group generated by reflections in the faces of P. (Note: not all the faces of Q, when $P \neq Q$.) At each edge of P, the product of two reflections in planes meeting at that edge is a rotation of order n when the dihedral angle is π/n.

Theorem 1.

(a) *G is generated by the edge rotations.*

(b) *G is presented by the union of the presentations for the vertex groups together with amalgamations of the cyclic edge groups between vertex groups.*

(c) *The relations at any single vertex of P are redundant in this presentation.*

(d) *The elements of finite order in G consist exactly of the edge rotations and their conjugates.*

(e) *The finite subgroups of G are subgroups of the vertex groups and their conjugates.*

Proof. Statements (a), (b), and (c) are done by Brunner, Lee and Wielenberg [Brunner-Lee-Wielenberg, 1985], for polyhedra with spherical or euclidean vertices. The result follows from applying Poincaré's Theorem to a particular fundamental polyhedron. The fundamental polyhedron is obtained by choosing an interior point of P, taking the cone from the interior point of P to each face, and joining to P the reflection of each of these cones across the corresponding face. This also works when there are hyperbolic vertices.

For statement (d), no elliptic element of G can have an axis which intersects the interior of the fundamental polyhedron or any of its images. Thus the edge rotations of P and their conjugates account for all elliptic elements of G.

Two elliptic elements which generate a finite subgroup must have a common fixed point in $\mathbb{H}^3$, hence they belong to a spherical vertex group of G.

□

4. Three-Manifold Subgroups

Theorem 2.

(a) *There is an algorithmic procedure to calculate presentations for all torsion-free finite index subgroups of G.*

(b) *The smallest index for a torsion-free subgroup is divisible by the least common multiple of the orders of the finite subgroups of G.*

Proof. The set of all torsion-free subgroups of G of index m are calculated as follows. First, find all transitive representations $\phi : G \to S_m$ such that no element of G of finite order has a representative which has a fixed point. There are finitely many conditions to be satisfied. For each representation the corresponding subgroup of G is $H = \{g \in G : \phi(g)(1) = 1\}$. Secondly, a presentation for H can be calculated by Todd-Coxeter coset enumeration and and Reidemeister-Schreier rewriting or by the c-cress method of Karrass, Solitar, and Pietrowski for polygonal products and graph amalgamation products. (See illustrations in [Brunner-Frame-Lee-Wielenberg, 1984].)

Since the number of cosets of each finite subgroup of G must divide the index of H in G, statement (b) follows.

Theorem 2

If the vertices of P are all of spherical or euclidean type, then the three-manifold $\mathbb{H}^3/H$, where H is a torsion-free finite index subgroup of G, has finite volume. If P has one or more hyperbolic vertices, then the Kleinian manifold $(\mathbb{H}^3 \cup \Omega)/H$ has boundary components which are parallel to totally geodesic surfaces in $\mathbb{H}^3/H$. Hence these groups are building blocks for *HNN* extensions and free products with amalgamation via the Klein-Maskit-Marden combination theorems [Marden, 1974].

5. Examples

Consider the labelled polyhedron in Figure 1. It can be realized with totally geodesic faces in $\mathbb{H}^3$. Of course, when $n \geq 6$ the faces at the $(2\ 3\ n)$ vertex do not meet at a point in $\mathbb{H}^3$. The vertex $(2\ 3\ n)$ is spherical for $n =$ 3, 4, 5; euclidean for $n = 6$; and hyperbolic for $n \geq 7$.

Rotations along the edges are the generators

$$a = \begin{pmatrix} 0 & -1 \\ 1 & 0 \end{pmatrix} \qquad b = \begin{pmatrix} 0 & 1 \\ 1 & 0 \end{pmatrix} \qquad c = \begin{pmatrix} 1 & -1 \\ 1 & 0 \end{pmatrix} \qquad d = \begin{pmatrix} 0 & 1 \\ 1 & -\omega \end{pmatrix}$$

where $\omega = 2i \cos\pi/n$. A presentation for the group is

$$\langle a,b\ ;\ a^2 = b^2 = (ab)^2 = 1\rangle$$

$$\langle b,c\ ;\ b^2 = c^3 = (bc)^2 = 1\rangle$$

$$\langle c,d\ ;\ c^3 = d^n = (cd)^2 = 1\rangle$$

$$\langle d,a\ ;\ d^n = a^2 = (da)^2 = 1\rangle$$

This presentation is represented as the polygonal product in Figure 2.

For $n = 3$, the group is $PSL\ [2, Z(i)]$.

For $n = 4$ it is $PGL\ [2, Z(i\sqrt{2})\}$.

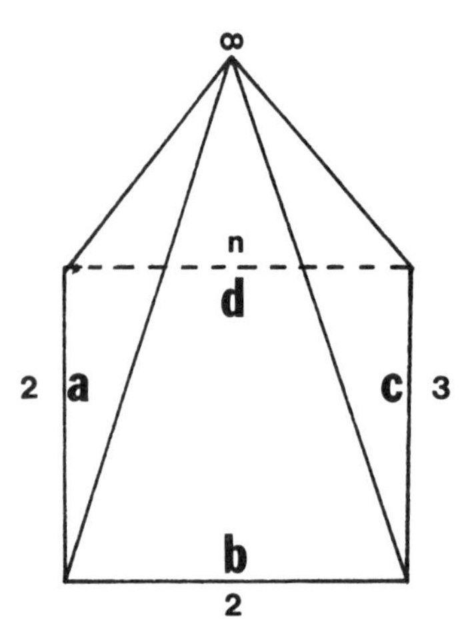

Figure 1.

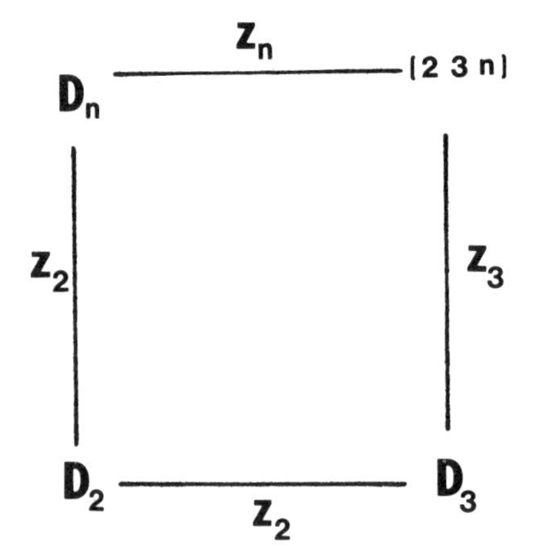

Figure 2.

For $n \geq 7$, the disk $|z - \frac{1+\omega}{2}| \leq \frac{1}{2}\sqrt{|\omega|^2 - 3}$ is stabilized by the Fuchsian group at that vertex. Conjugate the group by an inversion in the boundary circle of this disk. A free product with amalgamation of the group with the conjugate of itself, amalgamated along the Fuchsian subgroup, produces a finite volume group. Denote this finite volume group by G_n. Of course, G_n is itself a polyhedral group.

For $n = \infty$, G_n is still discrete but d has become a parabolic element. For values of ω between i and $2i$, but not equal to $2i\cos\pi/n$, G_n may not be discrete but is the holonomy group of a cone-manifold.

Since Borel [Borel, 1981] has shown that the set of covolumes of arithmetic subgroups of $PGL(2, \mathbb{C})$ is discrete, it follows that the sequence $\{G_n\}$ contains infinitely many non-arithmetic groups. This contrasts with the situation in our paper [Brunner-Lee-Wielenberg, 1985], where most of the examples are arithmetic.

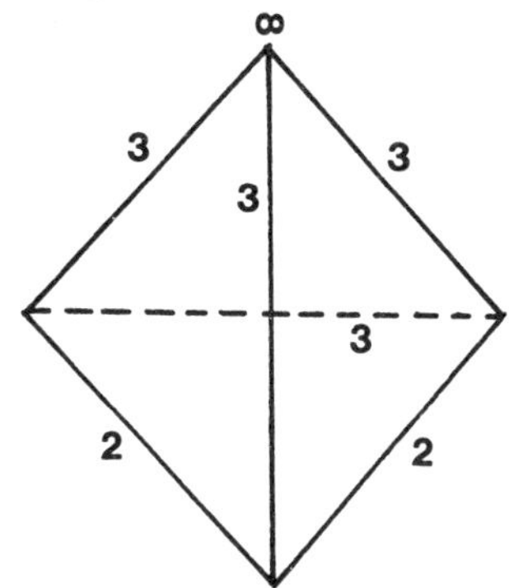

Figure 3.

Finally, we give some details of using the method described above, to discover the well-known but slightly elusive manifold with the same volume as the figure-eight knot complement. Both groups are subgroups of $\Gamma = PSL[2, Z(\omega)]$ where $\omega = \dfrac{-1+i\sqrt{3}}{2}$. The orbifold $\mathbb{H}^3/\Gamma$ is $S^3 - \{\infty\}$ with the singular set as shown in Figure 3.

The group Γ has a presentation as a triangular product.

$$\langle a, b \; ; \; a^2 = b^2 = (ab)^3 = 1 \rangle$$

$$\langle b, c \; ; \; b^2 = c^3 = (bc)^3 = 1 \rangle$$

$$\langle c, a \; ; \; c^3 = a^2 = (ca)^3 = 1 \rangle$$

$$a = \begin{pmatrix} 0 & -1 \\ 1 & 0 \end{pmatrix} \qquad b = \begin{pmatrix} 0 & -\omega^2 \\ \omega & 0 \end{pmatrix} \qquad c = \begin{pmatrix} 0 & -\omega \\ \omega^2 & 1 \end{pmatrix}$$

The finite subgroups of Γ have orders 2, 3, 6, and 12. A representation in S_{12} is

$$a = (1\ 2)(3\ 7)(4\ 8)(5\ 9)(6\ 10)(11\ 12)$$

$$b = (1\ 5)(2\ 11)(3\ 8)(4\ 10)(6\ 7)(9\ 12)$$

$$c = (1\ 3\ 4)(2\ 5\ 6)(8\ 11\ 9)(7\ 10\ 12)$$

The c-cress method now produces a presentation for the corresponding subgroup as

$$\langle x, y ; x^2 = y\ x\ y^3\ x\ y \rangle$$

where

$$x = \begin{pmatrix} -2-2\omega & 2+\omega \\ -1-\omega & 1+\omega \end{pmatrix} \quad \text{and } y = \begin{pmatrix} 0 & -\omega \\ \omega^2 & 1-\omega \end{pmatrix}$$

The first homology is $Z \oplus Z_5$. A peripheral subgroup is $\langle y^{-1}\ x,\ x\ y^{-1}\ x^{-1}\ y^{-1}\ x\ y \rangle$. These elements have a fixed point at $-\omega$. If the fixed point is moved to ∞, the peripheral subgroup has generators $\begin{pmatrix} 1 & 2\omega \\ 0 & 1 \end{pmatrix}$ and $\begin{pmatrix} 1 & 2 \\ 0 & 1 \end{pmatrix}$, respectively.

The manifold is given by a 5-surgery on one component of the Whitehead link complement, as shown in Figure 4.

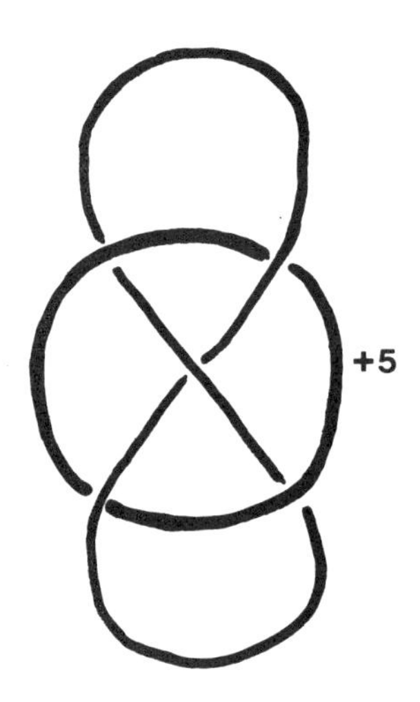

Figure 4.

References

Andreev, 1970.

E.M. Andreev, "On convex polyhedra of finite volume in Lobachevski spaces," *Math.USSR Sb.*, **12**, pp. 225-259, 1970.

Borel, 1981.

A. Borel, "Commensurability classes and volumes of hyperbolic 3-manifolds," *Annali Scuola Normale Superiore - Pisa*, **8**, pp. 1-33, 1981.

Brunner-Frame-Lee-Wielenberg, 1984.
A. M. Brunner, M. L. Frame, Y. W. Lee, and N. J. Wielenberg, "Classifying Torsion-Free subgroups of the Picard Group," *T.A.M.S.*, **282**, pp. 205-235, 1984.

Brunner-Lee-Wielenberg, 1985.
A.M. Brunner, Y.W. Lee, and N.J. Wielenberg, "Polyhedral groups and graph amalgamation products," *Topology and its Applications*, **20**, pp. 289-304, 1985.

Marden, 1974.
A. Marden, "The Geometry of Finitely Generated Kleinian Groups," *Annals of Mathematics*, **99**, pp. 383-462, 1974.

Thurston.
W. P. Thurston, *Lectures at the 1984 Durham Symposium on Kleinian groups, 3-Manifolds, and Hyperbolic Geometry.* unpublished

Thurston, 1979.
W. P. Thurston, *The Geometry and Topology of 3-Manifolds,* Princeton University Mathematics Department, 1979. Part of this material – plus additional material – will be published in book form by Princeton University Press

N.J.Wielenberg
University of Wisconsin-Parkside,
Kenosha,
Wisconsin 53141,
USA.